**Springer Proceedings in Mathematics & Statistics**

Volume 420

This book series features volumes composed of selected contributions from workshops and conferences in all areas of current research in mathematics and statistics, including data science, operations research and optimization. In addition to an overall evaluation of the interest, scientific quality, and timeliness of each proposal at the hands of the publisher, individual contributions are all refereed to the high quality standards of leading journals in the field. Thus, this series provides the research community with well-edited, authoritative reports on developments in the most exciting areas of mathematical and statistical research today.

Toka Diagana • Khalil Ezzinbi • Stanislas Ouaro
Editors

# Partial Differential Equations and Applications

## Colloquium in Honor of Hamidou Touré, Ouagadougou, Burkina Faso, November 5–9, 2018

Springer

*Editors*
Toka Diagana
Department of Mathematical Sciences
The University of Alabama in Huntsville
Huntsville, AL, USA

Khalil Ezzinbi
Faculty of Sciences Semlalia, Department of Mathematics
Cadi Ayyad University
Marrakesh, Morocco

Stanislas Ouaro
Ministry of National Education, Literacy and Promotion of National Languages
Ouagadougou, Burkina Faso

ISSN 2194-1009 ISSN 2194-1017 (electronic)
Springer Proceedings in Mathematics & Statistics
ISBN 978-3-031-27660-6 ISBN 978-3-031-27661-3 (eBook)
https://doi.org/10.1007/978-3-031-27661-3

Mathematics Subject Classification: 34K14, 35-XX, 35B25, 35B35, 35B65, 35Fxx, 35Q20, 39-XX, 39A23, 39A27, 49-XX, 58A10, 92D30



This Springer imprint is published by the registered company Springer Nature Switzerland AG
The registered company address is: Gewerbestrasse 11, 6330 Cham, Switzerland

# Preface

## Préface (Version française)

Le Professeur Hamidou TOURÉ est né le 14 octobre 1954 à Bobo Dioulasso dans l'actuelle Burkina Faso. Du 05 au 07 novembre 2018, alors qu'il a 64 ans et se prépare à faire valoir ses droits à la retraite dans moins d'un an, le Laboratoire de Mathématiques et Informatique (LAMI) de l'Université Joseph KI-ZERBO de Ouagadougou au Burkina Faso a décidé de l'honorer en organisant un colloque international avec pour thème «Équations aux dérivées partielles et applications». En effet, le Professeur Hamidou TOURÉ a, durant ses 37 ans de carrière universitaire formé beaucoup d'étudiants en mathématiques, contribué à beaucoup de travaux scientifiques et surtout a, avec deux de ses amis, les professeurs Mary Teuw NIANE du Sénégal et Iselkou Ould Ahmed IZID BIH de la Mauritanie, créé en mai 1999 au centre de Physique Théorique Abdoul Salam (ICTP) à Trieste en Italie, un réseau de recherche en mathématique dénommé Réseau EDP-Modélisation et Contrôle (EDP-MC), qui a contribué à la formation de plus d'une centaine de docteurs en mathématiques en Afrique et à la production de plus de 1000 publications scientifiques. Ce numéro spécial est une contribution des participants au colloque international à un hommage mérité au Professeur Hamidou TOURE.

## Préface (Version anglaise)

Prof. Hamidou TOURE was born on October 14, 1954 in present-day Burkina Faso's Bobo Dioulasso. The Laboratoire de Mathématiques et Informatique (LAMI) of the Joseph KI-ZERBO University in Ouagadougou, Burkina Faso, decided to honor him by hosting an International Colloquium titled "Partial Differential Equations and Applications" from November 5 to 7, 2018, when he was 64 years old and preparing to claim his retirement benefits in less than a year.

Throughout his 37-year university career, Prof. Hamidou TOURE has trained many students in mathematics, contributed to numerous scientific works, and, along with two of his friends, Prof. Mary Teuw NIANE from Senegal and Prof. Iselkou Ould Ahmed IZID BIH from Mauritania, established in May 1999 at the Abdoul Salam Theoretical Physics Centre (ICTP) in Trieste, Italy, a mathematical research network known as the "PDE-Modeling and Control Network", which has contributed to the training of over a hundred PhD students in mathematics in Africa and the production of over a thousand scientific publications.

The participants of the above-mentioned International Colloquium have put together these proceedings as their contribution to a fitting tribute to Prof. Hamidou TOURE, who has made major contributions to mathematics and related areas.

## Avant Propos (Version française)

Du 05 au 07 Novembre 2018, a eu lieu, à l'Université Joseph KI-ZERBO de Ouagadougou au Burkina Faso, un colloque international autour du thème : «Équations aux Dérivées Partielles et Applications» en honneur au Professeur Hamidou TOURE à l'occasion de ses 64 ans de vie et de 36 années consacrées à l'enseignement et au développement des mathématiques en Afrique. A l'issue du colloque, les communications originales non encore soumises à publication ont été soumises à publication dans Springer Nature. Les travaux examinés, évalués et acceptés font l'objet de ce numéro spécial dédié au Professeur Hamidou TOURE à l'occasion de ses 64 années.

## Avant Propos (Version anglaise)

From 05 to 07 November 2018, an international symposium was held at the Joseph KI-ZERBO University in Ouagadougou, Burkina Faso under the title: «Partial Differential Equations and Applications» in honour of Professor Hamidou TOURE on the occasion of his 64 years of life and 36 years devoted to the teaching and development of mathematics in Africa. At the end of the symposium, original papers not yet submitted for publication were submitted for publication in Springer Nature. The works reviewed, evaluated and accepted are the subject of this special issue dedicated to Professor Hamidou TOURE on the occasion of his 64 years.

Huntsville, AL, USA
Marrakesh, Morocco
Ouagadougou, Burkina Faso
November 2018

Toka Diagana
Khalil Ezzinbi
Stanislas Ouaro

# Contents

# Contributors

**Abdoul Aziz Kalifa Dianda** Unité de Formation et de Recherche en Sciences Exactes et Appliquées, Département de Mathématiques, Université Joseph KI-ZERBO, Ouagadougou, Burkina Faso

**Ngalla Djitte** Gaston Berger University, Saint Louis, Senegal

**Khalil Ezzinbi** Département de Mathématiques, Université Cadi Ayyad, Faculté des Sciences Semlalia, Marrakech, Marocco

**Aboudramane Guiro** Laboratoire de Mathématiques, d'Informatique et Applications (LaMIA), UFR, Sciences Exactes et Appliquées, Université Nazi BONI, Bobo Dioulasso, Burkina Faso

**Idrissa Ibrango** Laboratoire de Mathématiques, d'Informatique et Applications (LaMIA), Institut Universitaire de Technologie, Université Nazi BONI, Bobo Dioulasso, Burkina Faso

**Souleye Kane** Faculté des Sciences et Techniques, Université Cheikh Anta Diop de Dakar, Dakar, Senegal

**Kpè Kansie** LAboratoire de Mathématiques et Informatique (LA.M.I), UFR. Sciences et Techniques, Université Nazi Boni, Bobo-Dioulasso, Burkina Faso

**Blaise Koné** Laboratoire de Mathématiques et Informatique (LAMI), Institut Burkinabé des Arts et Métiers, Université Joseph KI-Zerbo, Ouagadougou, Burkina Faso

**Moussa El-KhalilL Kpoumiè** Département de Mathématiques et Informatique, Ecole de Géologie et d'exploitation minière, Université de N'gaoundéré, Meiganga, Cameroun

**Bila Adolphe Kyelem** Département de Mathématiques et Informatique, Université de Ouahigouya, Unité de Formation et de Recherche en Sciences et Technologies, Ouahigouya, Burkina Faso

**Mohamed Lazhar Tayeb** Department of Mathematics, Faculty of Sciences of Tunis, University of Tunis El-Manar, Tunis, Tunisia

**Daniel Y. Le Roux** Université de Lyon, CNRS, Université Lyon 1, Institut Camille Jordan, Villeurbanne Cedex, France

**Salifou Mboutngam** Département de Mathématiques, Ecole Normale Supérieure, Université de Maroua, Maroua, Cameroun

**Mariama Ndiaye** Gaston Berger University, Saint Louis, Senegal

**Evrad M. D. Ngom** CEA-MITIC, Université Gaston Berger, Saint-Louis, Sénégal

**Abdel Hamid Gamal Nsangou** Département de Mathématiques, Ecole Normale Supérieure, Université de Maroua, Maroua, Cameroun

**Stanislas Ouaro** LAboratoire de Mathématiques et Informatique (LA.M.I), UFR. Sciences Exactes et Appliquées, Université Ouaga I Pr Joseph KI ZERBO, Ouagadougou, Burkina Faso

**Dramane Ouedraogo** Centre Universitaire de Banfora, Université Nazi Boni, Dindéresso, Burkina Faso

**Harouna Ouedraogo** Département de Mathématiques, Université Joseph Ki-Zerbo, Ouagadougou, Burkina Faso

**Serigne Fallou Samb** Département de Mathématiques et Informatique, Université Cheikh Anta Diop de Dakar, Dakar, Senegal

**Rodrigue Sanou** Laboratoire de Mathématiques et Informatique (LAMI), UFR, Sciences et Techniques, Université Thomas SANKARA, Ouagadougou, Burkina Faso

**Diaraf Seck** Ecole Doctorale de Mathématiques et Informatique, Université Cheikh Anta Diop de Dakar, Dakar, Senegal

**Abdou Sène** CEA-MITIC, Université Virtuelle du Sénégal, Dakar-Fann, Sénégal

**Ali Traoré** LAboratoire de Mathématiques et Informatique (LAMI), Université Joseph KI-ZERBO, Ouagadougou, Burkina Faso

**Issa Zabsonre** Unité de Recherche et de Formation en Sciences Exactes et Appliquées, Département de Mathématiques, Université Joseph KI-ZERBO, Ouagadougou, Burkina Faso

# Chapter 1
# Existence and Uniqueness of Solution for Semi-linear Conservation Laws with Velocity Field in $L^\infty$

**Souleye Kane, Serigne Fallou Samb, and Diaraf Seck**

**Abstract** In this chapter, we extend results obtained in Besson and Pousin (Arch Ration Mech Analy 2:159–175, 2007) and Benmansour et al. (Discrete Contin Dyn Syst 29:1001–1030, 2011). By considering a semi-linear conservation law with velocity in $L^\infty$, we prove by fixed-point arguments existence and uniqueness result and even in a penalized situation.

**Keywords** Transport equations · Semi-linear PDE · Fixed-point methods · STILS method · Conservation laws · Advection–reaction · Finite-element method · Newton's method · Picard's iteration

## 1.1 Introduction

This chapter deals about semi-linear conservations laws with velocity field in $L^\infty$. Our goal is twofold. On the one hand, the focus is to propose a generalization of space–time integrated least-square (STILS) method introduced by O. Besson and J. Pousin in [1] for linear conservation laws to semi-linear ones. The STILS method has been widely studied in numerous linear cases. Our aim is to introduce a nonlinearity in the source term and look for theoretical methods to prove existence

S. Kane (✉)
Faculté des Sciences et Techniques, Université Cheikh Anta Diop de Dakar, Dakar, Senegal
e-mail: souleye@ucad.edu.sn

S. F. Samb
Département de Mathématiques et Informatique, Université Cheikh Anta Diop de Dakar, Dakar, Senegal
e-mail: serignefallou.samb@ucad.edu.sn

D. Seck
Ecole Doctorale de Mathématiques et Informatique, Université Cheikh Anta Diop de Dakar, Dakar, Senegal
e-mail: diaraf.seck@ucad.edu.sn

 
T. Diagana et al. (eds.), *Partial Differential Equations and Applications*,
Springer Proceedings in Mathematics & Statistics 420,
https://doi.org/10.1007/978-3-031-27661-3_1

and uniqueness results. For this, we shall propose methods combining variational and topological methods.

To reach this aim, we shall use two fixed-point theorems. The first one is the Banach's fixed-point theorem, and the second is due to Schauder. In this latter case, we shall need a penalization argument.

On the other hand, we endeavour to propose numerical methods to analyse semi-linear boundary-value problems. We shall use finite-element methods combined with Picard's iteration and Newton's methods.

Finite-element method is known to produce spurious oscillations and add diffusions in the orthogonal directions of integral curve when convection-dominated problem is solved, see [2] and references therein. To remedy it, the space–time integrated least-square method has been introduced in finite-element context by H. Nguyen and J. Reynen in [3] for solving advection–diffusion equation. And a time-marching approach of STILS has been proposed by O. Besson and G. De Montmollin in [4] for solving numerically linear transport equation using the finite-element method with $div(u) = 0$. To get discrete maximum principle and remove the oscillations produced by the STILS method, J. Pousin et al. in [5] added to the formulation a constraint of positivity and a penalization of the total variation.

Before presenting the organization of our work, let us point out that interesting works on the SILS method have been already realized. We quote some among them closely related to our theoretical works. In fact, it has been used by P.Azerad and O. Besson in [6] to give a coercive variational formulation to the transport equation with a free divergence $C^1$ regular velocity vector field. Existence and uniqueness of space–time least-square solution of linear conservation law with velocity field in $L^\infty$ is proved in [1] by O. Besson and J. Pousin. And in the same paper, these latter deduce a maximum principle result from Stampacchia's theorem and have established the comparison between the least-square solution and the renormalized solution of these equations.

This chapter is organized as follows. In the next section, we shall do the presentation of the problem with some useful mathematical tools for our study. The third section is devoted to the existence and uniqueness results. The main used arguments are fixed-point theorems (Banach–Picard's theorem, Schauder's Theorem). And in the last section, we propose two new numerical methods for computing the solution by using fixed-point algorithm.

## 1.2 Position of the Problem

### *1.2.1 Statement of the Aim and Functional Setting*

Let $\Omega \subset \mathbb{R}^d (d \in \mathbb{N}^*)$ be a domain with a Lipschitz boundary $\partial\Omega$ satisfying the cone property. Let us take $T > 0$, a set $Q = \Omega \times ]0, T[$, and consider an advection velocity $u : Q \to \mathbb{R}^d$ with the following regularity property:

$$u \in L^\infty(Q)^d \text{ with } div\,(u) \in L^\infty(Q).$$

Let $f : \mathbb{R} \longrightarrow \mathbb{R}$ be a function such that $f \in W^{1,\infty}(\mathbb{R})$. In some situations, we can consider $f$ as a $k$-Lipschitz, for $k$ small enough.

The first question we will look is to find a space–time least-square solution for the following boundary-value problem:

$$\begin{cases} \frac{\partial c}{\partial t} + div\,(uc) = f(c) \; in \; Q \\ c(x,0) = c_0(x) \text{ in } \Omega \\ c(x,t) = c_1(x,t) \text{ on } \Gamma_- \end{cases}, \tag{1.1}$$

where

$$\Gamma_- = \{x \in \partial\Omega : (n(x), u(x,t)) < 0; \;\; \forall t \in (0,T)\},$$

and $(.,.)$ is the inner product in $\mathbb{R}^d$, and $n(x)$ is the outer normal to $\partial\Omega$ at point $x$. For the sake of simplicity, one assumes that $\Gamma_-$ does not depend on the time t.

Let us consider

$$u \in L^\infty(Q)^d \text{ such that } div\,(u) \in L^\infty(Q),$$

and set $\widetilde{u} = (1, u_1, u_2, ...., u_d)$ and $\widetilde{n}(x,t)$ the outer normal to $\partial Q$ at $(x,t)$.

We shall use the notation $\mid E \mid$ to mean the Lebesgue measure of a set $E$ throughout this chapter. Let us recall that the space–time incoming flow boundary is given by

$$\partial Q_- = \big\{(x,t) \in \partial Q, (\widetilde{u}(x,t), \widetilde{n}(x,t)) < 0\big\} = \Omega \times \{0\} \cup \Gamma_- \times ]0,T[.$$

The incoming flow boundary condition in space–time is defined as follows:

$$c_b(x,t) = \begin{cases} c_0(x) \text{ if } t = 0 \\ c_1(x,t) \text{ on } \Gamma_- \end{cases}.$$

We introduce the following norm defined by:

1. $\|\phi\|^2 = \|\phi\|^2_{L^2(Q)} + \|\widetilde{div}(\widetilde{u}\phi)\|^2_{L^2(Q)} - \int_{\partial Q_-} \phi^2(\widetilde{u}, \widetilde{n})ds$ for all $\phi \in D(\overline{Q})$.

   where
2. $\widetilde{\nabla}\phi = \left(\frac{\partial \phi}{\partial t}, \frac{\partial \phi}{\partial x_1}, \frac{\partial \phi}{\partial x_1}, ..., \frac{\partial \phi}{\partial x_d}\right)$.
3. $div\,(\widetilde{u}\phi) = \frac{\partial \phi}{\partial t} + \sum_{i=1}^{d} \frac{\partial(\phi u)}{\partial x_i}$.
4. And the Sobolev space $H(u, Q) = \overline{D(\overline{Q})}^{\|.\|}$.

5. Note that if $u$ is regular enough for instance $u \in L^\infty(Q)^d$ with $div(u) \in L^\infty(Q)$, then $H(u, Q) \cap L^\infty = \{\phi \in L^2(Q); \widetilde{div}(\widetilde{u}\phi) \in L^2(Q), \phi_{/\partial Q_-} \in L^2(\partial Q_-; |(\widetilde{u}, \widetilde{n})|)\} \cap L^\infty$ for more details, see [1].

Before proceeding further, let us remind the following theorems that will be useful for our work and for their proofs, and we invite the reader to see [1].

**Theorem 1.2.1** *Let us consider $u \in L^\infty(Q)^d$ with $div(u) \in L^\infty(Q)$. Then the normal trace of $u$ $(\widetilde{u}, \widetilde{n}) \in L^\infty(\partial Q)$.* □

**Theorem 1.2.2** *Let $u \in L^\infty(Q)^d$ with $div(u) \in L^\infty(Q)$. Then there exists a linear continuous trace operator*

$$\begin{aligned} \gamma_{\widetilde{n}} : H(u, Q) &\longrightarrow L^2(\partial Q; (\widetilde{u}, \widetilde{n})) \\ \phi &\longmapsto \phi_{/\partial Q}, \end{aligned}$$

*which can be localized as*

$$\begin{aligned} \gamma_{\widetilde{n}\pm} : H(u, Q) &\longrightarrow L^2(\partial Q_\pm; (\widetilde{u}, \widetilde{n})) \\ \phi &\longmapsto \phi_{/\partial Q_\pm}. \end{aligned}$$

Finally, let us define the spaces

$$H_0(u, Q, \partial Q_-) = \{\phi \in H(u, Q), \phi = 0 \text{ on } \partial Q_-\} = H(u, Q) \cap Ker\gamma_{\widetilde{n}_-}$$

and

$$G_\pm = \gamma_{\widetilde{n}_\pm}(H(u, Q)).$$

Let us give the curved inequality still called curved Poincaré inequality, below that is fundamental and even is the precursor of existence of STILS solution. It has been introduced and proved in [6] for free divergence and extended in [1].

There exists $c_p > 0$ such that for any $\phi \in H(u, Q)$ :

$$\|\phi\|^2_{L^2(Q)} \le c_p^2\left(\|\widetilde{div}(\widetilde{u}\phi)\|^2_{L^2(Q)} - \int_{\partial Q_-} \phi^2(\widetilde{u}, \widetilde{n})\mathrm{d}s\right). \tag{1.2}$$

From the curved inequality, one deduces the following theorem.

**Theorem 1.2.3** *Let $u \in L^\infty(Q)^d$ with $div(u) \in L^\infty(Q)$. Then the semi-norm on $H(u, Q)$ defined by*

$$|\phi|^2_{1,u} = \|\widetilde{div}(\widetilde{u}\phi)\|^2_{L^2(Q)} - \int_{\partial Q_-} \phi^2(\widetilde{u}, \widetilde{n})\mathrm{d}s$$

*is a square of norm, equivalent to the norm defined on $H(u, Q)$.*

Thus $H(u, Q)$ can be equipped by the norm $| \, . \, |_{1,u}$ .

*Remark 1.2.1* In the free-divergence case, one gets that $c_p \leq 2T$ (see, for instance, [6] for additional information). □

### 1.2.2 Space–Time Least-Square and Linear Problem

In this section, we are going to recall the design and some proprieties of STILS method for solving the following linear conservation laws problem:

$$\begin{cases} \frac{\partial c}{\partial t} + div(uc) = f \text{ in } Q \\ c(x,0) = c_0(x) \text{ in } \Omega \\ c(x,t) = c_1(x,t) \text{ on } \Gamma_-. \end{cases} \tag{1.3}$$

The space–time least-square solution of (1.3) corresponds to a minimizer in

$$\{\phi \in H(u, Q); \gamma_{\widetilde{n}_-}(\phi) = c_b\}$$

of the following convex, $H(u, Q)$-coercive functional defined by

$$J(c) = \frac{1}{2}\Big( \int_Q (\widetilde{div}(\widetilde{u}c) - f)^2 \mathrm{d}x\mathrm{d}t - \int_{\partial Q_-} c^2(\widetilde{u}, \widetilde{n})\mathrm{d}s \Big). \tag{1.4}$$

The Gâteaux differential of $J$ yields

$$D[J(c)].\phi = \int_Q (\widetilde{div}(\widetilde{u}c) - f)\widetilde{div}(\widetilde{u}\phi)\mathrm{d}x\mathrm{d}t - \int_{\partial Q_-} c\phi(\widetilde{u}, \widetilde{n})\mathrm{d}s. \tag{1.5}$$

Thus, if $c_b \in G_-$, the space–time least-square formulation of (1.3) is expressed as follows:

$$\int_Q \widetilde{div}(\widetilde{u}c)\widetilde{div}(\widetilde{u}\phi)\mathrm{d}x\mathrm{d}t = \int_Q f\widetilde{div}(\widetilde{u}\phi)\mathrm{d}x\mathrm{d}t \ \forall \ \phi \in H_0(u, Q) \tag{1.6}$$

and

$$\gamma_{\widetilde{n}_-}(c) = c_b.$$

For more details, see [1, 5].

Thanks to Theorem 1.2.2, we can reduce the problem (1.6) in a homogeneous one in $\partial Q_-$. For $c_b \in G_-$, let $C_b \in H(u, Q)$ such that $\gamma_{\widetilde{n}_-}(C_b) = c_b$; then $\rho = c - C_b$ is the unique solution of

$$\int_Q \widetilde{div}(\widetilde{u}\rho)\widetilde{div}(\widetilde{u}\phi)\mathrm{d}x\mathrm{d}t = \int_Q (f - \widetilde{div}(\widetilde{u}C_b))\widetilde{div}(\widetilde{u}\phi)\mathrm{d}x\mathrm{d}t \ \forall \ \phi \in H_0(u, Q). \tag{1.7}$$

Finally, let us recall the following theorem proved in [1].

**Theorem 1.2.4** *For $u \in L^\infty(Q)^d$ with $div(u) \in L^\infty(Q)$, $c_b \in G_-$, and $f \in L^2(Q)$, the problem (1.7) has a unique solution. Moreover,*

$$| \rho |_{1,u} \leq \|f\|_{L^2(Q)} + \|\widetilde{div}(\widetilde{u}C_b)\|_{L^2(Q)},$$

*and the function $c = \rho + C_b$ is the space–time least-square solution of (1.3).* □

### 1.2.3 Space–Time Least-Square and Semi-linear Problem

This last subsection is devoted to introducing a variational formulation (1.1). Otherwise, our aim is to find $c \in H(u, Q)$ such that

$$\int_Q \widetilde{div}(\widetilde{u}c)\widetilde{div}(\widetilde{u}\phi)\mathrm{d}x\mathrm{d}t = \int_Q f(c)\widetilde{div}(\widetilde{u}\phi)\mathrm{d}x\mathrm{d}t \ \forall \ \phi \in H_0(u, Q, \partial Q_-) \tag{1.8}$$

and

$$\gamma_{\widetilde{n}_-}(c) = c_b. \tag{1.9}$$

It is important to stress that the above formulation is nonlinear. And we shall propose fixed-point methods to study it. Let us recall that there are at least three distinct classes of such abstract theorems that are useful for proving existence results in a wide family of partial differential equations. These classes are:

- Fixed-point theorems for strict contractions
- Fixed-point theorems for compact mappings
- Fixed-point theorems for order-preserving operators

We shall use in the following the first two types.

## 1.3 Existence and Qualitative Results

### 1.3.1 Existence and Uniqueness

In this section, we shall study the problem (1.8) by establishing and proving existence and uniqueness theorems for the STILS solution. These results are deduced thanks to the fixed-point theory, namely the Banach–Picard and Schauder theorems.

At first, in the case where $f$ is $k$-Lipschitz with $k$ is small enough that will be precised and by using the Banach–Picard fixed-point theorem [7], we have the following existence and uniqueness theorem of STILS solution.

**Theorem 1.3.1** *Let $u \in L^\infty(Q)$ with $div(u) \in L^\infty(Q)$, and $c_b \in G_-$, $f$ be $k$-Lipschitz in $\mathbb{R}$ with $k < \frac{1}{c_p}$. Then the problem (1.8)–(1.9) has a unique solution.* □

***Proof*** Let us consider

$$\mathbb{H} = \left\{\phi \in H(u, Q), \gamma_{\vec{n}_-}(\phi) = c_b\right\}.$$

For all $\rho \in \mathbb{H}, f(\rho) \in L^2(Q)$, then, by Theorem 1.2.4, there exists a unique element $c \in \mathbb{H}$ satisfying:

$$\int_Q \widetilde{div}(\widetilde{u}c)\widetilde{div}(\widetilde{u}\phi)\mathrm{d}x\mathrm{d}t = \int_Q f(\rho)\widetilde{div}(\widetilde{u}\phi)\mathrm{d}x\mathrm{d}t \tag{1.10}$$

for all $\phi \in H_0(u, Q, \partial Q_-)$.

Let us define

$$T : \mathbb{H} \to \mathbb{H} \tag{1.11}$$

such that

$$T(\rho) = c; \tag{1.12}$$

thus a solution of the nonlinear problem (1.8)–(1.9) is a fixed point of T.

Let $\rho_1, \rho_2 \in \mathbb{H}$ and $c_1 = T(\rho_1)$, $c_2 = (T\rho_2)$.

Since $c_1 - c_2 = 0$ on $\partial Q_-$,

$$\begin{aligned}| c_1 - c_2 |_{1,u}^2 = \|\widetilde{div}(\widetilde{u}(c_1 - c_2))\|_{L^2(Q)}^2 = \int_Q \widetilde{div}(\widetilde{u}(c_1 - c_2))\widetilde{div}(\widetilde{u}(c_1))\mathrm{d}x\mathrm{d}t \\ - \int_Q \widetilde{div}(\widetilde{u}(c_1 - c_2))\widetilde{div}(\widetilde{u}(c_2))\mathrm{d}x\mathrm{d}t.\end{aligned}$$

For $c_1 = T(\rho_1)$ and $c_2 = T(\rho_2)$, we have

$$\int_Q \widetilde{div}(\widetilde{u}c_1)\widetilde{div}(\widetilde{u}(c_1 - c_2))\mathrm{d}x\mathrm{d}t = \int_Q f(\rho_1)\widetilde{div}(\widetilde{u}(c_1 - c_2))\mathrm{d}x\mathrm{d}t$$

and

$$\int_Q \widetilde{div}(\widetilde{u}c_2)\widetilde{div}(\widetilde{u}(c_1 - c_2))\mathrm{d}x\mathrm{d}t = \int_Q f(\rho_2)\widetilde{div}(\widetilde{u}(c_1 - c_2))\mathrm{d}x\mathrm{d}t.$$

Then a computation yields

$$| c_1 - c_2 |_{1,u}^2 = \int_Q f(\rho_1)\widetilde{div}(\widetilde{u}(c_1 - c_2))\mathrm{d}x\mathrm{d}t - \int_Q f(\rho_2)\widetilde{div}(\widetilde{u}(c_1 - c_2))\mathrm{d}x\mathrm{d}t.$$

$$| c_1 - c_2 |_{1,u}^2 = \int_Q (f(\rho_1) - f(\rho_2))\widetilde{div}(\widetilde{u}(c_1 - c_2))\mathrm{d}x\mathrm{d}t.$$

By Young's inequality, we get

$$| c_1 - c_2 |_{1,u}^2 \leq \|f(\rho_1) - f(\rho_2)\|_{L^2(Q)} \|\widetilde{div}(\widetilde{u}(c_1 - c_2))\|_{L^2(Q)}.$$

Since $f$ is $k$-Lipschitz in $\mathbb{R}$ and $| c_1 - c_2 |_{1,u} = \|\widetilde{div}(\widetilde{u}(c_1 - c_2))\|_{L^2(Q)}$, we have

$$| c_1 - c_2 |_{1,u}^2 \leq k\|\rho_1 - \rho_2\|_{L^2(Q)}^2 \; | c_1 - c_2 |_{1,u},$$

and hence,

$$| c_1 - c_2 |_{1,u}^2 \leq kc_p \; | \rho_1 - \rho_2 |_{1,u} | c_1 - c_2 |_{1,u} .$$

Finally, we get

$$| T(\rho_1) - T(\rho_2) |_{1,u} \leq kc_p \; | \rho_1 - \rho_2 |_{1,u} .$$

Thus $T$ is a strict contraction, provided that $kc_p < 1$. The Banach's fixed-point theorem ensures the existence and uniqueness of $c \in \mathbb{H}$ with $T(c) = c$ that solves (1.8)–(1.9). □

*Remark 1.3.1* In the free-divergence case, the previous assumption gives $k < \frac{1}{2T}$; thus we get a solution for small times. But it cannot be extended because of the loss of continuity.

The constant $c_p$ is not optimal (see [1] for more details). And so, the condition $kc_p < 1$ could be improved. □

Now, let us state and prove the following technical lemmas that will be key steps in the building of the next existence theorem.

**Lemma 1.3.1** *There is a positive constant $C > 0$ such that for any $\phi \in D(\overline{Q})$ verifying $\phi = 0$ on $\partial Q_-$, we have $\|\widetilde{\nabla}\phi\|_{L^2(Q)^{d+1}} \leq C\|\widetilde{div}(\widetilde{u}\phi)\|_{L^2(Q)}$.* □

***Proof*** Let us suppose that the inequality is false. Then for any integer $n \in \mathbb{N}$, there is $\phi_n \in D(\overline{Q})$ such that:

$$\|\widetilde{\nabla}\phi_n\|_{L^2(Q)^{d+1}} > n\|\widetilde{div}(\widetilde{u}\phi_n)\|_{L^2(Q)}. \tag{1.13}$$

If $n$ is such that $\|\widetilde{\nabla}\phi_n\|_{L^2(Q)^{d+1}} = 0$, then $\|\widetilde{\nabla}\phi_n\|_{L^2(Q)^{d+1}} = n\|\widetilde{div}(\widetilde{u}\phi_n)\|_{L^2(Q)} = 0$, which is a contradiction with (1.13).

Now dividing (1.13) by $\|\widetilde{\nabla}\phi_n\|_{L^2(Q)^{d+1}}$, we have

$$\|\widetilde{\nabla}\frac{\phi_n}{\|\widetilde{\nabla}\phi_n\|}\|_{L^2(Q)^{d+1}} > n\|\widetilde{div}(\widetilde{u}\frac{\phi_n}{\|\widetilde{\nabla}\phi_n\|})\|_{L^2(Q)}. \tag{1.14}$$

Setting $\theta_n = \frac{\phi_n}{\|\widetilde{\nabla}\phi_n\|_{L^2(Q)^{d+1}}}$, we obtain

$$\|\widetilde{\nabla}\theta_n\|_{L^2(Q)^{d+1}} = 1 \tag{1.15}$$

and

$$\|\widetilde{div}(\widetilde{u}\theta_n)\|_{L^2(Q)} = \|\widetilde{div}\left(\widetilde{u}\frac{\phi_n}{\|\widetilde{\nabla}\phi_n\|}\right)\|_{L^2(Q)}. \tag{1.16}$$

Thanks to (1.15) and (1.16), the inequality (1.14) can be written as follows:

$$\|\widetilde{div}(\widetilde{u}\theta_n)\|_{L^2(Q)} < \frac{1}{n}. \tag{1.17}$$

By curved inequality (also named curved Poincaré inequality), we get existence of a positive constant $A > 0$ such that:

$$\|\theta_n\|_{L^2(Q)} \leq \sqrt{A}\|\widetilde{div}(\widetilde{u}\theta_n)\|_{L^2(Q)},$$

and then

$$\|\theta_n\|_{L^2(Q)} \leq \frac{\sqrt{A}}{n}. \tag{1.18}$$

This implies that

$$\theta_n \longrightarrow 0 \text{ in } L^2(Q). \tag{1.19}$$

From (1.15) and (1.18), one deduces that $(\theta_n)$ is bounded in $H^1(Q)$. Then there is a convex combination of the sequence $(\theta_n)$ that converges to $\theta^* \in H^1(Q)$ weakly, and so in $L^2(Q)$ too. Using (1.19), this convex combination converges to 0 in $L^2(Q)$. Thanks to the uniqueness of the limit, we have $\theta^* = 0$.

As a sum up, one sees that (1.13) yields existence of a sequence $(\theta_n)_n \subset D(\overline{Q}) \subset H^1(Q)$ satisfying:

$$\begin{cases} \theta_n \longrightarrow 0 \text{ weakly in } H^1(Q) \ (i) \\ \|\widetilde{\nabla}\theta_n\|_{L^2(Q)^{d+1}} = 1 \text{ for any } n \in \mathbb{N} \ (ii) \end{cases}. \tag{1.20}$$

(i) implies that $\widetilde{\nabla}\theta_n \longrightarrow 0$ weakly in $L^2(Q)$.

Let $\psi \in L^2(Q)^{d+1}$ such that $\|\psi\|_{L^2(Q)^{d+1}} = 1$.

We have $(\psi, \widetilde{\nabla}\theta_n) \longrightarrow 0$ in $\mathbb{R}$.

The translation of the definition of the limit allows us to write:

$\exists n_0 \in \mathbb{N}$ such that for any $n \geq n_0$, we have $|(\psi, \widetilde{\nabla}\theta_n)| < 1$.

Thus we get $\underset{\|\psi\|_{L^2(Q)^{d+1}}=1}{\text{Sup}} |(\psi, \widetilde{\nabla}\theta_n)| < 1$.

Hence, one deduces that $\|\widetilde{\nabla}\theta_n\|_{L^2(Q)^{d+1}} < 1$ for any $n \geq n_0$ : what is in contradiction with (ii). □

**Lemma 1.3.2** *Let $f : \mathbb{R} \longrightarrow \mathbb{R}$ be a k-Lipschitzian function.*

*For any $\rho \in H(u, Q)$, we have $f(\rho) \in H^1(Q)$. In addition, there exists a positive $C > 0$ such that*

$$\|\widetilde{\nabla} f(\rho)\|_{L^2(Q)^{d+1}} \leq C\|\widetilde{div}(\widetilde{u}\rho)\|_{L^2(Q)}.$$

***Proof*** Let $\rho \in H(u, Q)$; then there is a sequence $(\rho_n) \subset D(\overline{Q})$ that converges to $\rho$ in $H(u, Q)$.

Since $f$ is $k$-Lipschitzian, we get

$| f(\rho_n) | \leq k | \rho_n | + | f(0) |$ and $| f(\rho) | \leq k | \rho | + | f(0) |$.

Therefore, $(f(\rho_n)) \subset L^2(Q)$ and $f(\rho) \in L^2(Q)$. In addition: $\|f(\rho_n) - f(\rho)\|_{L^2(Q)} \leq k\|\rho_n - \rho\|_{L^2(Q)}$, and $\rho_n$ converges to $\rho$ in $L^2$; thus $f(\rho_n)$ converges to $f(\rho)$ in $L^2$. And in particular, any convex combination of $f(\rho_n)$ converges to $f(\rho)$ in $L^2$.

Now let us take $x, y$ in $Q$.

$$| f(\rho_n(x)) - f(\rho_n(y)) | \leq k | \rho_n(x) - \rho_n(y) | \tag{1.21}$$

$$| f(\rho_n(x)) - f(\rho_n(y)) | \leq k | \nabla\rho_n |_\infty | x - y | .$$

Under Rademacher's theorem, for any integer $n$, the function $f(\rho_n)$ is differentiable almost everywhere, and there is a positive constant depending on $n$, $C_n$ such that $| \frac{\partial f(\rho_n)}{\partial x_i} | \leq C_n$; then $\frac{\partial f(\rho_n)}{\partial x_i} \in L^2(Q)$ for any $i = 1, ..., d+1$.

Using again the inequality (1.21), one sees that

$$| \frac{\partial f(\rho_n)}{\partial x_i} | \leq k | \frac{\partial \rho_n}{\partial x_i} | \text{ for any } i = 1, ..., d+1,$$

and then

$$\|\widetilde{\nabla} f(\rho_n)\|_{L^2(Q)^{d+1}} \leq k\|\widetilde{\nabla}\rho_n\|_{L^2(Q)^{d+1}}.$$

By Lemma 1.3.1, we have $\|\widetilde{\nabla}\rho_n\|_{L^2(Q)^{d+1}} \leq C\|\widetilde{div}(\widetilde{u}\rho_n)\|_{L^2(Q)}$.
This yields

$$\|\widetilde{\nabla} f(\rho_n)\|_{L^2(Q)^{d+1}} \leq kC\|\widetilde{div}(\widetilde{u}\rho_n)\|_{L^2(Q)}. \tag{1.22}$$

Since $(\rho_n)$ converges to $\rho$ in $H(u, Q)$, we get $\|\widetilde{div}(\widetilde{u}\rho_n)\|_{L^2(Q)}$ converges to $\|\widetilde{div}(\widetilde{u}\rho)\|_{L^2(Q)}$. And we can conclude that $(f(\rho_n))$ is bounded in $H^1(Q)$.

And more, we have $(f(\rho_n))$ is bounded in $H^1(Q)$. Then there is $\theta \in H^1(Q)$ such that $(f(\rho_n))$ converges to $\theta$ weakly. Thanks to Mazur's lemma, there is a convex combination of the sequence $(f(\rho_n))$, denoted $\theta_n$ that strongly converges to $\theta$ in $H^1(Q)$and then in $L^2$. And the same convex combination converges to $f(\rho)$ in $L^2(Q)$.

Under uniqueness in $L^2(Q)$, we have $f(\rho) = \theta$ but $\theta \in H^1(Q)$. This ensures us that $f(\rho) \in H^1(Q)$.

Passing to the limit, the inequality (1.22) yields

$$\|\widetilde{\nabla} f(\rho)\|_{L^2(Q)^{d+1}} \leq kC\|\widetilde{div}(\widetilde{u}\rho)\|_{L^2(Q)}.$$

□

**Lemma 1.3.3** *Let $f : \mathbb{R} \longrightarrow \mathbb{R}$ be a k-Lipschitzian function, $C$ be a bounded subset of $H(u, Q)$, and $(\rho_n)$, $(c_n)$ be sequences in $C$. Denoting by $c$ the weak limit of $(c_n)$ in $H_0(u, Q)$. We have*

$$\int_Q f(\rho_n)\widetilde{div}(\widetilde{u}(c_n - c))\mathrm{d}x\mathrm{d}t \longrightarrow 0.$$

***Proof*** Since $C \subset H(u, Q)$, $(\rho_n)$, $(c_n)$ are sequences of $C$, there are $M > 0$ and $c \in H(u, Q)$ such that

$$\|\widetilde{div}(\widetilde{u}\rho_n)\|_{L^2(Q)} \leq M \tag{1.23}$$

and

$$c_n \rightharpoonup c \text{ faiblement dans } H(u, Q). \tag{1.24}$$

Using the curved Poincaré inequality (1.23), we have

$$\|\rho_n\|_{L^2(Q)} \leq M\sqrt{A} \tag{1.25}$$

$$| f(\rho_n) |\leq k | \rho_n | + | f(0) | . \tag{1.26}$$

Thus (1.25)–(1.26) yield a constant $C_{(1.27)}$ such that:

$$\|f(\rho_n)\|_{L^2(Q)} \leq C_{(1.27)}. \tag{1.27}$$

In another way, by Lemma 1.3.2, there exists a constant $C > 0$ such that

$$\|\widetilde{\nabla} f(\rho_n)\|_{L^2(Q)^{d+1}} \leq C\|\widetilde{div}(\widetilde{u}\rho_n)\|_{L^2(Q)}. \tag{1.28}$$

From (1.23) and (1.28), we have the following estimation

$$\|\widetilde{\nabla} f(\rho_n)\|_{L^2(Q)^{d+1}} \leq CM. \tag{1.29}$$

Relations (1.27) and (1.29) imply that the sequence $(f(\rho_n))$ is bounded in $H^1(Q)$. Then, by Rellich's theorem, even if it means extracting a subsequence, there is $F \in L^2(Q)$ such that

$$f(\rho_n) \longrightarrow F \text{ strongly in } L^2(Q). \tag{1.30}$$

From (1.24) and (1.30), we get

$$\langle f(\rho_n), \widetilde{div}(\widetilde{u}(c_n - c))\rangle \longrightarrow (F, 0) = 0. \tag{1.31}$$

Finally, we have

$$\int_Q f(\rho_n)\widetilde{div}(\widetilde{u}(c_n - c))\mathrm{d}x\mathrm{d}t \longrightarrow 0.$$

□

Having at hands these lemmas and using fixed Schauder's theorem, we can proceed further to get existence and uniqueness results.

**Theorem 1.3.2** *Let $u \in L^\infty(Q)$ with $div(u) \in L^\infty(Q)$ and $c_b \in G_-$, $f \in W^{1,\infty}(\mathbb{R})$. Then the problem (1.8)–(1.9) has a solution in $H_0(u, Q, \partial Q_-)$.*

***Proof*** Since $c_b \in G_-$ changing the source term if necessary, we shall assume that $c_b = 0$ on $\partial Q_-$.

Existence.

The proof is relied mainly on the Schauder's fixed theorem.

Step 1: We first have to choose a bounded subset $\mathbb{X}$ of $H_0(u, Q, \partial Q_-)$ and a mapping $T : \mathbb{X} \to \mathbb{X}$. To achieve this aim, for all $\rho \in V$, under Lemma 1.3.2, or since $f \in W^{1,\infty}(\mathbb{R})$, we have $f(\rho) \in L^2(Q)$. Then by Theorem 1.2.4, there exists a function $c \in H_0(u, Q, \partial Q_-)$ such that

$$\int_Q \widetilde{div}(\widetilde{u}c)\widetilde{div}(\widetilde{u}\phi)\mathrm{d}x\mathrm{d}t = \int_Q f(\rho)\widetilde{div}(\widetilde{u}\phi)\mathrm{d}x\mathrm{d}t \text{ for all } \phi \in H_0(u, Q, \partial Q_-).$$

Moreover, $| c |_{1,v} \leq \|f(\rho)\|_{L^2(Q)}$.

Since $f \in W^{1,\infty}(\mathbb{R})$, we have $\mid c \mid_{1,u} \leq \mid f \mid_{L^\infty} \mid Q \mid^{\frac{1}{2}}$ .
Let us define $T : H_0(u, Q, \partial Q_-) \to H_0(u, Q, \partial Q_-)$ such that $c = T(\rho)$.
Solving (1.39) is equivalent to show the existence of fixed-point theorem of $T$.
Let us proceed further and choose a convex set $\mathbb{X}$ as follows:

$$\mathbb{X} = \{\phi \in H_0(u, Q, \partial Q_-), \mid \phi \mid_{1,u} \leq M\}$$

when $M$ is to be precised later.

$$\mid T\rho \mid_{1,u} = \mid c \mid_{1,u} \leq \mid f \mid_{L^\infty} \mid Q \mid^{\frac{1}{2}}, \text{ for all } \rho \in \mathbb{X}.$$

Thus, choosing $M = \mid f \mid_{L^\infty} \mid Q \mid^{\frac{1}{2}}$, the following inclusion yields

$$T(H_0(u, Q, \partial Q_-)) \subset \mathbb{X}$$

and then

$$T(\mathbb{X}) \subset \mathbb{X}.$$

So we will consider $T : \mathbb{X} \to \mathbb{X}$.
Step 2: Thus T is continuous. □

***Proof of Step 2*** Then a computation yields

$$\mid c_1 - c_2 \mid^2_{1,u} = \int_Q f(\rho_1)\widetilde{div}(\widetilde{u}(c_1 - c_2))\mathrm{d}x\mathrm{d}t - \int_Q f(\rho_2)\widetilde{div}(\widetilde{u}(c_1 - c_2))\mathrm{d}x\mathrm{d}t$$

$$\mid c_1 - c_2 \mid^2_{1,u} = \int_Q (f(\rho_1) - f(\rho_2))\widetilde{div}(\widetilde{u}(c_1 - c_2))\mathrm{d}x\mathrm{d}t.$$

By Young's inequality, we get

$$\mid c_1 - c_2 \mid^2_{1,u} \leq \|f(\rho_1) - f(\rho_2)\|_{L^2(Q)} \|\widetilde{div}(\widetilde{u}(c_1 - c_2))\|_{L^2(Q)}.$$

Since $f \in W^{1,\infty}(\mathbb{R})$, we have

$$\|f(\rho_1) - f(\rho_2)\|_{L^2(Q)} \leq \mid f' \mid_{L^\infty} \mid \|\rho_1 - \rho_2\|^2_{L^2(Q)},$$

and hence,

$$\mid c_1 - c_2 \mid^2_{1,u} \leq \mid f' \mid_{L^\infty} c_p \mid \rho_1 - \rho_2 \mid_{1,u} \mid c_1 - c_2 \mid_{1,u};$$

finally, we get

$$\mid T\rho_1 - T\rho_2 \mid_{1,u} \leq \mid f' \mid_{L^\infty} c_p \mid \rho_1 - \rho_2 \mid_{1,u} .$$

Thus $T$ is Lipschitz so continuous. □

Step 3: $\mathbb{X}$ is a subset convex, closed in $H_0(u, Q, \partial Q_-)$, and $T(\mathbb{X})$ compact in $L^2(Q)$.

***Proof of Step 3*** It is clear that $\mathbb{X}$ is convex and closed in $H_0(u, Q, \partial Q_-)$.

Let $(c_n)$ be sequences in $T(\mathbb{X})$; then there exists $(\rho_n)$ sequence in $H_0(u, Q, \partial Q_-)$ such that

$$\int_Q \widetilde{div}(\widetilde{u}c_n)\widetilde{div}(\widetilde{u}\phi)\mathrm{d}x\mathrm{d}t = \int_Q f(\rho_n)\widetilde{div}(\widetilde{u}\phi)\mathrm{d}x\mathrm{d}t \ \forall \phi \in H_0(v, Q, \partial Q_-). \tag{1.32}$$

Since $(c_n)$ bounded in $H_0(u, Q, \partial Q_-)$, then there exists $c \in H_0(u, Q, \partial Q_-)$ such that

$$c_n \rightharpoonup c \text{ weakly in } H_0(u, Q, \partial Q_-),$$

then $\widetilde{div}(\widetilde{u}(c_n - c)) \rightharpoonup 0$ weakly in $L^2(Q)$, and in particular,

$$\int_Q \widetilde{div}(\widetilde{u}(c_n - c))\widetilde{div}(\widetilde{u}(c))\mathrm{d}x\mathrm{d}t \longrightarrow 0 \tag{1.33}$$

$$\begin{aligned} |\, c_n - c\,|_{1,u}^2 = \|\widetilde{div}(\widetilde{u}(c_n - c))\|_{L^2(Q)}^2 &= \int_Q \widetilde{div}(\widetilde{u}(c_n - c))\widetilde{div}(\widetilde{u}(c_n))\mathrm{d}x\mathrm{d}t \\ &- \int_Q \widetilde{div}(\widetilde{u}(c_n - c))\widetilde{div}(\widetilde{u}(c))\mathrm{d}x\mathrm{d}t. \end{aligned} \tag{1.34}$$

Using (1.32), we have

$$|\, c_n - c\,|_{1,u}^2 = \int_Q f(\rho_n)\widetilde{div}(\widetilde{u}(c_n - c))\mathrm{d}x\mathrm{d}t - \int_Q \widetilde{div}(\widetilde{u}(c_n - c))\widetilde{div}(\widetilde{u}(c))\mathrm{d}x\mathrm{d}t.$$

And by Lemma 1.3.3, even if it means extracting a subsequence, we have

$$\int_Q f(\rho_n)\widetilde{div}(\widetilde{u}(c_n - c))\mathrm{d}x\mathrm{d}t \longrightarrow 0. \tag{1.35}$$

Equations (1.33) and (1.35) imply that

$$|\, c_n - c\,|_{1,u}^2 \longrightarrow 0.$$

□

Since $\mathbb{X}$ is convex, closed in $H_0(u, Q, \partial Q_-)$, and $T : \mathbb{X} \to \mathbb{X}$ continuous, $T(\mathbb{X})$ is relatively compact in $H_0(u, Q, \partial Q_-)$. By Schauder's theorem, T has a fixed point.

## 1.4 Existence and Uniqueness Result for the Penalization Version

Let us consider the space

$$(i)\ \mathbb{V} = H_0(u, Q, \partial Q_-) \cap H^1(Q),$$

where $H^1(Q)$ is the usual Sobolev spaces, with the norm

$$(ii)\ \|\phi\|_{\mathbb{V}}^2 = \|\phi\|_{L^2(Q)}^2 + \|\widetilde{div}(\widetilde{u}\phi)\|_{L^2(Q)}^2 + \|\widetilde{\nabla}\phi\|_{L^2(Q)}^2.$$

From the curved inequality (1.2), one deduces that the following semi-norm

$$\mid \phi \mid_{\mathbb{V}} = (\|\widetilde{div}(\widetilde{u}\phi)\|_{L^2(Q)}^2 + \|\widetilde{\nabla}\phi\|_{L^2(Q)}^2)^{\frac{1}{2}}$$

becomes a norm, equivalent to the norm given on $\mathbb{V}$. And the space $\mathbb{V}$ will be equipped with the norm $|\ .\ |_{\mathbb{V}}$ .

For any $\lambda \in \mathbb{R}_+$ and $f \in L^2(Q)$, we are going to study the following optimization problem:

$$\rho_\lambda = \underset{c\in\mathbb{V}}{\text{Argmin}}\ J(c) + \lambda\|\widetilde{\nabla}c\|_{L^2(Q)}^2 = \underset{c\in\mathbb{V}}{\text{Argmin}}\ J_\lambda(c), \tag{1.36}$$

where

$$J(c) = \frac{1}{2}\left(\int_Q (\widetilde{div}(\widetilde{u}c) - f)^2 \mathrm{d}x\mathrm{d}t\right).$$

**Proposition 1.4.1** *For any non-negative real number $\lambda$ and $f \in L^2(Q)$, the problem (1.36) has a unique solution.*

*Moreover, for any $\lambda \geq 1$, there exists $\alpha := \alpha(\lambda)$ such that $\mid c \mid_{\mathbb{V}} \leq \alpha \| f \|_{L^2(Q)}$.*

***Proof*** Since $J_\lambda$ is strictly convex and Gâteaux-differentiable, we have to show that there is a function $c \in \mathbb{V}$ such that $DJ_\lambda(c).\phi = 0$ for all $\phi \in \mathbb{V}$.

An easy computation gives

$$DJ_\lambda(c).\phi = \int_Q (\widetilde{div}(\widetilde{u}c) - f)\widetilde{div}(\widetilde{u}\phi)\mathrm{d}x\mathrm{d}t + \lambda \int_Q \widetilde{\nabla}c\widetilde{\nabla}\phi \mathrm{d}x\mathrm{d}t. \tag{1.37}$$

And we obtain the following weak formulation:

$$\int_Q \widetilde{div}(\widetilde{u}c)\widetilde{div}(\widetilde{u}\widetilde{\nabla}\phi)\mathrm{d}x\mathrm{d}t + \lambda \int_Q \widetilde{\nabla}c\widetilde{\nabla}\phi \mathrm{d}x\mathrm{d}t = \int_Q f\widetilde{div}(\widetilde{u}\phi)\mathrm{d}x\mathrm{d}t \tag{1.38}$$

for all $\phi \in \mathbb{V}$.

Let us now consider the bilinear form $a_\lambda(.,.) : \mathbb{V} \times \mathbb{V} \to \mathbb{R}$ defined for all $\phi$ , $\psi \in \mathbb{V}$ by

$$a_\lambda(\phi, \psi) = \int_Q \widetilde{div}(\widetilde{u}\phi)\widetilde{div}(\widetilde{u}\psi)\mathrm{d}x\mathrm{d}t + \lambda \int_Q \widetilde{\nabla}\phi\widetilde{\nabla}\psi\mathrm{d}x\mathrm{d}t$$

and the linear form $L : \mathbb{V} \to \mathbb{R}$ defined for all $\phi \in \mathbb{V}$ by:

$$L(\phi) = \int_Q f\widetilde{div}(\widetilde{u}\phi)\mathrm{d}x\mathrm{d}t.$$

Thus the expression (1.36) can be written as follows: find $c \in \mathbb{V}$ such that

$$a_\lambda(c, \phi) = L(\phi) \text{ for all } \phi \in \mathbb{V}.$$

Taking $m = \min(\lambda, 1) > 0$, we have

$$a_\lambda(\phi, \phi) = \int_Q \widetilde{div}(\widetilde{u}\phi)^2\mathrm{d}x\mathrm{d}t + \lambda \int_Q | \widetilde{\nabla}\phi |^2 \mathrm{d}x\mathrm{d}t \geq m \mid \phi \mid_{\mathbb{V}}^2 .$$

Then $a(.,.)_\lambda$ is $\mathbb{V}$ elliptic on the one hand.
On the other hand, by using Holder's inequality, we have

$$| a_\lambda(\phi, \psi) | \leq \|\widetilde{div}(\widetilde{u}\phi)\|_{L^2(Q)}\|\widetilde{div}(\widetilde{u}\psi)\|_{L^2(Q)} + \lambda\|\nabla\phi\|_{L^2(Q)}\|\nabla\psi\|_{L^2(Q)}.$$

And the following estimate holds

$$\begin{aligned} | a_\lambda(\phi, \psi) | \leq \max(\lambda, 1)(&\|\widetilde{div}(\widetilde{u}\phi)\|_{L^2(Q)}\|\widetilde{div}(\widetilde{u}\psi)\|_{L^2(Q)} \\ &+ \|\nabla\phi\|_{L^2(Q)}\|\nabla\psi\|_{L^2(Q)}). \end{aligned}$$

By taking $C = \max(\lambda, 1)$ and using Cauchy–Schwarz's inequality in $\mathbb{R}^2$, we have

$$| a_\lambda(\phi, \psi) | \leq C \mid \phi \mid_{\mathbb{V}} \mid \psi \mid_{\mathbb{V}}, \text{ for all } \phi, \psi \in \mathbb{V}.$$

And we conclude that $a_\lambda(.,.)$ is continuous.
Let us now prove that $L$ is continuous.

$$| L(\phi) | \leq \|f\|_{L^2(Q)}\|\widetilde{div}(\widetilde{u}\phi)\|_{L^2(Q)}$$

so,

$$| L(\phi) | \leq \|f\|_{L^2(Q)} \mid \phi \mid_{\mathbb{V}} .$$

Since $L$ is linear with respect to $\phi$, we get its continuity.

Hence by the Lax–Milgram's theorem, there is a unique solution of (1.36) that satisfies

$$\min(1,\lambda) \mid c_\lambda \mid_{\mathbb{V}}^2 \leq a_\lambda(c_\lambda, c_\lambda) = \mid L(c_\lambda) \mid \leq \|f\|_{L^2(Q)} \mid c_\lambda \mid_{\mathbb{V}} .$$

So for $\lambda \geq 1$, we get the desired result $\mid c \mid_{\mathbb{V}} \leq \|f\|_{L^2(Q)}$.

□

**Theorem 1.4.1** *Let* $\lambda > 1$ *and* $f \in W^{1,\infty}(\mathbb{R})$. *Then there exists function* $c_\lambda \in \mathbb{V}$ *such that*

$$\int_Q \widetilde{div}(\widetilde{u}c_\lambda)\widetilde{div}(\widetilde{u}\phi)\mathrm{d}x\mathrm{d}t + \lambda \int_Q \widetilde{\nabla} c_\lambda \widetilde{\nabla}\phi \mathrm{d}x\mathrm{d}t = \int_Q f(c_\lambda)\widetilde{div}(\widetilde{u}\phi)\mathrm{d}x\mathrm{d}t \text{ for all } \phi \in \mathbb{V} \tag{1.39}$$

*for all* $\phi \in \mathbb{V}$.

*The solution is unique if* $\lambda > 2T^2 \mid f' \mid_{L^\infty(\mathbb{R})}^2 \|\widetilde{u}\|_{L^2(Q)}^2$ *and* $div(u) = 0$. □

***Proof***

**A-Existence:**

The proof is relied mainly on the Schauder's fixed theorem.

Step 1: We first have to choose a bounded subset $\mathbb{X}$ of $\mathbb{V}$ and a mapping $T : \mathbb{X} \to \mathbb{X}$. To achieve this aim, for all $\rho \in V$, since $f \in W^{1,\infty}(Q)$, $f(\rho) \in L^2(Q)$, then by Proposition 1.4.1, there exists a function $c_\lambda \in \mathbb{V}$ such that

$$\int_Q \widetilde{div}(\widetilde{u}c_\lambda)\widetilde{div}(\widetilde{u}\phi)\mathrm{d}x\mathrm{d}t + \int_Q \widetilde{\nabla} c_\lambda \widetilde{\nabla}\phi \mathrm{d}x\mathrm{d}t = \int_Q f(\rho)\widetilde{div}(\widetilde{u}\phi)\mathrm{d}x\mathrm{d}t \text{ for all } \phi \in \mathbb{V}.$$

Moreover, $\mid c_\lambda \mid_{\mathbb{V}} \leq \|f(\rho)\|_{L^2(Q)}$.

Since $f \in W^{1,\infty}(\mathbb{R})$, we have $\mid c_\lambda \mid_{\mathbb{V}} \leq \mid f \mid_{L^\infty} \mid Q \mid^{\frac{1}{2}}$ .

Let us define $T : \mathbb{V} \to \mathbb{V}$ such that $c_\lambda = T(\rho)$.

Solving (1.39) is equivalent to showing the existence of fixed-point theorem of T.

Let us proceed further and choose a convex set $\mathbb{X}$ as follows:

$$\phi \in \mathbb{V}, \mid \phi \mid_{\mathbb{V}} \leq M,$$

when $M$ is to be precised later. And

$$\mid T\rho \mid_{\mathbb{V}} = \mid c_\lambda \mid_{\mathbb{V}} \leq \mid f \mid_{L^\infty} \mid Q \mid^{\frac{1}{2}}, \text{ for all } \rho \in \mathbb{X}.$$

Thus, choosing $M = \mid f \mid_{L^\infty} \mid Q \mid^{\frac{1}{2}}$, the following inclusion yields

$$T(\mathbb{V}) \subset \mathbb{X}$$

and then

$$T(\mathbb{X}) \subset \mathbb{X}.$$

So we will consider $T : \mathbb{X} \to \mathbb{X}$.
Step 2: T is continuous for all $\lambda \geq 1$. □

***Proof of Step 2*** T can be written as composition of following application:

$$L^2(Q) \longrightarrow L^2(Q) \longrightarrow \mathbb{V} \hookrightarrow L^2(Q)$$

$$\rho \longmapsto \widetilde{f}(\rho) = f \circ \rho \longmapsto T(\rho) \hookrightarrow T(\rho).$$

By Caratheodory theorem, $\rho \longmapsto \widetilde{f}(\rho) = f \circ \rho$ is continuous from $L^2(Q)$ into $L^2(Q)$. And Lax–Milgram's lemma gives the continuity of $f \circ \rho \longmapsto T(\rho)$ from $L^2(Q)$ into $\mathbb{V}$.Using the curved inequality (1.2), it is easy to see that the injection $\rho \in V \longmapsto \rho \in L^2(Q)$ is also continuous.

Then T is continuous

□

Step 3: $\mathbb{X}$ is a subset, convex, and compact in $L^2(Q)$.

***Proof of Step 3***

$$\|\phi\|^2_{H^1(Q)} = \|\phi\|^2_{L^2(Q)} + \|\widetilde{\nabla}\phi\|^2_{L^2(Q)} \quad \forall \phi \in H^1(Q).$$

By the inequality (1.2), we have

$$\|\phi\|^2_{H^1(Q)} \leq (1+c_p^2)(\|\widetilde{div}(\widetilde{u}\phi)\|^2_{L^2(Q)} + \|\widetilde{\nabla}\phi\|^2_{L^2(Q)}) = (1+c_p^2) \mid \phi \mid_{\mathbb{V}} \quad \forall \phi \in \mathbb{V}.$$

Then $\mathbb{X}$ that is bounded in $\mathbb{V}$ is bounded in $H^1(Q)$. And by Rellich's theorem, we know that $H^1(Q) \subset L^2(Q)$ with compact injection so $\mathbb{X}$ is relatively compact in $L^2(Q)$.

Moreover, $\mathbb{X}$ is closed in $L^2(Q)$.

In fact, let $x_n$ be a sequence in $\mathbb{X}$ with $x_n \longrightarrow x \in L^2(Q)$; then $x_n$ is bounded in $\mathbb{V}$, which is a reflexive Banach space; then there is a subsequence $x_{nk}$ that converges in the weak topology $\sigma(\mathbb{V}, \mathbb{V}^*)$ to $x^* \in \mathbb{V}$.

$\mathbb{X}$ is convex closed in the strong topology, then $\mathbb{X}$ is convex closed in the weak topology (see [8],Theorem 3.2), so we have $x^* \in \mathbb{X}$.

And from Mazur's theorem, there are a convex combination of $x_{nk}$, themselves elements of $\mathbb{X}$ which converge strongly towards $x^* \in \mathbb{X}$.

But these same convex combinations converge towards $x \in \mathbb{X}$ in $L^2(Q)$. By uniqueness of the limit in $L^2(Q)$, we have $x = x^*$.

Furthermore,

$$| v |_{\mathbb{V}} \leq \liminf | x_{nk} |_{\mathbb{V}} \leq M \text{ a.e } x \in \mathbb{X};$$

therefore, $\mathbb{X}$ is closed in $L^2(Q)$.

Since $\mathbb{X}$ is relatively compact and closed in $L^2(Q)$, then it is compact in $L^2(Q)$. □

Since $\mathbb{X}$ is convex, compact in $L^2(Q)$, and $T : \mathbb{X} \to \mathbb{X}$ continuous, from Schauder's fixed-point theorem, T has a fixed point.

**B-Uniqueness:**

Let $\rho_\lambda$ and $\overline{\rho_\lambda}$ be two solutions of 1.39, and we have

$$\int_Q | \widetilde{div}(\widetilde{u}(\rho_\lambda - \overline{\rho_\lambda})) |^2 \,\mathrm{d}x\mathrm{d}t + \lambda \int_Q | \widetilde{\nabla}(\rho_\lambda - \overline{\rho_\lambda}) |^2 \,\mathrm{d}x\mathrm{d}t$$
$$= \int_Q (f(\rho_\lambda) - f(\overline{\rho_\lambda}))\widetilde{div}(\widetilde{u}(\rho_\lambda - \overline{\rho_\lambda}))\mathrm{d}x\mathrm{d}t.$$

By Young's inequality, we have

$$2\|\widetilde{div}(\widetilde{u}(\rho_\lambda - \overline{\rho_\lambda}))\|^2_{L^2(Q)} + 2\lambda\|\widetilde{\nabla}(\rho_\lambda - \overline{\rho_\lambda})\|^2_{L^2(Q)} \leq \|(f(\rho_\lambda) - f(\overline{\rho_\lambda}))\|^2_{L^2(Q)} + \|\widetilde{div}(\widetilde{u}(\rho_\lambda - \overline{\rho_\lambda}))\|^2_{L^2(Q)}.$$

Since $f \in W^{1,\infty}(\mathbb{R})$, we have

$$\|f(\rho_\lambda) - f(\overline{\rho_\lambda})\|_{L^2(Q)} \leq | f' |_{L^\infty(\mathbb{R})} \|\rho_\lambda - \overline{\rho_\lambda}\|_{L^2(Q)},$$

and it follows that

$$2\|\widetilde{div}(\widetilde{u}(\rho_\lambda - \overline{\rho_\lambda}))\|^2_{L^2(Q)} + 2\lambda\|\widetilde{\nabla}(\rho_\lambda - \overline{\rho_\lambda})\|^2_{L^2(Q)} \leq | f' |^2_{L^\infty(\mathbb{R})} \|\rho_\lambda - \overline{\rho_\lambda}\|^2_{L^2(Q)} + \|\widetilde{div}(\widetilde{u}(\rho_\lambda - \overline{\rho_\lambda}))\|^2_{L^2(Q)}.$$

Since $div(u) = 0$, Remark 1.2.1 yields

$$\|\rho_\lambda - \overline{\rho_\lambda}\|^2_{L^2(Q)} \leq 4T^2\|(\widetilde{u}, \widetilde{\nabla}(\rho_\lambda - \overline{\rho_\lambda}))\|^2_{L^2(Q)}.$$

And then, we have

$$2\|\widetilde{div}(\widetilde{u}(\rho_\lambda - \overline{\rho_\lambda}))\|^2_{L^2(Q)} + 2\lambda\|\widetilde{\nabla}(\rho_\lambda - \overline{\rho_\lambda})\|^2_{L^2(Q)}$$
$$\leq 4T^2 | f' |^2_{L^\infty(\mathbb{R})} \|(\widetilde{u}, \widetilde{\nabla}(\rho_\lambda - \overline{\rho_\lambda}))\|^2_{L^2(Q)} + \|\widetilde{div}(\widetilde{u}(\rho_\lambda - \overline{\rho_\lambda}))\|^2_{L^2(Q)}.$$

By using Cauchy–Schwarz's inequality, we have

$$
\begin{aligned}
&2\|\widetilde{div}(\widetilde{u}(\rho_\lambda - \overline{\rho_\lambda}))\|^2_{L^2(Q)} + 2\lambda\|\widetilde{\nabla}(\rho_\lambda - \overline{\rho_\lambda})\|^2_{L^2(Q)} \\
&\quad \leq 4T^2 \mid f' \mid^2_{L^\infty(\mathbb{R})} \|\widetilde{u}\|^2_{L^2(Q)}\|\widetilde{\nabla}(\rho_\lambda - \overline{\rho_\lambda})\|^2_{L^2(Q)} + \|\widetilde{div}(\widetilde{u}(\rho_\lambda - \overline{\rho_\lambda}))\|^2_{L^2(Q)}
\end{aligned}
$$

$$
\|\widetilde{div}(\widetilde{u}(\rho_\lambda - \overline{\rho_\lambda}))\|^2_{L^2(Q)} + (2\lambda - 4T^2 \mid f' \mid^2_{L^\infty(\mathbb{R})} \|\widetilde{u}\|^2_{L^2(Q)})\|\widetilde{\nabla}(\rho_\lambda - \overline{\rho_\lambda})\|^2_{L^2(Q)} \leq 0.
$$

Thus $\rho_\lambda = \overline{\rho_\lambda}$ provided that $\lambda > 2T^2 \mid f' \mid^2_{L^\infty(\mathbb{R})} \|\widetilde{u}\|^2_{L^2(Q)}$.

## 1.5 Numerical Study and Simulations

In this section, two numerical methods are presented for computing the solution of semi-linear conservation law problem (1.10). The first consists in using Picard's iteration or Newton-adaptive for the linearization of the semi-linear problem. These linearized problems are discretized by using discontinuous Galerkin's method of the STILS formulation (1.6) and continuous finite-element method for the penalization version (1.38). Moreover, a posteriori error bounds are established when Newton iteration is used.

In the sequel, we shall assume that the function $f$ is $k$-Lipschitz; then by Rademacher's theorem (see [9] for more details), $f$ is differentiable almost everywhere.

### 1.5.1 *A Finite-Element Method for Semi-linear Conservations Laws*

Let us assume that the problem (1.8)–(1.9) admits a unique solution $c \in H^{k+1}(Q) \cap H(u, Q)$. In order to provide numerical approximation for computing the solution of (1.8)–(1.9) after linearization, we shall use a simple finite-element approximation that can be derived from the use of discontinuous Galerkin's approximations of the space–time least-square formulation. This method is introduced in [10] for linear hyperbolic problem and [11] for Poisson problem.

Let $\mathcal{T}_h$ be a regular partition of the domain $Q$ more precisely a triangulation in which each element is a polygon (respectively, polyhedra) in two dimensions (respectively, in three dimensions). For $k \geq 1$, we consider the discontinuous finite-element space (see [10] )

$$
\mathcal{V}_h = \{\phi \in L^2(Q), \phi \mid T \in Q_k(T) \; \forall T \in \mathcal{T}_h\}, \tag{1.40}
$$

where $Q_k(T)$ is the space of linear polynomials of degree k in each variable on $T$ and

$$\mathcal{V} = \{\phi \in L^2(Q), \phi \mid T \in H^{k+1}(T) \cap H(u, T) \ \forall T \in \mathcal{T}_h\}. \tag{1.41}$$

It is easy to remark that $\mathcal{V}$ contains $\mathcal{V}_h$ and $H^{k+1}(Q) \cap H(u, Q)$. Let $\mathcal{E}_h$ be the set of all edges for $d = 1$ or flat face for $d = 2$ and $\mathcal{E}_h^0 = \mathcal{E}_h \backslash \partial Q_-$. For $T \in \mathcal{T}_h$, let us denote by $h_K$ the diameter of $K$ and $\rho_K$ the supremum of the diameters of the inscribed spheres of $K$, $h = \max h_T$ the mesh size of $\mathcal{T}_h$. Let us suppose that $\mathcal{T}_h$ is shape regular, and also there exists two non-negative constants $C_{(1.42)}^{(1)}$ and $C_{(1.42)}^{(2)}$ such that

$$C_{(1.42)}^{(1)} \leq \frac{h_T}{h_e} \leq C_{(1.42)}^{(2)} \ \forall \ T \in \mathcal{T}_h \ \forall \ e \subset T. \tag{1.42}$$

Moreover, for $T \in \mathcal{T}_h$, we introduce the following notations:

$$\mathcal{E}_h(T) = \{E \in \mathcal{E}_h \, ; \ E \subset \partial T\}.$$

For $\phi \in \mathcal{V}_h$ and $e \in \mathcal{E}_h$ with $e = \partial T_1 \cap \partial T_2$, $T_1, T_2 \in \mathcal{T}_h$, let we define $[\phi]$ the jump of $\phi$ across $e \in \mathcal{E}_h^0$ as following:

$$[\phi] = \phi \mid_{\partial T_1} \widetilde{n_1} + \phi \mid_{\partial T_1} \widetilde{n_2}$$

and also

$$[(\widetilde{u}, \widetilde{n})\phi] = (\widetilde{u}, \widetilde{n_1})\phi \mid_{\partial T_1} + (\widetilde{u}, \widetilde{n_2})\phi \mid_{\partial T_2},$$

where $\widetilde{n_1}$ and $\widetilde{n_2}$ denote the unit outward vectors on $\partial T_1$ and $\partial T_2$, respectively. For $e \in \partial Q_-$, $[\phi] = \phi$ and $[(\widetilde{u}, \widetilde{n})\phi] = (\widetilde{u}, \widetilde{n})\phi$.

By considering the following bilinear form in $\mathcal{V} \times \mathcal{V}$

$$\mathcal{A}(c, \phi) = \sum_{T \in \mathcal{T}_h} \int_T \widetilde{div}(\widetilde{u}c)\widetilde{div}(\widetilde{u}\phi)\mathrm{d}x\mathrm{d}t + \sum_{e \in \mathcal{E}_h^0} \int_e h_e^{-1}[(\widetilde{u}, \widetilde{n})c][(\widetilde{u}, \widetilde{n})\phi]\mathrm{d}s. \tag{1.43}$$

Since $c \in \mathcal{V}$, then

$$\mathcal{A}(c, \phi) = \sum_{T \in \mathcal{T}_h} \int_T f(c)\widetilde{div}(\widetilde{u}\phi)\mathrm{d}x\mathrm{d}t + \sum_{e \in \partial Q_-} \int_e h_e^{-1}[(\widetilde{u}, \widetilde{n})c_b][(\widetilde{u}, \widetilde{n})\phi]\mathrm{d}s \ \forall \phi \in \mathcal{V}_h. \tag{1.44}$$

The corresponding approximation of (1.44) is called in ([10]) simple finite-element methods. It is easy to see that the bilinear form

$$\|\phi\|_{DG}^2 = \mathcal{A}(\phi, \phi) + |\phi|_{\mathcal{T}_h, k+1}$$

defines a norm in $\mathcal{V}$. Moreover, we have where

$$|\rho|_{\mathcal{T}_h,k+1} = \sum_{T\in\mathcal{T}_h} |\rho|^2_{k+1,T} \tag{1.45}$$

$$\mathcal{A}(\phi,\psi) \le \|\phi\|_{DG}\|\psi\|_{DG} \forall\ \psi\ ,\phi\in\mathcal{V}. \tag{1.46}$$

As in [12], we shall use the following abbreviation $x \preceq y$ for signifying $x \le Cy$ for some constant $C > 0$ independent to the mesh size $h$ and $\lambda$. Let $P_h$ be the $L^2$ projection onto $\mathcal{V}_h$; we have the following results, see [13] for more details.

There exists a constant $C_{(1.47)} > 0$ such that for all $\rho \in \mathcal{V}$

$$\|\widetilde{\nabla}(\rho - P_h\rho)\|_{0,T} \le C_{(1.47)} h^k |\rho|_{k+1,T} \tag{1.47}$$

for all $T \in \mathcal{T}_h$ and

$$\|\rho - P_h\rho\|_{0,T} \le C_{(1.47)} h^{k+1} |\rho|_{k+1,T}. \tag{1.48}$$

It is also proved in [14] that there exists a constant $C_{(1.49)}$ independent of the mesh size $h$ such that for any $T \in \mathcal{T}_h$ and $e \subset \partial T$, we have

$$\|\rho\|_e^2 \le C_{(1.49)}(h^{-1}\|\rho\|_T^2 + h\|\widetilde{\nabla}\rho\|_T^2). \tag{1.49}$$

Finally, we deduce the following approximation lemma.

**Lemma 1.5.1** *For all $\rho \in \mathcal{V}$,*

$$\|\rho - P_h\rho\|_{DG} \preceq h^k \|\rho\|_{\mathcal{T}_h,k+1}\ \forall\ T \in \mathcal{T}_h. \tag{1.50}$$

***Proof***

$$\|\rho - P_h\rho\|_{DG}^2 = \sum_{T\in\mathcal{T}_h}\int_T \widetilde{div}(\widetilde{u}(\rho-P_h\rho))^2 \mathrm{d}x\mathrm{d}t + \sum_{e\in\mathcal{E}_h}\int_e h_e^{-1}\|[(\widetilde{u},\widetilde{n})(\rho-P_h\rho)]\|^2 \mathrm{d}s.$$

By theorem (1.2.1), $(\widetilde{u},\widetilde{n}) \in L^\infty(\partial T)$, then it follows from (1.49) and (1.42)

$$\int_e h_e^{-1}\|[(\widetilde{u},\widetilde{n})(\rho - P_h\rho)]\|^2 \mathrm{d}s \le C^{(2)}_{(1.42)} h^{-1}\|(\widetilde{u},\widetilde{n})\|_{L^\infty(e)}\int_e \|[(\rho - P_h\rho)]\|^2 \mathrm{d}s. \tag{1.51}$$

This and (1.49) yield

$$\int_e \|[(\widetilde{u},\widetilde{n})(\rho - P_h\rho)]\|^2 \mathrm{d}s \le 4C^{(2)}_{(1.42)}\|(\widetilde{u},\widetilde{n})\|_{L^\infty(e)} C_{(1.49)}(h^{-2}\|\rho - P_h\rho\|_T^2 + \|\widetilde{\nabla}(\rho - P_h\rho)\|_T^2). \tag{1.52}$$

And from (1.47) and (1.48), it follows

$$\int_e \|[(\widetilde{u}, \widetilde{n})(\rho - P_h\rho)]\|^2 ds \le c_{(1.53)} h^{2k} |\rho|^2_{k+1,T}, \tag{1.53}$$

where

$$c_{(1.53)} = C_{(1.49)} C_{(1.47)} C^{(2)}_{(1.42)} \|(\widetilde{u}, \widetilde{n})\|_{L^\infty(e)}. \tag{1.54}$$

We also have from triangular inequality

$$\|\widetilde{div}(\rho - P_h\rho)\|_T \le \|(\widetilde{u}\widetilde{\nabla}(\rho - P_h\rho))\|_T + \|div(\widetilde{u})(\rho - P_h\rho)\|_T. \tag{1.55}$$

Since $\widetilde{u} \in L^\infty(T)$ and $div(\widetilde{u}) \in L^\infty(T)$, we get from (1.47)–(1.48)

$$\|\widetilde{div}(\rho - P_h\rho)\|_T \le C_{(1.47)} \alpha_{(u,1.57)} (h^k |\rho|_{k+1,T}), \tag{1.56}$$

where

$$\alpha_{(u,1.57)} = \max\{\|div(\widetilde{u})\|_{L^\infty(T)}, |Q| \|\widetilde{u}\|_{L^\infty(T)}\}. \tag{1.57}$$

From (1.53) and (1.56), we get the result. □

#### 1.5.1.1 A Finite-Element Method and Picard's Iteration

Let $f$ be a $k$-Lipschitz function in $\mathbb{R}$ with $k < \frac{1}{c_p}$. In this case, the solution $c^h$ can be computed by using the Picard iteration of some linear problem. The Picard iteration in this context is given by the following scheme:

**Algorithm**

- Start STILS-MT1 with some given $C^0$.
- Compute $c^h_{n+1}$ from $c^h_n$ such that

$$\begin{aligned}\mathcal{A}(c^h_{n+1}, \phi_h) &= \sum_{T \in \mathcal{T}_h} \int_T f(c^h_n) \widetilde{div}(\widetilde{u}\phi_h) dx dt \\ &+ \sum_{e \in \partial Q_-} \int_e h_e^{-1} [(\widetilde{u}, \widetilde{n}) c_b][(\widetilde{u}, \widetilde{n})\phi_h] ds \ \forall \phi_h \in \mathcal{V}_h.\end{aligned} \tag{1.58}$$

#### 1.5.1.2 A Finite-Element Method and Newton's Method

We suppose that the problem (1.8)–(1.9) has a unique solution $\mathcal{V} = H^2(Q) \cap H(u, Q)$. Recalling (1.58), we can write (1.8)–(1.9) as follows:

$$\text{find } c \in \mathcal{V} \text{ such that } F(c) = 0, \tag{1.59}$$

where

$$F : \mathcal{V} \longrightarrow \mathcal{V}^*$$

$$\begin{aligned}\langle F(c), \phi\rangle_{\mathcal{V}^*,\mathcal{V}} &= \mathcal{A}(c, \phi) - \sum_{T\in\mathcal{T}_h} \int_T f(c)\widetilde{div}(\widetilde{u}\phi)\mathrm{d}x\mathrm{d}t \\ &- \sum_{e\in\partial Q_-} \int_e h_e^{-1}[(\widetilde{u}, \widetilde{n})c_b][(\widetilde{u}, \widetilde{n})\phi]\mathrm{d}s \ \forall\phi \in \mathcal{V}.\end{aligned} \tag{1.60}$$

Given some initial guess $c^0$, the classical Newton–Raphson's method for solving equation (1.59), when $F$ is differentiable and consists in generating a sequence of approximation that converges in the quadratic sense, to the exact solution as follows:

$$\begin{cases} c_0 \in \mathbb{V} \\ c_{n+1} = c_n - F^{'}(c_n)^{-1}.F(c_n) \ \forall n \in \mathbb{N}^*. \end{cases} \tag{1.61}$$

This method is known to produce a chaotic behaviour when $c_0$ is far to the desired root, see, for instance, (see[15]) for more details. In order to remedy the chaotic behaviour, the following Newton damping method is proposed (see[16]). In that case, (1.61) is written as

$$\begin{cases} C_0 \in \mathbb{V} \\ c_{n+1} = c_n - \delta t F^{'}(c_n)^{-1}.F(c_n) \ \forall n \in \mathbb{N}^*. \end{cases} \tag{1.62}$$

We shall use adaptive Newton–Galerkin's method; more precisely, the damping parameter $\delta t$ in (1.62) may be adjusted and adapted in each iteration. For illustration of the choice of $\delta t$, let us define the Newton–Raphson's transform as follows:

$$\rho \longmapsto N_F(\rho) := -F^{'}(\rho)^{-1}.F(\rho).$$

By (1.62), we have

$$\frac{c_{n+1} - c_n}{\delta_n} = N_F(c_n).$$

And we remark that (1.62) may be seen as a forward Euler scheme of the following ordinary differential equation:

$$\frac{\mathrm{d}}{\mathrm{d}s}\rho(s) = N_F(\rho(s)) \ \forall s \ \rho(0) = c_0. \tag{1.63}$$

If $c_n \in \mathcal{V}$ for all $n \geq 1$ and $F$ is enough smooth, for instance, $F^{'}(\rho)^{-1}.F(\rho)$ exists for all $\rho \in \mathcal{V}$, then we obtain the solution of (1.63) satisfies

$$F(\rho(t)) = F(\rho(0))\exp(-t), \ \ \forall\, t \geq 0.$$

It is easy to see that $F(\rho(t)) \longrightarrow 0$ as $t \longrightarrow 0$.

The adaptive Newton–Raphson (see [17]) consists in choosing the damping parameter $\delta t_n$ so that the discrete forward Euler's solution of (1.62) stays reasonably close to the continuous solution of (1.63). Finally, we obtain the following algorithm, see [12].

**Algorithm**

Fix a tolerance $\epsilon$:

(i) Start the Newton iteration with some initial guess $c_0 \in \mathbb{V}$.
(ii) In each iteration step $n = 1, 2, ...,$ compute

$$\delta t_n = \min\left(\sqrt{\frac{2\epsilon}{\|N_F(c_n)\|_{\mathbb{V}}}}, 1\right). \tag{1.64}$$

(iii) Compute $c_{n+1}$ from (1.62) and go (ii).

In the sequel, we suppose that $f^{'}(c_n)$ exists for all $n \geq 1$, thus the sequels in 1.62 are well-defined, and we have

$$\begin{aligned}\beta(c, \rho, \phi) =: \langle F^{'}(c)\rho, \phi\rangle_{\mathcal{V}^*,\mathcal{V}} &= \mathcal{A}(\rho, \phi)\\ &- \sum_{T\in\mathcal{T}_h}\int_T f^{'}(c)\rho(\widetilde{u}\phi)\mathrm{d}x\mathrm{d}t \text{ for all } \phi \in \mathcal{V}.\end{aligned} \tag{1.65}$$

Let us define

$$L(c, \phi) := \langle F(c), \phi\rangle_{\mathcal{V}^*,\mathcal{V}}$$

with the previous notation (1.62) can be written as follows: given $c_n \in \mathcal{V}$, find $c_{n+1} \in \mathcal{V}$ such that

$$\beta(c_n, c_{n+1}, \phi) = \beta(c_n, c_n, \phi) - \delta t_n L(c_n, \phi) \text{ for all } \phi \in \mathcal{V}. \tag{1.66}$$

Let us now consider the following finite-element approximation: find $c_{n+1}^h \in \mathcal{V}$ from $c_n^h \in \mathcal{V}_h$ such that

$$\beta(c_n^h, c_{n+1}^h, \phi) = \beta(c_n^h, c_n^h, \phi) - \delta t_n L(c_n^h, \phi) \text{ for all } \phi \in \mathcal{V}_h. \tag{1.67}$$

By introducing the following notation:

$$c_{n+1}^{(\delta t_n, h)} := c_{n+1}^h - (1 - \delta t_n) c_n^h \tag{1.68}$$

and

$$f^{\delta t_n}(c_{n+1}^h) := \delta t_n f(c_n^h) + f'(c_n^h)(c_{n+1}^h - c_n^h), \tag{1.69}$$

we have, from (1.66),

$$\sum_{T \in \mathcal{T}_h} \int_T \widetilde{div}(\widetilde{u} c_{n+1}^{(\delta t_n, h)}) \widetilde{div}(\widetilde{u}\phi) \mathrm{d}x \mathrm{d}t + \sum_{e \in \mathcal{E}_h^0} \int_e h_e^{-1} [(\widetilde{u}, \widetilde{n}) c_{n+1}^{(\delta t_n, h)}][(\widetilde{u}, \widetilde{n})\phi] \mathrm{d}s$$

$$= \sum_{T \in \mathcal{T}_h} \int_T f^{\delta t_n}(c_{n+1}^h) \widetilde{div}(\widetilde{u}\phi) \mathrm{d}x \mathrm{d}t$$

$$+ \sum_{e \in \partial Q_-} \int_e \delta t_n h_e^{-1} [(\widetilde{u}, \widetilde{n}) c_b][(\widetilde{u}, \widetilde{n})\phi] \mathrm{d}s \ \forall \phi \in \mathcal{V}_h.$$

Let us define the following quantities:

$$\alpha_T = \|\widetilde{div}(\widetilde{u} c_{n+1}^{(\delta t_n)}) - f(c_{n+1}^{(\delta t_n, h)})\|_{0,T} \text{ and } \beta_T = \|f^{\delta t_n}(c_{n+1}^h) - f(c_{n+1}^{\delta t_n, h})\|_{0,T} \tag{1.70}$$

$$\alpha_e = \|[(\widetilde{u}, \widetilde{n}) c_{n+1}^{(\delta t_n)}]\|_{0,e} \text{ and } \beta_e = \|[(\widetilde{u}, \widetilde{n}) c_b]\|_{0,e}. \tag{1.71}$$

We also have the following result expressed by an inequality.

**Theorem 1.5.1**

$$\|F(c_{n+1}^{(\delta t_n, h)})\|_{\mathcal{V}^*} \preceq h^k \max\left(\left(\sum_{T \in \mathcal{T}_h} \beta_T^2\right)^{\frac{1}{2}}, \max\left(\left(\sum_{T \in \mathcal{T}_h} \alpha_T^2\right)^{\frac{1}{2}}, \left(\sum_{e \in \mathcal{E}_h^0} h_e^{-1} \alpha_e^2\right)^{\frac{1}{2}} + \left(\sum_{e \in \mathcal{E}_h^0} h_e^{-1} \beta_e^2\right)^{\frac{1}{2}}\right)\right). \tag{1.72}$$

***Proof***

$$\langle F(c), \phi \rangle_{\mathcal{V}^*, \mathcal{V}} = \langle F(c), \phi - P_h \phi \rangle_{\mathcal{V}^*, \mathcal{V}} + \langle F(c), P_h \phi \rangle_{\mathcal{V}^*, \mathcal{V}}.$$

Since $P_h \in \mathcal{V}$, from (1.66), it follows

$$\langle F(c_{n+1}^{(\delta t_n,h)}), P_h\phi\rangle_{\mathcal{V}^*,\mathcal{V}} = \sum_{T\in\mathcal{T}_h}\int_T (f^{\delta t_n}(c_{n+1}^h) - f(c_{n+1}^{\delta t_n,h}))\widetilde{div}(\widetilde{u}P_h\phi)\mathrm{d}x\mathrm{d}t,$$

and it follows, from Cauchy–Schwarz inequality in $L_2(T)$ and $R^q$ with $q = \dim(\mathcal{T}_h)$,

$$\langle F(c_{n+1}^{(\delta t_n,h)}), P_h\phi\rangle_{\mathcal{V}^*,\mathcal{V}} \le \sum_{T\in\mathcal{T}_h} \|f^{\delta t_n}(c_{n+1}^h) - f(c_{n+1}^{\delta t_n,h})\|_{0,T}\|P_h\phi\|_{0,T}$$

$$|\langle F(c_{n+1}^{(\delta t_n,h)}), P_h\phi\rangle_{\mathcal{V}^*,\mathcal{V}}| \le \Big(\sum_{T\in\mathcal{T}_h} \|f^{\delta t_n}(c_{n+1}^h) - f(c_{n+1}^{\delta t_n,h})\|_{0,T}^2\Big)^{\frac{1}{2}}\Big(\sum_{T\in\mathcal{T}_h}\|P_h\phi\|_{0,T}^2\Big)^{\frac{1}{2}}.$$

Since $P_h$ satisfies $\sum_{T\in\mathcal{T}_h}\|P_h\phi\|_{0,T}^2 \le \sum_{T\in\mathcal{T}_h}\|\phi\|_{0,T}^2$ $\forall\phi\in L^2(Q)$ (see [8]), we have

$$|\langle F(c_{n+1}^{(\delta t_n,h)}), P_h\phi\rangle_{\mathcal{V}^*,\mathcal{V}}| \le \Big(\sum_{T\in\mathcal{T}_h}\beta_T^2\Big)^{\frac{1}{2}}\Big(\sum_{T\in\mathcal{T}_h}\|\phi\|_{0,T}^2\Big)^{\frac{1}{2}}. \tag{1.73}$$

And using Lemma 1.5.1, we have

$$|\langle F(c_{n+1}^{(\delta t_n,h)}), P_h\phi\rangle_{\mathcal{V}^*,\mathcal{V}}| \le \Big(\sum_{T\in\mathcal{T}_h}\beta_T^2\Big)^{\frac{1}{2}} h^k\|\phi\|_{\mathcal{T}_h,k+1}. \tag{1.74}$$

$$\begin{aligned}\langle F(c_{n+1}^{(\delta t_n,h)}), \phi - P_h\phi\rangle_{\mathcal{V}^*,\mathcal{V}} = &\sum_{T\in\mathcal{T}_h}\int_T (\widetilde{div}(\widetilde{u}c_{n+1}^{(\delta t_n)}) - f(c_{n+1}^{(\delta t_n,h)}))\\ &\widetilde{div}(\widetilde{u}(\phi - P_h\phi))\mathrm{d}x\mathrm{d}t\\ &+\sum_{e\in\mathcal{E}_h^0}\int_e h_e^{-1}[(\widetilde{u},\widetilde{n})c_{n+1}^{(\delta t_n)}][(\widetilde{u},\widetilde{n})(\phi - P_h\phi)]\mathrm{d}s\\ &-\sum_{e\in\partial Q_-}\int_e h_e^{-1}[(\widetilde{u},\widetilde{n})c_b][(\widetilde{u},\widetilde{n})(\phi - P_h\phi)]\mathrm{d}s\end{aligned}$$

$$\sum_{T\in\mathcal{T}_h}\int_T \left(\widetilde{div}\left(\widetilde{u}c_{n+1}^{(\delta t_n)}\right) - f\left(c_{n+1}^{(\delta t_n,h)}\right)\right)\widetilde{div}(\widetilde{u}(\phi - P_h\phi))\mathrm{d}x\mathrm{d}t$$

$$\leq \left(\sum_{T\in\mathcal{T}_h}\alpha_T^2\right)^{\frac{1}{2}}\left(\sum_{T\in\mathcal{T}_h}\|\widetilde{div}(\widetilde{u}(\phi - P_h\phi)\|_{0,T}^2)^{\frac{1}{2}}\right) \tag{1.75}$$

□

$$\sum_{e\in\mathcal{E}_h^0}\int_e h_e^{-1}[(\widetilde{u},\widetilde{n})c_{n+1}^{(\delta t_n)}big][(\widetilde{u},\widetilde{n})(\phi - P_h\phi)]\mathrm{d}s$$

$$\leq \left(\sum_{e\in\mathcal{E}_h^0}h_e^{-1}\alpha_e^2\right)^{\frac{1}{2}}\left(\sum_{e\in\mathcal{E}_h^0}h_e^{-1}\|[(\widetilde{u},\widetilde{n})(\phi - P_h\phi)]\|_{0,e}^2\right)^{\frac{1}{2}} \tag{1.76}$$

$$\sum_{e\in\partial Q_-}\int_e h_e^{-1}[(\widetilde{u},\widetilde{n})c_b][(\widetilde{u},\widetilde{n})(\phi - P_h\phi)]\mathrm{d}s$$

$$\leq \left(\sum_{e\in\mathcal{E}_h^0}h_e^{-1}\beta_e^2\right)^{\frac{1}{2}}\left(\sum_{e\in\mathcal{E}_h^0}h_e^{-1}\|[(\widetilde{u},\widetilde{n})(\phi - P_h\phi)]\|_{0,e}^2\right)^{\frac{1}{2}}. \tag{1.77}$$

The inequalities (1.75)–(1.77) yield

$$|\langle F\left(c_{n+1}^{(\delta t_n,h)}\right), \phi - P_h\phi\rangle_{\mathcal{V}^*,\mathcal{V}}| \leq \max\left(\sum_{T\in\mathcal{T}_h}\alpha_T^2\right)^{\frac{1}{2}}, \left(\sum_{e\in\mathcal{E}_h^0}h_e^{-1}\alpha_e^2\right)^{\frac{1}{2}}$$

$$+\left(\sum_{e\in\mathcal{E}_h^0}h_e^{-1}\beta_e^2\right)^{\frac{1}{2}}\|\phi - P_h\phi\|_{DG}. \tag{1.78}$$

Thus, it follows from Lemma 1.5.1

$$|\langle F\left(c_{n+1}^{(\delta t_n,h)}\right), \phi - P_h\phi\rangle_{\mathcal{V}^*,\mathcal{V}}| \preceq \max\left(\left(\sum_{T\in\mathcal{T}_h}\alpha_T^2\right)^{\frac{1}{2}}, \left(\sum_{e\in\mathcal{E}_h^0}h_e^{-1}\alpha_e^2\right)^{\frac{1}{2}}\right.$$

$$\left.+\left(\sum_{e\in\mathcal{E}_h^0}h_e^{-1}\beta_e^2\right)^{\frac{1}{2}}\right)h^k\|\phi\|_{\mathcal{T}_h,k+1}. \tag{1.79}$$

From (1.74) and (1.79), we deduce

$$
\begin{aligned}
&|\langle F(c_{n+1}^{(\delta t_n,h)}), \phi\rangle_{\mathcal{V}^*,\mathcal{V}}| \\
&\preceq \max\left(\left(\sum_{T\in\mathcal{T}_h} \beta_T^2\right)^{\frac{1}{2}}, \max\left(\left(\sum_{T\in\mathcal{T}_h} \alpha_T^2\right)^{\frac{1}{2}}, \left(\sum_{e\in\mathcal{E}_h^0} h_e^{-1}\alpha_e^2\right)^{\frac{1}{2}}\right.\right. \\
&\quad \left.\left. + \left(\sum_{e\in\mathcal{E}_h^0} h_e^{-1}\beta_e^2\right)^{\frac{1}{2}}\right)\right) h^k \|\phi\|_{\mathcal{T}_h,k+1}. \qquad (1.80)
\end{aligned}
$$

□

## 1.5.2 STILS for Semi-linear Conservations Laws

In the following, we assume that the problem (1.39) admits a unique solution $c \in \mathbb{V} := H_0(u, Q) \cap H^{k+1}(Q)$, and we will omit the dependency of the function according to the parameter $\lambda$. Our aims are to give numerical methods for solving problem (1.39) based on the classical finite-element approximation of STILS formulation and establish a posteriori estimations. For this, we shall first consider first a finite-element approximation based on quadrilateral mesh by starting with the following finite-dimensional spaces:

$$
V(\widehat{K}) = \{\phi \in C^0(Q), \phi \mid \widehat{K} \in \widehat{Q_k}\}, \qquad (1.81)
$$

where $\widehat{K}$ is the so-called reference element and $\widehat{Q_k}$ is the space of polynomials of degree at most k in each variable, separately defined in $\widehat{K}$. Let $S$ be a class of invertible affine mapping defined on $\widehat{K}$ into $\mathbb{R}^{d+1}$. For $K = F_K(\widehat{K})$ with $F_K \in S$, the finite-element space can be defined by composition with the inverse of $F_K$ as follows:

$$
V(K) = \{\rho : K \to \mathbb{R} : \rho = \widehat{\rho} \circ F_K \text{ for some } \rho \in V(\widehat{K})\}. \qquad (1.82)
$$

Let $\mathcal{T}_h$ be a triangulation of $Q$ such that each of its element is the transformation of $\widehat{K}$ with some mapping in S. Thus we get the classical finite-element approximation

$$
V_h = \{\rho : Q \to \mathbb{R} : \rho \mid K \in V(K) \text{ for all } K\}.
$$

In order to obtain the CFL condition stability of STILS-MT1 (see [4]), we shall consider a strict rectangular mesh. Let

$$
\Pi : V \longrightarrow V_h \text{ such that } \Pi q = q \text{ for all } q \in Q_k \qquad (1.83)
$$

be a linear operator.

Let us recall the following approximation result proved in [18], pp 103, Corollary 4.4.2.

**Lemma 1.5.2** *Let us suppose that $d \leq 2$ and $k \geq 1$; then, there exist C such that, for all $0 \leq m \leq k+1$, $c \in H^{k+1}(K)$, the following inequality holds:*

$$| c - \Pi c |_{m,K} \leq \frac{h^{k+1}}{\rho^m} C_{\Pi,Q} | c |_{k+1,K} . \tag{1.84}$$

The above lemma provides us the existence of $C > 0$ such that the following inequalities hold for any $c \in H(u, T) \cap H^{k+1}(T)$,

$$\|c - \Pi c\|_{0,T} \leq C h^{k+1} |c|_{k+1,T} \ \forall\ T \in \mathcal{T}_h, \tag{1.85}$$

and

$$\|\widetilde{\nabla}(c - \Pi c)\|_{0,T} \leq C h^k |c|_{k+1,T} \ \forall\ T \in \mathcal{T}_h. \tag{1.86}$$

As in the proof of Lemma 1.5.1, there is a non-negative constant $C_u$ such that:

$$\|\widetilde{div}(\widetilde{u}(c - \Pi c))\|_{0,T} \leq C_u h^k |c|_{k+1,T} \ \forall\ T \in \mathcal{T}_h. \tag{1.87}$$

#### 1.5.2.1 STILS and Picard's Iteration

In this section, we suppose that $f$ is $k$-Lipschitz with $k < \frac{1}{c_p}$. Then the mapping T defined by (1.11)–(1.12) is a strict contraction, and thus, we shall use Picard's iteration algorithm for the linearization of (1.10)–(1.9).

The Picard's iteration in this context is given by following scheme:

**Algorithm**

- Start STILS-MT1 with some given $C^0$.
- Find $c^h_{n+1} \in \mathbb{V}_h$ from $c^h_n$ by the formula

$$\begin{aligned} &\int_Q \widetilde{div}(\widetilde{u} c^h_{n+1}) \widetilde{div}(\widetilde{u}\phi_h) \mathrm{d}x \mathrm{d}t + \lambda \int_Q \widetilde{\nabla} c^h_{n+1} \widetilde{\nabla}\phi \mathrm{d}x \mathrm{d}t \\ &= \int_Q (f(c^h_n) - \widetilde{div}(\widetilde{u} C_b)) \widetilde{div}(\widetilde{u}\phi_h) \mathrm{d}x \mathrm{d}t \ \forall\ \phi_h \in \mathbb{V}_h. \end{aligned} \tag{1.88}$$

### 1.5.3 STILS Adaptive Newton Method

Since the problem (1.39) has a unique solution, $\mathbb{V} = H^{k+1}(Q) \cap H_0(u, Q)$. Then the problem can be written as follows:

$$\text{find } c \in \mathbb{V} \quad \text{such} \quad \text{that } F_\lambda(c) = 0, \tag{1.89}$$

where

$$F_\lambda : \mathbb{V} \longrightarrow \mathcal{V}^*$$

such that

$$\langle F_\lambda(c), \phi\rangle_{\mathcal{V}^*,\mathcal{V}} := \int_Q \widetilde{div}(\widetilde{u}c)\widetilde{div}(\widetilde{u}\phi)\mathrm{d}x\mathrm{d}t - \lambda \int_Q \widetilde{\nabla}c\widetilde{\nabla}\phi\mathrm{d}x\mathrm{d}t$$
$$- \int_Q f(c)\widetilde{div}(\widetilde{u}\phi)\mathrm{d}x\mathrm{d}t \ \forall\ \phi \in \mathbb{V}. \tag{1.90}$$

Since $f$ is differentiable, then $F_\lambda$ is differentiable, and we have

$$\beta_\lambda(c, \rho, \phi) =: \langle F_\lambda^{'}(c)\rho, \phi\rangle_{\mathcal{V}^*,\mathcal{V}} = \int_Q \widetilde{div}(\widetilde{u}\rho)\widetilde{div}(\widetilde{u}\phi)\mathrm{d}x\mathrm{d}t - \lambda \int_Q \widetilde{\nabla}\rho\widetilde{\nabla}\phi\mathrm{d}x\mathrm{d}t$$
$$- \int_Q f^{'}(c)\rho\widetilde{div}(\widetilde{u}\phi)\mathrm{d}x\mathrm{d}t \ \forall\ \phi_h \in \mathbb{V}. \tag{1.91}$$

Let us also define the following linear form in $\mathbb{V}$

$$\lambda(\rho, \phi) = \langle F_\lambda(\rho), \phi\rangle_{\mathcal{V}^*,\mathcal{V}}. \tag{1.92}$$

We assume that $F$ is invertible, and inserting (1.91) and (1.92) in (1.62), we get

$$\beta_\lambda(c_n, c_{n+1}, \phi) = \beta_\lambda(c_n, c_n, \phi) - \delta t_n L_\lambda(c_n, \phi) \text{ for all } \phi \in \mathcal{V}. \tag{1.93}$$

Let $c_n^h$ be the finite-element approximation of $c_n$ (1.66). We obtain the following FEM adaptive Newton:

$$\beta_\lambda(c_n^h, c_{n+1}^h, \phi) = \beta_\lambda(c_n^h, c_n^h, \phi) - \delta t_n L_\lambda(c_n^h, \phi) \text{ for all } \phi \in \mathcal{V}_h. \tag{1.94}$$

By introducing the following notation:

$$c_{n+1}^{(\delta t_n, h)} := c_h^{n+1} - (1 - \delta t_n)c_h^n \tag{1.95}$$

and

$$f^{\delta t_n}(c^h_{n+1}) := \delta t_n f(c^h_n) + f'(c^h_n)(c^h_{n+1} - c^h_n), \tag{1.96}$$

it follows from (1.94) the following result

$$\begin{aligned}\int_Q \widetilde{div}\big(\widetilde{u} c_{n+1}^{(\delta t_n,h)}\big)\widetilde{div}(\widetilde{u}\phi)\mathrm{d}x\mathrm{d}t + \lambda \int_Q \widetilde{\nabla} c_{n+1}^{(\delta t_n,h)} \widetilde{\nabla}\phi \mathrm{d}x\mathrm{d}t \\ = \int_Q f^{\delta t_n}(c^h_{n+1})\widetilde{div}(\widetilde{u}\phi)\mathrm{d}x\mathrm{d}t \text{ for all } \phi \in \mathbb{V}_h.\end{aligned} \tag{1.97}$$

We also get the following result.

**Theorem 1.5.2**

$$\| F_\lambda\big(c_{n+1}^{(\delta t_n,h)}\big)\|_{\mathbb{V}^*} \preceq h^k\Bigg(\Big(\sum_{T\in\mathcal{T}_h} \alpha_T^2\Big)^{\frac{1}{2}} + \Big(\sum_{T\in\mathcal{T}_h} \beta_T^2\Big)^{\frac{1}{2}} + \lambda\Big(\sum_{T\in\mathcal{T}_h} \gamma_T^2\Big)^{\frac{1}{2}}\Bigg), \tag{1.98}$$

*where*

$$\begin{aligned}\alpha_T &= \big\|\big(\big(\widetilde{div}\big(\widetilde{u} c_{n+1}^{(\delta t_n,h)}\big)\big) - f^{\delta t_n}(c^h_{n+1})\big)\big\|_{0,T},\\ \beta_T &= \big\| f^{\delta t_n}(c^h_{n+1}) - f\big(c_{n+1}^{(h,\delta t_n)}\big)\big\|_{0,T} \textit{ and } \gamma_T = \big\|\widetilde{\nabla} c_{n+1}^{(\delta t_n,h)}\big\|_{0,T}.\end{aligned}$$

***Proof***

$$\begin{aligned}\big\langle F_\lambda\big(c_{n+1}^{(\delta t_n,h)}\big),\phi\big\rangle = \int_Q \big(\widetilde{div}\big(\widetilde{u} c_{n+1}^{(\delta t_n,h)}\big) - f\big(c_{n+1}^{(h,\delta t_n)}\big)\big)\widetilde{div}(\widetilde{u}\phi)\mathrm{d}x\mathrm{d}t \\ + \lambda \int_Q \widetilde{\nabla} c_{n+1}^{(\delta t_n,h)} \widetilde{\nabla}\phi \mathrm{d}x\mathrm{d}t.\end{aligned} \tag{1.99}$$

By adding and subtracting $\phi_h = \Pi\phi$ in (1.99) and using (1.97), the following result holds

$$\begin{aligned}\big\langle F_\lambda\big(c_{n+1}^{(\delta t_n,h)}\big),\phi\big\rangle = &\int_Q \big(\widetilde{div}\big(\widetilde{u} c_{n+1}^{(\delta t_n,h)}\big) - f\big(c_{n+1}^{(h,\delta t_n)}\big)\big)\widetilde{div}(\widetilde{u}(\phi-\phi_h))\mathrm{d}x\mathrm{d}t \\ &+\lambda \int_Q \widetilde{\nabla} c_{n+1}^{(\delta t_n,h)} \widetilde{\nabla}(\phi-\phi_h)\mathrm{d}x\mathrm{d}t + \int_Q \big(f^{\delta t_n}(c^h_{n+1}) \\ &- f\big(c_{n+1}^{(h,\delta t_n)}\big)\big)\widetilde{div}(\widetilde{u}\phi_h)\mathrm{d}x\mathrm{d}t\end{aligned}$$

$$\int_Q \big((\widetilde{div}(\widetilde{u}c_{n+1}^{(\delta t_n,h)})) - f^{\delta t_n}(c_{n+1}^h)\big)\widetilde{div}(\widetilde{u}(\phi-\phi_h))\mathrm{d}x\mathrm{d}t$$

$$\leq \sum_{T\in\mathcal{T}_h} \|\big((\widetilde{div}(\widetilde{u}c_{n+1}^{(\delta t_n,h)})) - f^{\delta t_n}(c_{n+1}^h)\big)\|_{0,T}\|\widetilde{div}(\widetilde{u}(\phi-\phi_h))\|_{0,T}. \tag{1.100}$$

Applying Cauchy–Schwarz inequality leads to

$$\int_Q \big((\widetilde{div}(\widetilde{u}c_{n+1}^{(\delta t_n,h)})) - f^{\delta t_n}(c_{n+1}^h)\big)\widetilde{div}(\widetilde{u}(\phi-\phi_h))\mathrm{d}x\mathrm{d}t$$

$$\leq \Big(\sum_{T\in\mathcal{T}_h} \|\big(\widetilde{div}(\widetilde{u}c_{n+1}^{(\delta t_n,h)}) - f^{\delta t_n}(c_{n+1}^h)\big)\|_{0,T}^2\Big)^{\frac{1}{2}}\Big(\sum_{T\in\mathcal{T}_h}\|\widetilde{div}(\widetilde{u}(\phi-\phi_h))\|_{0,T}^2\Big)^{\frac{1}{2}}, \tag{1.101}$$

and recalling (1.87), it follows

$$\int_Q \big((\widetilde{div}(\widetilde{u}c_{n+1}^{(\delta t_n,h)})) - f^{\delta t_n}(c_{n+1}^h)\big)\widetilde{div}(\widetilde{u}(\phi-\phi_h))\mathrm{d}x\mathrm{d}t$$

$$\leq C_u\Big(\sum_{T\in\mathcal{T}_h} \|\big((\widetilde{div}(\widetilde{u}c_{n+1}^{(\delta t_n,h)})) - f^{\delta t_n}(c_{n+1}^h)\big)\|_{0,T}^2\Big)^{\frac{1}{2}} h^k|\phi|_{k+1,Q}. \tag{1.102}$$

Thus,

$$\int_Q \big(f^{\delta t_n}(c_{n+1}^h) - f(c_{n+1}^{(h,\delta t_n)})\big)\widetilde{div}(\widetilde{u}\phi_h)\mathrm{d}x\mathrm{d}t \preceq \Big(\sum_{T\in\mathcal{T}_h}\alpha_T^2\Big)^{\frac{1}{2}} h^k|\phi|_{k+1,Q}. \tag{1.103}$$

It also follows

$$\lambda\int_Q \widetilde{\nabla}c_{n+1}^{(\delta t_n,h)}\widetilde{\nabla}(\phi-\phi_h)\mathrm{d}x\mathrm{d}t \leq \Big(\sum_{T\in\mathcal{T}_h}\|\widetilde{\nabla}c_{n+1}^{(\delta t_n,h)}\|_{0,T}\|\widetilde{\nabla}(\phi-\phi_h)\|_{0,T}\Big).$$

Using inequality (1.85), the following estimation holds

$$\lambda\int_Q \widetilde{\nabla}c_{n+1}^{(\delta t_n,h)}\widetilde{\nabla}(\phi-\phi_h)\mathrm{d}x\mathrm{d}t \preceq \lambda h^k\Big(\sum_{T\in\mathcal{T}_h}\gamma_T^2\Big)^{\frac{1}{2}}|\phi|_{k+1,Q}. \tag{1.104}$$

It follows from (1.102), (1.103), and (1.104).

$$\langle F_\lambda(c_{n+1}^{(\delta t_n,h)}),\phi\rangle \preceq h^k\Big(\Big(\sum_{T\in\mathcal{T}_h}\alpha_T^2\Big)^{\frac{1}{2}} + \Big(\sum_{T\in\mathcal{T}_h}\beta_T^2\Big)^{\frac{1}{2}} + \lambda\Big(\sum_{T\in\mathcal{T}_h}\gamma_T^2\Big)^{\frac{1}{2}}\Big)|\phi|_{k+1,Q}.$$

Furthermore, $\| F_\lambda(c_{n+1}^{(\delta t_n,h)})\|_{\mathbb{V}^*,\mathbb{V}} = \sup\langle F_\lambda(c_{n+1}^{(\delta t_n,h)}), \phi\rangle$ and $|\phi|_{k+1,Q} \preceq \|\phi\|_{\mathbb{V}}$, and then we get the results. □

Since $\delta t_n = 1$, if the adaptive Newton converges, $\| F_\lambda(c_{n+1}^{(\delta t_n,h)})\|_{\mathbb{V}^*}$ is a reasonable approximation; moreover, under certain conditions on f, we can show that $\|c - c_{n+1}^{(\delta t_n,h)}\|_{\mathbb{V}}$ is equivalent to $\| F_\lambda(c_{n+1}^{(\delta t_n,h)})\|_{\mathbb{V}^*}$.

### 1.5.4 Numerical Experiment

Let we consider the following one-dimension hyperbolic conservations laws with linear convection and stiff source terms (see [19]).

$$f(s) = -\mu s(s-1)\left(s - \frac{1}{2}\right),$$

and initial data

$$c_0(x) = \begin{cases} 1 \text{ if } x \leq 0.3 \\ 0 \text{ if } x > 0.3 \end{cases}.$$

The exact solution approaches the following waves solution $\omega(x - t)$ with

$$\omega(x) = \begin{cases} 0 \text{ if } c_0(x) < \frac{1}{2} \\ \frac{1}{2} \text{ if } c_0(x) = \frac{1}{2} \\ 1 \text{ if } c_0(x) > \frac{1}{2} \end{cases}.$$

*Example 1.5.1* We first choose $\mu$ such that $T$ is a contraction for instance $\mu = \frac{1}{7}$, and we will compute the solution of (1.8)–(1.9) by using Picard iteration and simple finite-element method and (1.39) by Picard iteration and STILS-MT. The mesh size of the space is $\frac{1}{60}$ and the times step $\frac{1}{65}$, which give $60 \times 65$ element in space–time. The solution is presented at $t = \frac{1}{4}$ in Fig. 1.1. □

*Example 1.5.2* Let we choose now $\mu = 7$ and compute the solution of simple finite-element method and STILS-MT1 with penalization $\lambda = \frac{5}{12}$ and using Newton–Raphson iteration for the semi-linearity. The mesh size of the space is $\frac{1}{20}$ and the times step $\frac{1}{25}$, which give $20 \times 25$ element in space–time. The solution is presented at $t = \frac{1}{4}$. □

Both numerical methods can be used to tame the spurious oscillations produced by STILS-MT and classical finite-element methods when advection problem is solved. In the case of simple finite-element methods, we have spurious diffusion for this semi-linear conservation; on the other hand, the same fact can be obtained when

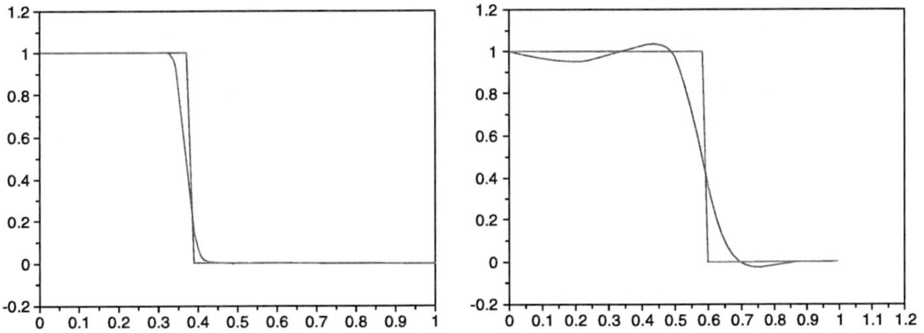


**Fig. 1.1** Left: Picard's iteration with STILS-MT1 with penalization $\lambda = \frac{5}{12}$. Right: Picard's iteration with simple finite-element method

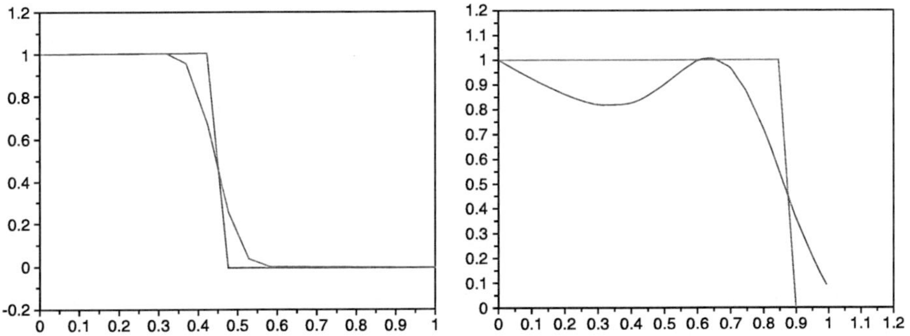


**Fig. 1.2** Left Newton-adaptative–Raphson's iteration with STILS-MT1 with penalization $\lambda = \frac{5}{12}$, right Newton-adaptative–Raphson's iteration with simple finite-element method

penalization version is used, but it can be controlled by the parameter $\lambda$. Moreover, STILS-MT cannot be used for simple finite element and that gives an important time calculation. We can clearly see that STILS-MT with penalization provides effective methods for solving semi-linear conservation law numerically (Fig. 1.2).

## References

1. O. Besson, J. Pousin, Solutions for Linear Conservation Laws with Velocity Fields in $L^\infty$. Arch. Ration. Mech. Analy. **2**, 159–175 (2007)
2. L.P. Franca, G. Hauke, A. Masud, Revisiting finite elements methods for the advective-diffusive equation. Comput. Meths. Appl. Mech. Engrg. **195**, 1560–1572 (2006)
3. H. Nguyen, J. Reynen, A space-time integrated least squares finite element scheme for advection-diffusion equation. Comput. Meths. Appl. Mech. Engrg. **42**, 331–342 (1984)
4. O. Besson, G. de Montmolling, Space-time integred least squares: a time marching. In. J. Num. Methods Fluids **44**, 525–543 (2004)

5. K. Benmansour, E. Bretin, L. Piffet, J. Pousin, Discrete Maximum principle for a space-time least squares formulation of the transport equation with finite element.. Discrete Cont. Dyn. Syst. **29**(3), 1001–1030 (2011)
6. P. Azerad , O. Besson, Inégalité de Poincaré courbe pour le traitement variationnel de l'équation de transport. C. R. Acad. Sci. Math, Paris **322**, 721–727 (1996)
7. L.C. Evans, *Partial Differential Equations*. Graduate Studies in Mathematics, vol. 19 (AMS, Providence, 1998)
8. H. Brezis, *Analyse fonctionnelle, Théorie et Applications* (Collection Sciences Sup, Dunod, 2005)
9. H. Rademacher, Uber partielle und totale differenzierbarkeit von Funktion mehrerer Variabeln und Uber die Transformation der Doppelintegrale. Math. Annalen. **79**, 340–369 (1919)
10. L. Mu, X. Ye, A simple finite element methods for linear hyperbolic problem. Comput. Appl. Math. **330**, 330–339 (2017)
11. X. Hu, L. Mu, X. Ye, A simple finite element methods of the Cauchy problem for Poisson equation. Numer. Analy. Modell. **14**, 591–603 (2017)
12. M. Amrein, T.P. Wihler, An adaptive Newton-method based on a dynamical systems approach. Commun. Nonlinear Sci. Numer. Simul. **19**, 2958–2973 (2014)
13. B. Riviere, *Discontinuous Galerkin Methods for Solving Elliptic and Parabolic Equations: Theory and Implementation* (SIAM 2008)
14. J. Wang, X. Ye, A weak *A weak Galerkin mixed finite element method for second-order ellipticproblems*. Math. Comp. **83**, 2101–2126 (2014)
15. H.O. Peitgen, P.H. Richter, *The Beauty of Fractals* (Springer, Berlin, 1986)
16. J.L. Varona, Graphic and numerical comparison between iterative methods. Math. Intell. **24**, 37–46 (2002)
17. H.R. Schneebeli, T.P. Wihler, The Newton-Raphson method and adaptive ODE solvers, Fractals. Complex Geom. Patterns Scaling Nat. Soc. **19**, 87–99 (2011)
18. P.A. Raviart, J.M. Thomas, *Introduction a l'analyse numérique des équations aux dérivées partielles* (Elsevier, Masson, 1992)
19. R.J. Leveque, H.C. Yee, A stady of numerical methods for hyperbolic conservation laws with stiff source terms. J. Comput. Phys. **86**, 187–210 (1990)

# Chapter 2
# Structural Stability of $p(x)$-Laplace Problems with Robin-Type Boundary Condition

**Kpè Kansie and Stanislas Ouaro**

**Abstract** In this chapter, a continuous dependence result on coefficients of solutions of the nonlinear nonhomogeneous Robin boundary-value problems involving the $p(x)$-Laplace operator is established.

**Keywords** Generalized Lebesgue and Sobolev spaces · Leray–Lions operator · Weak solution · Renormalized solution · Thermorheological fluids · Continuous dependence · Young measures · Robin-type boundary condition

## 2.1 Introduction

Our work has for goal to study the convergence of sequences of solutions of degenerate elliptic problems with variable coercivity and growth exponents $p_n$ of the form

$$(Pb_n):\begin{cases} b(u_n) - \operatorname{div} a_n(x, \nabla u_n) = f_n & \text{in } \Omega, \\ a_n(x, \nabla u_n).\eta = -|u_n|^{p_n(x)-2}u_n & \text{on } \partial\Omega, \end{cases}$$

where $\Omega$ is an open bounded domain of $\mathbb{R}^N$ ($N \geq 3$) with smooth boundary $\partial\Omega$ and $\eta$ is the outer unit normal to $\partial\Omega$. Here, $b : \mathbb{R} \longrightarrow \mathbb{R}$ is a continuous, onto, and nondecreasing function such that $b(0) = 0$; $(a_n(x, \xi))_{n\in\mathbb{N}}$ is a family of applications that verify the classical Leray–Lions hypotheses but with a variable summability exponent $p_n(x)$ converging in measure to some exponent $p$ such that $1 < p_- \leq$

K. Kansie
Laboratoire de Mathématiques et Informatique (LA.M.I), UFR. Sciences et Techniques, Université Nazi Boni, Bobo-Dioulasso, Burkina Faso

S. Ouaro (✉)
Laboratoire de Mathématiques et Informatique (LA.M.I), UFR. Sciences Exactes et Appliquées, Université Ouaga I Pr Joseph KI ZERBO, Ouagadougou, Burkina Faso
e-mail: souaro@univ-ouaga.bf

 
T. Diagana et al. (eds.), *Partial Differential Equations and Applications*,
Springer Proceedings in Mathematics & Statistics 420,
https://doi.org/10.1007/978-3-031-27661-3_2

$p_n(.) \leq p_+ < \infty$ and $(f_n)_{n\in\mathbb{N}} \subset L^1(\Omega)$. We show that the limit of this sequence of solutions is the solution of the following problem associated to the exponent $p$:

$$(Pb):\begin{cases} b(u) - \operatorname{div} a(x, \nabla u) = f & \text{in } \Omega, \\ a(x, \nabla u).\eta = -|u|^{p(x)-2}u & \text{on } \partial\Omega. \end{cases} \tag{2.1}$$

Andreianov, Bendahmane and Ouaro (see [1]) studied the structural stability of weak and renormalized solutions $u_n$ of the following nonlinear homogeneous Dirichlet boundary-value problem:

$$\begin{cases} b(u_n) - \operatorname{div} a_n(x, \nabla u_n) = f_n \text{ in } \Omega, \\ u_n = 0 \qquad\qquad\qquad\qquad \text{on } \partial\Omega, \end{cases} \tag{2.2}$$

where $(a_n(x,\xi))_{n\in\mathbb{N}}$ verifies the classical Leray–Lions hypotheses with the variable exponents $p_n(x)$ such that $1 < p_- \leq p_n(.) \leq p_+ < \infty$. Since the exponent $p_n$, and thus the underlying function space for the solution $u_n$, varies with $n$, the convergence of weak solutions $u_n$ requires some involved assumptions on the convergence of the sequence $f_n$ of the source terms. To bypass this difficulty, they used the technique of renormalized solutions. Indeed, the study of convergence of renormalized solutions of the problem (2.2) permits them to deduce convergence results for the weak solutions under much simpler assumptions on $(f_n)_{n\in\mathbb{N}}$, in particular the weak $L^1$ convergence of $f_n$ to a limit $f$ sufficiently regular so that to allow for the existence of a weak solution. Moreover, the structural stability results permit them to deduce also new existence results of solutions.

This chapter is related to Robin-type boundary condition, so we cannot work in the space $W_0^{1,p(.)}(\Omega)$, but in the space $W^{1,p(.)}(\Omega)$. Therefore, Poincaré inequality does not apply. Nevertheless, we use in this chapter a Poincaré–Sobolev-type inequality. The technique of Young measures (see [10, 12, 14]) is essential for the convergence of the sequence of gradients of solutions.

Problems with variable exponents $p(x)$ and $p_n(x)$ were arisen and studied by Zhikov in the pioneering paper [23]. The study of problems involving variable exponent has received considerable attention in recent years due to the fact that they can model various phenomena that arise in the study of elastic mechanics, electrorheological, and thermorheological fluids (see [6, 18–20]) or image restoration (see [5, 13]).

Let us give the outline of the paper. In Sect. 2.2, we introduce some preliminary results. In Sect. 2.3, we prove the existence and uniqueness of the renormalized solution of (2.1) with $L^1$-data $f$. In Sect. 2.4, we tackle the question of continuous dependence for renormalized solutions.

## 2.2 Preliminaries

In this section, we do some assumptions on the model problem (2.1) and give some preliminary results.

$$p : \overline{\Omega} \longrightarrow \mathbb{R} \text{ is a continuous function such that } 1 < p_- \leq p_+ < \infty, \tag{2.3}$$

where $p_- := \inf_{x\in\Omega} p(x)$ and $p_+ := \sup_{x\in\Omega} p(x)$.

$$\begin{cases} b : \mathbb{R} \longrightarrow \mathbb{R} \text{ is a continuous, non-decreasing,} \\ \text{and onto function such that } b(0) = 0. \end{cases} \tag{2.4}$$

$a(.,.) : \Omega \times \mathbb{R}^N \longrightarrow \mathbb{R}^N$ is a Carathéodory function with

$$a(x, 0) = 0 \text{ for a.e. } x \in \Omega, \tag{2.5}$$

satisfying, for a.e. $x \in \Omega$, the strict monotonicity assumption

$$(a(x, \xi) - a(x, \eta)).(\xi - \eta) > 0 \text{ for all } \xi, \eta \in \mathbb{R}^N, \xi \neq \eta, \tag{2.6}$$

and the following growth and coercivity assumptions in $\xi$:

$$|a(x, \xi)| \leq C_1(\mathcal{M}(x) + |\xi|^{p(x)-1}), \tag{2.7}$$

$$a(x, \xi).\xi \geq C_2|\xi|^{p(x)}, \tag{2.8}$$

where $C_1$ and $C_2$ are the positive constants, and $\mathcal{M}$ is a non-negative function in $L^{p'(.)}(\Omega)$ with $1/p(x) + 1/p'(x) = 1$.

*Remark 2.2.1* The condition (2.5) is a consequence of the continuity and the coercivity of $a(.,.)$. □

For the given exponent $p$, we denote by $p'$ its conjugate exponent such that $1/p(x)+ 1/p'(x) = 1$ and by $p^*$ its optimal Sobolev embedding exponent such that

$$p^* := \begin{cases} Np/(N-p) & \text{if } p < N, \\ \text{any real value} & \text{if } p = N, \\ \infty & \text{if } p > N. \end{cases}$$

For any given $k > 0$, we define the truncation function $T_k : \mathbb{R} \longrightarrow \mathbb{R}$ by

$$T_k(r) = \max(\min(r, k), -k).$$

We put

$$sign(z) = \begin{cases} 1 & \text{if } z > 0, \\ 0 & \text{if } z = 0, \\ -1 & \text{if } z < 0. \end{cases}$$

The truncation function $T_k$ has so the following properties:

$$|T_k(z)| = \min(|z|, k), \quad \lim_{k \to \infty} T_k(z) = z \text{ and } \lim_{k \to 0} \frac{1}{k} T_k(z) = sign(z).$$

For a Lebesgue measurable set $A \subset \Omega$, $\chi_A$ denotes its characteristic function, and $meas(A)$ denotes its Lebesgue measure. Also, we denote by $meas_{N-1}(B)$ or $\mu(B)$ the Lebesgue measure of $B \subset \partial\Omega$.

Let $u : \Omega \longrightarrow \mathbb{R}$ be a function and $k \in \mathbb{R}$, and we write $\{|u| \leq k\}$ for the set $\{x \in \Omega : |u(x)| \leq k\}$ (respectively, $\geq, =, <, >$).

We will also need to truncate vector-valued functions with the help of the following maps:

$$\text{for } m > 0,\, h_m : \mathbb{R}^N \longrightarrow \mathbb{R}^N,\, h_m(\lambda) = \begin{cases} \lambda & \text{if } |\lambda| \leq m, \\ \dfrac{m}{|\lambda|}\lambda & \text{if } |\lambda| > m. \end{cases} \tag{2.9}$$

We have the following property (see [1, Lemma 2.1]).

**Lemma 2.2.1** *Let $h_m(.)$ be defined by (2.9) and $a(x,.)$ be monotone in the sense (2.6). Then, for all $\lambda \in \mathbb{R}^N$, the map $m \longmapsto a(x, h_m(\lambda)).h_m(\lambda)$ is non-decreasing and converges to $a(x,\lambda).\lambda$ as $m \to \infty$.* □

The exponent $p(.)$ appearing in (2.7) and (2.8) depends on the spatial variable $x$ and then requires so to work with Lebesgue and Sobolev spaces with variable exponents.

We define the Lebesgue space with variable exponent $L^{p(.)}(\Omega)$ as the set of all measurable functions $u : \Omega \longrightarrow \mathbb{R}$ for which the convex modular

$$\rho_{p(.)}(u) := \int_\Omega |u|^{p(x)} dx$$

is finite. If the exponent is bounded, i.e., if $p_+ < \infty$, then the expression

$$\|u\|_{p(.)} := \|u\|_{L^{p(.)}(\Omega)} := \inf\left\{\lambda > 0 : \rho_{p(.)}\left(\frac{f}{\lambda}\right) \leq 1\right\}$$

defines a norm in $L^{p(.)}(\Omega)$, called the Luxembourg norm. The space $\left(L^{p(.)}(\Omega), \|.\|_{p(.)}\right)$ is a separable Banach space. Moreover, if $1 < p_- \leq p_+ < \infty$, then $L^{p(.)}(\Omega)$ is uniformly convex, hence reflexive, and its dual space is isomorphic to $L^{p'(.)}(\Omega)$.

Moreover, we have the Hölder-type inequality

$$\left| \int_{\Omega} uv \, dx \right| \leq \left( \frac{1}{p_-} + \frac{1}{(p')_-} \right) \|u\|_{p(.)} \|v\|_{p'(.)} , \tag{2.10}$$

for all $u \in L^{p(.)}(\Omega)$ and $v \in L^{p'(.)}(\Omega)$.

$W^{1,p(.)}(\Omega)$ denotes the space of all functions $u \in L^{p(.)}(\Omega)$ such that its gradient $\nabla u$, taken in the sense of distributions, belongs to $(L^{p(.)}(\Omega))^N$. This space is a Banach space equipped with the following norm:

$$\|u\|_{1,p(.)} := \|u\|_{W^{1,p(.)}(\Omega)} := \|u\|_{p(.)} + \|\nabla u\|_{p(.)} .$$

The space $\left(W^{1,p(.)}(\Omega), \|.\|_{1,p(.)}\right)$ is a separable and reflexive Banach space; for more details on the generalized Lebesgue and Sobolev spaces, see [8, 15] and the references therein.

In manipulating the generalized Lebesgue and Sobolev spaces, the following lemma (cf.[11]) permits to pass from norm to convex modular and vice versa.

**Lemma 2.2.2** *If $u_n, u \in L^{p(.)}(\Omega)$ and $p_+ < \infty$, then the following properties hold:*

(*i*) $\rho_{p(.)}\left(u/\|u\|_{p(.)}\right) = 1$, *if* $u \neq 0$.
(*ii*) $\rho_{p(.)}(u) < 1$ (*respectively* $= 1; > 1$) $\Longleftrightarrow \|u\|_{p(.)} < 1$ (*respectively* $= 1; > 1$).
(*iii*) $\rho_{p(.)}(u) \leq 1 \Longrightarrow \|u\|_{p(.)}^{p_+} \leq \rho_{p(.)}(u) \leq \|u\|_{p(.)}^{p_-}$.
(*iv*) $\rho_{p(.)}(u) \geq 1 \Longrightarrow \|u\|_{p(.)}^{p_-} \leq \rho_{p(.)}(u) \leq \|u\|_{p(.)}^{p_+}$.
(*v*) $\|u_n\|_{p(.)} \to 0$ (*respectively* $\to \infty$) $\Longleftrightarrow \rho_{p(.)}(u_n) \to 0$ (*respectively* $\to \infty$). □

For a measurable function $u : \Omega \longrightarrow \mathbb{R}$, we introduce the function

$$\rho_{1,p(.)}(u) := \int_{\Omega} |u|^{p(x)} dx + \int_{\Omega} |\nabla u|^{p(x)} dx.$$

Then, we have the following lemma (see [21]).

**Lemma 2.2.3** *If $u \in W^{1,p(.)}(\Omega)$, then the following properties are true:*

(*i*) $\rho_{1,p(.)}(u) < 1$ *(respectively* $= 1; > 1$*)* $\Longleftrightarrow \|u\|_{1,p(.)} < 1$ *(respectively* $= 1; > 1$*)*.
(*ii*) $\rho_{1,p(.)}(u) \leq 1 \Longrightarrow \|u\|_{1,p(.)}^{p_+} \leq \rho_{1,p(.)}(u) \leq \|u\|_{1,p(.)}^{p_-}$.
(*iii*) $\rho_{1,p(.)}(u) \geq 1 \Longrightarrow \|u\|_{1,p(.)}^{p_-} \leq \rho_{1,p(.)}(u) \leq \|u\|_{1,p(.)}^{p_+}$.
(*iv*) $\|u_n\|_{1,p(.)} \to 0$ *(respectively* $\to \infty$*)* $\Longleftrightarrow \rho_{1,p(.)}(u_n) \to 0$ *(respectively* $\to \infty$*)*. □

One has below, imbedding result between Lebesgue and Sobolev spaces with variable exponent.

**Proposition 2.2.1 (See [8, 11])** *Let $p, q \in C(\overline{\Omega})$ with $p_- > 1$. Assume that $q(x) < p^*(x)$ for all $x \in \overline{\Omega}$. Then, there is a compact imbedding $W^{1,p(.)}(\Omega) \hookrightarrow L^{q(.)}(\Omega)$. In particular, there is a compact imbedding $W^{1,p(.)}(\Omega) \hookrightarrow L^{p(.)}(\Omega)$.* □

Put

$$p^{\partial}(x) := (p(x))^{\partial} := \begin{cases} \dfrac{(N-1)p(x)}{N-p(x)} & \text{if } p(x) < N, \\ \infty & \text{if } p(x) \geq N; \end{cases} \tag{2.11}$$

then, one also has the following imbedding result.

**Proposition 2.2.2 (See [22])** *Let $p \in C(\overline{\Omega})$ with $p_- > 1$. If $q \in C(\partial\Omega)$ satisfies the condition*

$$1 \leq q(x) < p^{\partial}(x) \ \forall x \in \partial\Omega,$$

*then there is a compact imbedding $W^{1,p(.)}(\Omega) \hookrightarrow L^{q(.)}(\partial\Omega)$. In particular, there is a compact embedding $W^{1,p(.)}(\Omega) \hookrightarrow L^{p(.)}(\partial\Omega)$.* □

For any $u \in W^{1,p(.)}(\Omega)$, we denote by $\tau(u)$ the trace of $u$ on $\partial\Omega$ in the usual sense. Proposition 2.2.2 means that, for every $1 \leq p \leq \infty$, the trace operator

$$\tau : W^{1,p(.)}(\Omega) \to L^{p(.)}(\partial\Omega), u \mapsto \tau(u) = u_{|\partial\Omega},$$

is compact.

The following result (corollary of Lebesgue-dominated convergence theorem) is very powerful to prove strong convergence results.

**Lemma 2.2.4 (Lebesgue Generalized Convergence Theorem)** *Let $(f_n)_{n\in\mathbb{N}}$ be a sequence of measurable functions and $f$ a measurable function such that $f_n \to f$ a.e. in $\Omega$. Let $(g_n)_{n\in\mathbb{N}} \subset L^1(\Omega)$ such that for all $n \in \mathbb{N}$, $|f_n| \leq g_n$ a.e. in $\Omega$ and $g_n \to g$ in $L^1(\Omega)$. Then,*

$$\int_{\Omega} f_n \, dx \longrightarrow \int_{\Omega} f \, dx.$$

We also recall a Poincaré-type inequality and a Poincaré–Sobolev-type inequality (see [17]).

**Lemma 2.2.5** *There exists $C_1 > 0$ such that for all $u \in W^{1,1}(\Omega)$, one has*

$$\int_{\Omega} |u| \, dx \leq C_1 \left( \int_{\Omega} |\nabla u| \, dx + \int_{\partial\Omega} |u| \, d\sigma \right),$$

*and there exists $C_2 > 0$ such that for all $u \in W^{1,q}(\Omega)$, $1 < q < N$, one has*

$$\left(\int_\Omega |u|^{q^*}\,dx\right)^{q/q^*} \le C_2\left(\int_\Omega |\nabla u|^q\,dx + \left(\int_{\partial\Omega} |u|\,d\sigma\right)^q\right),$$

*where* $1/q^* = 1/q - 1/N$. □

For the applications we have in mind, we will need the following theorem in which the results of $(ii)$ and $(iii)$ express measure convergence of some sequences.

**Theorem 2.2.1 (Young Measures and Nonlinear weak-* Convergence, cf. [1])**

$(i)$ *Let* $\Omega \subset \mathbb{R}^N$, $N \in \mathbb{N}$, *and* $(v_n)_{n\in\mathbb{N}}$ *an equi-integrable sequence on* $\Omega$ *of functions to values in* $\mathbb{R}^d$, $d \in \mathbb{N}$. *Then, there exist a subsequence* $(v_{n_k})_{k\in\mathbb{N}}$ *and a parametrized family* $(\nu_x)_{x\in\Omega}$ *of probability measures on* $\mathbb{R}^d$, *weakly measurable in* $x$ *with respect to the Lebesgue measure on* $\Omega$, *such that for all Carathéodory function* $F : \Omega \times \mathbb{R}^d \to \mathbb{R}^t$, $t \in \mathbb{N}$, *we have*

$$\lim_{k\to\infty} \int_\Omega F(x, v_{n_k}(x))\,dx = \int_\Omega \int_{\mathbb{R}^d} F(x,\lambda)\,d\nu_x(\lambda)dx, \tag{2.12}$$

*whenever the sequence* $(F(., v_n(.)))_{n\in\mathbb{N}}$ *is equi-integrable on* $\Omega$. *In particular,*

$$v(x) := \int_{\mathbb{R}^d} \lambda d\nu_x(\lambda) \tag{2.13}$$

□

*is the weak limit of the sequence* $(v_{n_k})_{k\in\mathbb{N}}$ *in* $L^1(\Omega)$, *as* $k \to \infty$.

*The family* $(\nu_x)_{x\in\Omega}$ *is called the Young measure generated by the subsequence* $(v_{n_k})_{k\in\mathbb{N}}$.

$(ii)$ *If* $\Omega$ *is of finite measure, and* $(\nu_x)_{x\in\Omega}$ *is the Young measure generated by a sequence* $(v_n)_{n\in\mathbb{N}}$, *then*

$$\left(\nu_x = \delta_{v(x)} a.e.\, x \in \Omega\right) \Longleftrightarrow (v_n \text{ converges in measure on } \Omega \text{ to } v \text{ as } n \to \infty).$$

$(iii)$ *If* $\Omega$ *is of finite measure,* $(u_n)_{n\in\mathbb{N}}$ *generates a Dirac Young measure* $\left(\delta_{u(x)}\right)_{x\in\Omega}$ *on* $\mathbb{R}^{d_1}$, *and* $(v_n)_{n\in\mathbb{N}}$ *generates a Young measure* $(\nu_x)_{x\in\Omega}$ *on* $\mathbb{R}^{d_2}$, *then the sequence* $((u_n, v_n))_{n\in\mathbb{N}}$ *generates the Young measure* $\left(\delta_{u(x)} \otimes \nu_x\right)_{x\in\Omega}$ *on* $\mathbb{R}^{d_1+d_2}$. □

Whenever a sequence $(v_n)_{n\in\mathbb{N}}$ generates a Young measure $(\nu_x)_{x\in\Omega}$, following the terminology of [9], we will say that $(v_n)_{n\in\mathbb{N}}$ nonlinear weak-* converges, and $(\nu_x)_{x\in\Omega}$ is the nonlinear weak-* limit of the sequence $(v_n)_{n\in\mathbb{N}}$. In the case $(v_n)_{n\in\mathbb{N}}$ possesses a nonlinear weak-* convergent subsequence, we will say that it is nonlinear weak-* compact. Theorem 2.2.1–$(i)$ thus means that any equi-integrable sequence of measurable functions is nonlinear weak-* compact on $\Omega$.

## 2.3 Renormalized Solution

In this part, we define the notion of associated renormalized solution to the problem (2.1) and prove an existence and uniqueness result of renormalized solution for $L^1$–data $f$.

Let us put

$$\mathcal{T}^{1,p(.)}(\Omega) := \{u : \Omega \longrightarrow \mathbb{R} \text{ measurable such that } T_k(u) \in W^{1,p(.)}(\Omega), \text{ for any } k > 0\}.$$

Then, we define $\mathcal{T}_{tr}^{1,p(.)}(\Omega)$ as the set of functions $u \in \mathcal{T}^{1,p(.)}(\Omega)$ such that there exists a sequence $(u_n)_{n\in\mathbb{N}} \subset W^{1,p(.)}(\Omega)$ satisfying the following conditions:

$(C_1)$ $u_n \longrightarrow u$ a.e. in $\Omega$.
$(C_2)$ $\nabla T_k(u_n) \longrightarrow \nabla T_k(u)$ in $L^1(\Omega)$ for any $k > 0$.
$(C_3)$ There exists a measurable function $v$ on $\partial\Omega$ such that $u_n \rightarrow v$ on $\partial\Omega$.

The function $v$ is the trace of $u$ in the generalized sense. In the sequel, the trace of $u \in \mathcal{T}_{tr}^{1,p(.)}(\Omega)$ on $\partial\Omega$ will be denoted by $tr(u)$. If $u \in W^{1,p(.)}(\Omega)$, then $tr(u)$ coincides with $\tau(u)$ in the usual sense. Moreover, for $u \in \mathcal{T}_{tr}^{1,p(.)}(\Omega)$ and for every $k > 0$, $\tau(T_k(u)) = T_k(tr(u))$, and if $\phi \in W^{1,p(.)}(\Omega) \cap L^\infty(\Omega)$, then $(u - \phi) \in \mathcal{T}_{tr}^{1,p(.)}(\Omega)$ and $tr(u - \phi) = tr(u) - tr(\phi)$ (see [2, 3] for more details).

*Remark 2.3.1* We will use the same notation $u$ for $u$ and its trace when there is not an inconvenience. □

The following proposition (see, e.g., [4]) is useful because it allows us to give a sense to the definition of the renormalized solution for problem (2.1) (see Definition 2.3.1 below).

**Proposition 2.3.1** *Let $u \in \mathcal{T}^{1,p(.)}(\Omega)$. Then, there exists a unique measurable function $v : \Omega \longrightarrow \mathbb{R}^N$ such that*

$$\nabla T_k(u) = v\chi_{\{|u|<k\}}, \textit{ for all } k > 0,$$

*where $\chi_E$ is the characteristic function of a measurable set $E$. The function $v$ is a generalized gradient and is denoted by $\nabla u$ (weak gradient of $u$). If, moreover, $u$ belongs to $W^{1,p(.)}(\Omega)$, then $v$ belongs to $\left(L^{p(.)}(\Omega)\right)^N$ and coincides with the standard distributional gradient of $u$.* □

Let us also set

$$\mathbb{S} := \{S \in W^{1,\infty}(\mathbb{R}) \text{ such that } suppS \text{ is compact}\}.$$

The following function,

$$\text{for } k > 0,\ S_k : z \in \Omega \longmapsto \begin{cases} 1 & \text{if } |z| \leq k - 1, \\ k - |z| & \text{if } k - 1 \leq |z| \leq k, \\ 0 & \text{if } |z| \geq k, \end{cases} \tag{2.14}$$

is an example of function in $\mathbb{S}$ that will be used a lot in the sequel. Note that this function is non-negative with $suppS_k = [-k, k]$, and $suppS'_k$ is contained in $[-k, -k+1] \cup [k-1, k]$ and that the sequences $S_k$ and $S'_k$ are uniformly bounded by one.

Now, we give the definition of renormalized solution for problem (2.1) under the assumptions (2.3)–(2.8).

**Definition 2.3.1** A measurable function $u : \Omega \to \mathbb{R}$ is a renormalized solution of problem (2.1) if:

For all $k > 0$, $T_k(u) \in W^{1,p(.)}(\Omega)$;
there exists $v \in L^{p(.)-1}(\partial\Omega)$ such that for a.e. $k > 0$, one has $T_k(v) = \tau(T_k(u))$;
$b(u) \in L^1(\Omega)$;

$$\lim_{k\to\infty} \int_{\{k<|u|<k+1\}} a(x, \nabla u).\nabla u dx = 0, \tag{2.15}$$

and, for all $S \in \mathbb{S}$ and for all $\phi \in W^{1,p(.)}(\Omega) \cap L^\infty(\Omega)$,

$$\int_\Omega S(u)a(x, \nabla u).\nabla\phi dx + \int_\Omega S'(u)a(x, \nabla u).(\nabla u)\phi dx + \int_\Omega b(u)S(u)\phi dx$$
$$+ \int_{\partial\Omega} S(u)|u|^{p(x)-2}u\phi d\sigma = \int_\Omega f S(u)\phi dx, \tag{2.16}$$

where $d\sigma$ is the surface measure on $\partial\Omega$. □

*Remark 2.3.2* All the integrals in (2.15) and (2.16) make sense. For the third and fifth integrals of (2.16), it is clear. We focus our attention on the other integrals. As the support of $S$ is compact, we can write $suppS \subset [-k, k]$ with $k > 0$. So, one has $S(u)|u|^{p(x)-2}u = S(u)|T_k(u)|^{p(x)-2}T_k(u)$, and as $T_k(u) \in W^{1,p(.)}(\Omega) \hookrightarrow L^{p(.)}(\partial\Omega)$, then $T_k(u) \in L^{p(.)}(\partial\Omega)$, and so $|T_k(u)|^{p(x)-2}T_k(u) \in L^{p'(.)}(\partial\Omega)$. Also, as $\phi \in W^{1,p(.)}(\Omega) \hookrightarrow L^{p(.)}(\partial\Omega)$, then we have $S(u)|T_k(u)|^{p(x)-2}T_k(u)\phi \in L^1(\partial\Omega)$ by Hölder-type inequality, and the fourth integral of (2.16) makes sense. Moreover, as $suppS \subset [-k, k]$ and thanks to Proposition 2.3.1, we can replace the terms $\nabla u$ by $\nabla T_k(u)$ in Eq. (2.16). Since $T_k(u) \in W^{1,p(.)}(\Omega)$, then, by the growth assumption (2.7), the term $S(u)a(x, \nabla u)$ is in $L^{p'(.)}(\Omega)$, and so, the terms $S(u)a(x, \nabla u).\nabla\phi$ and $S'(u)a(x, \nabla u).\nabla u$ both lie in $L^1(\Omega)$ by Hölder-type inequality. Thus, all the terms in (2.16) make sense.

For the integral in (2.15), one can replace $\nabla u$ by $\nabla T_{k+1}(u)$ thanks to Proposition 2.3.1, and so, by Hölder-type inequality, $\chi_{\{k<|u|<k+1\}}a(x, \nabla u).\nabla u \in L^1(\Omega)$. Hence, the integral in (2.15) makes sense.

Notice that we do not require explicitly that $u \in \mathcal{T}^{1,p(.)}_{tr}(\Omega)$ in the definition above because we can replace it by the technical hypothesis:

"There exists $v \in L^{p(.)-1}(\partial\Omega)$ such that for a.e. $k > 0$, one has $T_k(v) = \tau(T_k(u))$"

for the analytic interpretation of the trace of $u$ such that $T_k(u) \in W^{1,p(.)}(\Omega)$ (see [4]). □

The following theorem guarantees the existence of the sequence of renormalized solutions associated to problems $(Pb_n)$.

**Theorem 2.3.1** *Assume (2.3)–(2.8). Then, there exists at least one renormalized solution u of the elliptic equation (2.1).* □

For the proof of existence of solution, we recall the definition of the weak solution of the elliptic equation (2.1) for data $f \in L^{\infty}(\Omega)$.

**Definition 2.3.2 (See [16, Definition 3.1])** Let $f \in L^{\infty}(\Omega)$; a measurable function $u : \Omega \to \mathbb{R}$ is a weak solution of (2.1) if $u \in W^{1,p(.)}(\Omega)$, $b(u) \in L^{\infty}(\Omega)$, $|u|^{p(x)-2}u \in L^{\infty}(\partial\Omega)$, and

$$\int_{\Omega} a(x, \nabla u).\nabla\phi\, dx + \int_{\Omega} b(u)\phi\, dx + \int_{\partial\Omega} |u|^{p(x)-2}u\phi\, d\sigma = \int_{\Omega} f\phi\, dx, \qquad (2.17)$$

for all $\phi \in W^{1,p(.)}(\Omega)$. □

### 2.3.1 *Proof of Theorem 2.3.1*

The proof of existence of a renormalized solution of (2.1) is done in three steps: first, we introduce approximating problems for which existence of weak solutions $u_n$ is obvious; second, we establish some basic convergence results; third, we prove that these approximate solutions $u_n$ tend, as $n$ goes to infinity, to a measurable function $u$ that is a renormalized solution of the problem (2.1).

#### 2.3.1.1 Approximate Solutions

***Proof of Theorem 2.3.1*** Let $f_n = T_n(f)$ for all $n \in \mathbb{N}$; let us consider the approximate problems

$$\begin{cases} b(u_n) - \operatorname{div} a(x, \nabla u_n) = f_n & \text{in } \Omega, \\ a(x, \nabla u_n).\eta = -|u_n|^{p(x)-2}u_n & \text{on } \partial\Omega. \end{cases} \qquad (2.18)$$

One has $f_n \in L^{\infty}(\Omega)$, so according to Theorem 3.2 in [16], the problem (2.18) admits a weak solution $u_n$, i.e., $u_n \in W^{1,p(.)}(\Omega)$, $b(u_n) \in L^{\infty}(\Omega)$, $|u_n|^{p(x)-2}u_n \in L^{\infty}(\partial\Omega)$, and

$$\int_{\Omega} a(x, \nabla u_n).\nabla\phi\, dx + \int_{\Omega} b(u_n)\phi\, dx + \int_{\partial\Omega} |u_n|^{p(x)-2}u_n\phi\, d\sigma = \int_{\Omega} f_n\phi\, dx, \qquad (2.19)$$

for all $\phi \in W^{1,p(.)}(\Omega)$.

We are going to prove that the sequence of these approximated solutions $u_n$ converges to a measurable function $u$ that is a renormalized solution of the limit problem (2.1). Note that the sequence of the terms $f_n$ converges strongly to $f \in L^1(\Omega)$. Moreover, we have

$$\|f_n\|_{L^1(\Omega)} \leq \|f\|_{L^1(\Omega)}, \text{ for all } n \in \mathbb{N}.$$

#### 2.3.1.2 Convergence Results

The following proposition (see the propositions 4.9 and 4.12, see also the relation (4.70) in [16]) will be useful in the sequel.

**Proposition 2.3.2** *Assume that (2.3)–(2.8) hold true, and let $u_n$ be the weak solution of (3.6); then:*

$$(i)\ \begin{cases} \text{The sequence } (u_n)_{n\in\mathbb{N}} \text{ is a Cauchy sequence in measure. In particular,} \\ \text{there exist a measurable function } u \text{ and a subsequence still denoted } (u_n)_{n\in\mathbb{N}} \\ \text{such that } u_n \longrightarrow u \text{ in measure and } u_n \longrightarrow u \text{ a.e. in } \Omega. \end{cases}$$

*(ii) For all $k > 0$, $\nabla T_k(u_n)$ converges to $\nabla T_k(u)$ in $(L^1(\Omega))^N$.*
*(iii) For all $k > 0$, $a(x, \nabla T_k(u_n))$ converges strongly to $a(x, \nabla T_k(u))$ in $(L^1(\Omega))^N$ and weakly in $(L^{p'(.)}(\Omega))^N$.*
*(iv) $u_n$ converges a.e. to some function $v$ on $\partial\Omega$.*
*(v) $|u_n|^{p(x)-2}u_n$ converges strongly to $|u|^{p(x)-2}u$ in $L^1(\partial\Omega)$.* □

*Remark 2.3.3* For any $k > 0$, the sequence $(T_k(u_n))_{n\in\mathbb{N}}$ is uniformly bounded in $W^{1,p(.)}(\Omega)$ (see [16, Lemma 4.8]). Then, we can assume, up to a subsequence, that

$$T_k(u_n) \rightharpoonup T_k(u) \text{ in } W^{1,p(.)}(\Omega),$$

and by the compact imbedding of $W^{1,p(.)}(\Omega)$ in $L^{p(.)}(\Omega)$ and in $L^{p(.)}(\partial\Omega)$, we have

$$T_k(u_n) \to T_k(u) \text{ strongly in } L^{p(.)}(\Omega)$$

and

$$T_k(u_n) \to T_k(u) \text{ strongly in } L^{p(.)}(\partial\Omega).$$

The function $v$ in Proposition 2.3.2–(iv) is defined on $\partial\Omega$ by

$$v(x) := T_k(u(x)) \text{ if } x \in \partial\Omega \text{ with } |T_k(u(x))| < k,$$

moreover, $T_k(u_n) \longrightarrow T_k(u)$ a.e. on $\partial\Omega$, and we have so $v(x) = u(x)$ a.e. on $\partial\Omega$.

We also have the following convergence result.

**Lemma 2.3.1** *For all* $k > 0$, *the sequence* $a(x, \nabla T_k(u_n)).\nabla T_k(u_n)$ *converges strongly to* $a(x, \nabla T_k(u)).\nabla T_k(u)$ *in* $(L^1(\Omega))^N$. □

***Proof*** We use Vitali's theorem to get this strong convergence in $L^1(\Omega)$.

By Proposition 2.3.2, one has

$$a(x, \nabla T_k(u_n)).\nabla T_k(u_n) \longrightarrow a(x, \nabla T_k(u)).\nabla T_k(u) \text{ a.e. in } \Omega.$$

Moreover, by Hölder-type inequality, we get, for $E \subset \Omega$,

$$\int_E a(x, \nabla T_k(u_n)).\nabla T_k(u_n)\, dx \le 2 \left\| a(x, \nabla T_k(u_n)) \right\|_{L^{p'(.)}(\Omega)} \left\| \nabla T_k(u_n)\chi_E \right\|_{L^{p(.)}(\Omega)} .$$

But, the sequence $\big(a(x, \nabla T_k(u_n))\big)_{n\in\mathbb{N}}$ is bounded in $L^{p'(.)}(\Omega)$ because it converges weakly in $L^{p'(.)}(\Omega)$ and $\big(|\nabla T_k(u_n)|^{p(x)}\big)_{n\in\mathbb{N}}$ is equi-integrable in $\Omega$ because $(\nabla T_k(u_n))_{n\in\mathbb{N}}$ converges weakly in $L^{p(.)}(\Omega)$. So,

$$\lim_{meas(E)} \int_E |\nabla T_k(u_n)|^{p(x)}\, dx = 0.$$

Therefore, by Lemma 2.2.2, $\|\nabla T_k(u_n)\chi_E\|_{L^{p(.)}(\Omega)} \to 0$ as $meas(E) \to 0$. Hence, one has $a(x, \nabla T_k(u_n)).\nabla T_k(u_n)$ that is equi-integrable in $\Omega$, and so, by Vitali's theorem, one has the result. □

#### 2.3.1.3 Existence of Renormalized Solution

**Lemma 2.3.2** *The function u verifies the renormalized formulation (2.16).* □

***Proof*** Let $\phi \in W^{1,p(.)}(\Omega) \cap L^\infty(\Omega)$ and $S \in \mathbb{S}$. We take $S(u_n)\phi$ as test function in (2.17) to get

$$\int_\Omega S'(u_n)a(x, \nabla u_n).(\nabla u_n)\phi\, dx + \int_\Omega S(u_n)a(x, \nabla u_n).\nabla\phi\, dx$$
$$+ \int_\Omega b(u_n)S(u_n)\phi\, dx + \int_{\partial\Omega} |u_n|^{p(x)-2}u_n S(u_n)\phi\, d\sigma = \int_\Omega f_n S(u_n)\phi\, dx. \tag{2.20}$$

As $suppS \subset (-k, k)$ for some real number $k > 0$, $\nabla u_n$ can be replaced by $\nabla T_k(u_n)$ in (2.20), and we get

$$\int_\Omega S'(u_n)a(x,\nabla T_k(u_n)).\nabla T_k(u_n)\phi\,dx+\int_\Omega S(u_n)a(x,\nabla T_k(u_n)).\nabla\phi\,dx$$
$$+\int_\Omega b(u_n)S(u_n)\phi\,dx+\int_{\partial\Omega}|u_n|^{p(x)-2}u_nS(u_n)\phi\,d\sigma=\int_\Omega f_nS(u_n)\phi\,dx. \tag{2.21}$$

By definition, the function $S$ is continuous and $suppS$ is compact, so the sequences $S(u_n)$ and $|u_n|^{p(x)-2}u_nS(u_n)$ are bounded. The function $b$ is continuous, non-decreasing, and $b(0)=0$, so the sequence $b(u_n)S(u_n)$ is bounded. Moreover, the sequences $b(u_n)S(u_n)$ and $|u_n|^{p(x)-2}u_nS(u_n)$ converge almost everywhere, respectively, to $b(u)S(u)$ and $|u|^{p(x)-2}uS(u)$, respectively, in $\Omega$ and on $\partial\Omega$. Thus, by Lebesgue-dominated convergence theorem, the sequences $b(u_n)S(u_n)$ and $|u_n|^{p(x)-2}u_nS(u_n)$ converge to $b(u)S(u)$ and $|u|^{p(x)-2}uS(u)$, respectively, strongly in $L^1(\Omega)$ and in $L^1(\partial\Omega)$. One has so

$$\lim_{n\to\infty}\int_\Omega b(u_n)S(u_n)\phi\,dx=\int_\Omega b(u)S(u)\phi\,dx$$

and

$$\lim_{n\to\infty}\int_{\partial\Omega}|u_n|^{p(x)-2}u_nS(u_n)\phi\,d\sigma=\int_{\partial\Omega}|u|^{p(x)-2}uS(u)\phi\,d\sigma.$$

By Proposition 2.3.2-$(iii)$, one has $a(x,\nabla T_k(u_n))$ that converges weakly to $a(x,\nabla T_k(u))$ in $L^{p'(.)}(\Omega)$, and as $S(u_n)\nabla\phi$ converges strongly to $S(u)\nabla\phi$ in $L^{p(.)}(\Omega)$, we deduce that

$$\lim_{n\to\infty}\int_\Omega S(u_n)a(x,\nabla T_k(u_n)).\nabla\phi\,dx=\int_\Omega S(u)a(x,\nabla T_k(u)).\nabla\phi\,dx.$$

By Lemma 2.3.1, $a(x,\nabla T_k(u_n)).\nabla T_k(u_n)$ converges strongly to $a(x,\nabla T_k(u)).\nabla T_k(u)$ in $L^1(\Omega)$. So,

$$\lim_{n\to\infty}\int_\Omega S'(u_n)a(x,\nabla T_k(u_n)).\nabla T_k(u_n)\phi\,dx=\int_\Omega S'(u)a(x,\nabla T_k(u)).\nabla T_k(u)\phi\,dx.$$

Now, we are interested in the right-hand side of (2.21). Since $f_n=T_n(f)$ converges strongly to $f$ in $L^1(\Omega)$, then we conclude that

$$\lim_{n\to\infty}\int_\Omega f_nS(u_n)\phi\,dx=\int_\Omega fS(u)\phi\,dx.$$

Thus, passing to the limit in (2.21), we get that $u$ verifies equality (2.16). □

**Lemma 2.3.3** *The function $u$ respects the estimate (2.15).* □

***Proof*** Let us take $\phi = T_{k+1}(u_n) - T_k(u_n)$ as test function in (2.17) to get

$$\int_\Omega a(x, \nabla u_n).\nabla(T_{k+1}(u_n) - T_k(u_n))\, dx + \int_\Omega b(u_n)(T_{k+1}(u_n) - T_k(u_n))\, dx$$
$$+ \int_{\partial\Omega} |u_n|^{p(x)-2} u_n (T_{k+1}(u_n) - T_k(u_n))\, d\sigma = \int_\Omega f_n (T_{k+1}(u_n) - T_k(u_n))\, dx. \tag{2.22}$$

One has $T_{k+1}(z) - T_k(z) = \begin{cases} sign(z) & \text{if } |z| > k+1, \\ 0 & \text{if } |z| < k, \\ z - ksign(z) & \text{if } k \leq |z| \leq k+1. \end{cases}$

The test function $T_{k+1}(u_n) - T_k(u_n)$ has a support contained in the set $\{|u_n| \geq k\}$, is bounded by one and has the same sign that $u_n$ which has the same sign that $b(u_n)$ as $b$ is non-decreasing and $b(0) = 0$. So, $u_n(T_{k+1}(u_n) - T_k(u_n))$ and $b(u_n)(T_{k+1}(u_n) - T_k(u_n))$ are non-negative. We also have $\nabla(T_{k+1}(u_n) - T_k(u_n)) = \nabla u_n \chi_{\{k<|u_n|<k+1\}}$, and so, the equality (2.22) gives

$$\int_{\{k<|u_n|<k+1\}} a(x, \nabla u_n).\nabla u_n\, dx \leq \int_{\{|u_n|\geq k\}} |f_n|\, dx. \tag{2.23}$$

The sequence $(f_n)_{n\in\mathbb{N}}$ is equi-integrable on $\Omega$ as it converges strongly in $L^1(\Omega)$. It is sufficient to prove that $meas(\{|u_n| \geq k\})$ converges to zero as $k$ goes to infinity uniformly in $n$. For that, we take $T_k(u_n)$ as test function in the weak formulation (2.17) to get

$$\int_\Omega a(x, \nabla T_k(u_n)).\nabla T_k(u_n)\, dx + \int_\Omega b(u_n) T_k(u_n) dx + \int_{\partial\Omega} |u_n|^{p(x)-2} u_n T_k(u_n) d\sigma$$
$$= \int_\Omega f_n T_k(u_n) dx. \tag{2.24}$$

As $a(x, \nabla T_k(u_n)).\nabla T_k(u_n)$ is positive by (2.6) and as $b(u_n)T_k(u_n)$ and $u_n T_k(u_n)$ are positive since $b$ and $T_k$ are non-decreasing and $b(0) = T_k(0) = 0$, so, from (2.24), we get

$$\int_\Omega b(u_n) T_k(u_n)\, dx \leq \int_\Omega f_n T_k(u_n)\, dx, \tag{2.25}$$

which gives

$$\int_{\{|u_n|\geq k\}} b(u_n) T_k(u_n)\, dx \leq \int_\Omega f_n T_k(u_n)\, dx. \tag{2.26}$$

Since $b$ is non-decreasing and $b(0) = 0$, then $|b(u_n)| \geq \min(b(k), |b(-k)|)$ on $\{|u_n| \geq k\}$, and (2.26) gives

$$\min(b(k), |b(-k)|) \int_{\{|u_n| \geq k\}} |T_k(u_n)|\, dx, \leq \int_\Omega f_n T_k(u_n)\, dx,$$

which becomes

$$k \min(b(k), |b(-k)|) meas(\{|u_n| \geq k\}) \leq k \, \|f\|_{L^1(\Omega)},$$

since $|T_k(u_n)| = k$ on $\{|u_n| \geq k\}$ and $\|f_n\|_{L^1(\Omega)} \leq \|f\|_{L^1(\Omega)}$.

Thus,

$$meas(\{|u_n| \geq k\}) \leq \frac{\|f\|_{L^1(\Omega)}}{\min(b(k), |b(-k)|)} \longrightarrow 0, \text{ as } k \longrightarrow \infty, \tag{2.27}$$

since $b$ is non-decreasing and onto and so has an infinity limit at infinity.

Hence, by (2.27) and the equi-integrability of $f_n$, the right-hand side of (2.23) tends to zero uniformly in $n$ as $k \to \infty$. So, by the monotonicity (2.6), one has

$$\lim_{k \to \infty} \sup_n \int_{\{k < |u_n| < k+1\}} a(x, \nabla u_n).\nabla u_n dx = 0$$

or

$$\lim_{k \to \infty} \lim_{n \to \infty} \int_\Omega a(x, \nabla T_{k+1}(u_n)).\nabla T_{k+1}(u_n) \chi_{\{k < |u_n| < k+1\}} dx = 0. \tag{2.28}$$

Let

$$D_{n,k} := a(x, \nabla T_{k+1}(u_n)).\nabla T_{k+1}(u_n).$$

According to Lemma 2.3.1, $D_{n,k} \to a(x, \nabla T_{k+1}(u)).\nabla T_{k+1}(u)$ strongly in $L^1(\Omega)$. Moreover, as $u_n$ converges a.e. to $u$ by Proposition 2.3.2, so by the continuity of $\chi_{(k,k+1)\cup(-k-1,-k)}(.)$ on the image of $\Omega$ by $u(.)$, we conclude that, as $n \to \infty$,

$$\begin{aligned}\chi_{\{k < |u_n| < k+1\}} &= \chi_{(k,k+1)\cup(-k-1,-k)}(u_n) \to \chi_{(k,k+1)\cup(-k-1,-k)}(u) \\ &= \chi_{\{k < |u| < k+1\}} \text{ a.e. in } \Omega.\end{aligned}$$

Indeed, $\chi_{(k,k+1)\cup(-k-1,-k)}(.)$ is continuous if $meas\ (\{|u| = k\}) = 0$ for a.e. $k > 0$. But, for all $n$, one has

$$\left\{|T_k(u)| \geq k - \frac{1}{2}\right\} \subset \{|u_n| \geq k - 1\} \cup \left\{|T_k(u_n) - T_k(u)| > \frac{1}{2}\right\},$$

and so

$$meas\left(\left\{|T_k(u)| \geq k - \frac{1}{2}\right\}\right) \leq meas\left(\{|u_n| \geq k-1\}\right) + meas\left(\left\{|T_k(u_n) - T_k(u)| > \frac{1}{2}\right\}\right).$$

From (2.27) and as $T_k(u_n)$ converges to $T_k(u)$ in measure in $\Omega$, one gets, as $n \to \infty$,

$$meas\left(\{|u| = k\}\right) \leq meas\left(\left\{|T_k(u)| \geq k - \frac{1}{2}\right\}\right) \leq 0 \Longrightarrow meas\left(\{|u| = k\}\right) = 0.$$

Now, one has

$$\begin{cases} D_{n,k} \to a(x, \nabla T_{k+1}(u)).\nabla T_{k+1}(u) \text{ strongly in } L^1(\Omega), \\ D_{n,k}\chi_{\{k<|u_n|<k+1\}} \to a(x, \nabla T_{k+1}(u)).\nabla T_{k+1}(u)\chi_{\{k<|u|<k+1\}} \text{ a.e. in } \Omega, \text{ and} \\ \left|D_{n,k}\chi_{\{k<|u_n|<k+1\}}\right| \leq D_{n,k} \in L^1(\Omega) \text{ a.e. in } \Omega, \text{ for all } n \in \mathbb{N}. \end{cases}$$

So, by the Lebesgue generalized convergence theorem, we can write

$$\lim_{n\to\infty} \int_\Omega D_{n,k}\chi_{\{k<|u_n|<k+1\}}\,dx = \int_\Omega a(x, \nabla T_{k+1}(u)).\nabla T_{k+1}(u)\chi_{\{k<|u|<k+1\}}\,dx. \tag{2.29}$$

Now, coming back to the equality (2.28), we get the equality

$$\lim_{k\to\infty} \int_{\{k<|u|<k+1\}} a(x, \nabla u).\nabla u dx = 0, \tag{2.30}$$

which proves Lemma 2.3.3. □

**Lemma 2.3.4** *The function u is a renormalized solution of (2.1).* □

***Proof*** From Proposition 2.3.2 and Remark 2.3.3 and results of [2, proof of Theorem 3.1], one has $T_k(u) \in W^{1,p(.)}(\Omega)$, and there exists a function $v \in L^1(\partial\Omega)$ such that $u_n \to v$ a.e. on $\partial\Omega$ and $T_k(v) = \tau(T_k(u))$ a.e. on $\partial\Omega$ for all $k > 0$. Now, taking $\phi = sign(u_n)$ as test function in the weak formulation (2.17) for $u_n$, we get

$$\int_\Omega b(u_n)sign(u_n)\,dx + \int_{\partial\Omega} |u_n|^{p(x)-2}u_n sign(u_n)\,d\sigma = \int_\Omega f_n sign(u_n)\,dx,$$

which implies

$$\int_{\partial\Omega} |u_n|^{p(x)-2}|u_n|\,d\sigma \leq \|f\|_{L^1(\Omega)}. \tag{2.31}$$

By Fatou's lemma, as $n \to \infty$, (2.31) gives

$$\int_{\partial\Omega} |u|^{p(x)-2}|u|\, d\sigma \le \|f\|_{L^1(\Omega)},$$

which means that $|u|^{p(x)-2}u \in L^1(\partial\Omega)$, i.e., $v \in L^{p(x)-1}(\partial\Omega)$.

Now, we prove that $b(u) \in L^1(\Omega)$. By (2.26), one has

$$\begin{aligned}\int_\Omega |b(u_n)|\, dx &= \int_{\{|u_n|<k\}} |b(u_n)|\, dx + \int_{\{|u_n|\ge k\}} |b(u_n)|\, dx \\ &\le \max(b(k), |b(-k)|)\text{meas}(\Omega) + \|f\|_{L^1(\Omega)}.\end{aligned}$$

Therefore, $\|b(u_n)\|_{L^1(\Omega)}$ is uniformly bounded. One also has, by the continuity of $b$, $b(u_n) \longrightarrow b(u)$ a.e. in $\Omega$. So, Fatou's lemma gives us

$$\begin{aligned}\int_\Omega |b(u)|\, dx &\le \liminf_{n\to\infty} \int_\Omega |b(u_n)|\, dx \\ &\le \max(b(k), |b(-k)|)\text{meas}(\Omega) + \|f\|_{L^1(\Omega)}.\end{aligned}$$

Hence, $b(u) \in L^1(\Omega)$. Also, thanks to lemmas 2.3.2 and 2.3.3, we conclude that $u$ is a renormalized solution to the problem (2.1).

This ends the proof of Theorem 2.3.1. □

Now, let us go to the uniqueness of the solution of problem (2.1).

### 2.3.2 Uniqueness of Renormalized Solution

**Theorem 2.3.2** *Assume (2.3)–(2.8), $f \in L^1(\Omega)$. Then, there is uniqueness of the renormalized solution to the problem (2.1).* □

***Proof*** Let $k, h > 0$ and $u_1$ and $u_2$ be two renormalized solutions of problem (2.1) associated to the same data $f \in L^1(\Omega)$. As $T_h(u_2) \in W^{1,p(.)}(\Omega) \cap L^\infty(\Omega)$, then one has $T_k(u_1 - T_h(u_2)) \in W^{1,p(.)}(\Omega) \cap L^\infty(\Omega)$ that can be taken as test function in (2.16) for $u_1$. Similarly, we can take $T_k(u_2 - T_h(u_1))$ as test function in (2.16) for $u_2$. Upon addition, we get

$$\begin{aligned}&\int_{\{|u_1-T_h(u_2)|\le k\}} S_M(u_1)a(x, \nabla u_1).\nabla(u_1 - T_h(u_2))\, dx \\ &+ \int_{\{|u_2-T_h(u_1)|\le k\}} S_M(u_2)a(x, \nabla u_2).\nabla(u_2 - T_h(u_1))\, dx \\ &+ \int_\Omega S'_M(u_1)a(x, \nabla u_1).(\nabla u_1)T_k(u_1 - T_h(u_2))\, dx\end{aligned}$$

$$+\int_\Omega S'_M(u_2)a(x,\nabla u_2).(\nabla u_2)T_k(u_2-T_h(u_1))\,dx$$
$$+\int_\Omega S_M(u_1)b(u_1)T_k(u_1-T_h(u_2))\,dx+\int_\Omega S_M(u_2)b(u_2)T_k(u_2-T_h(u_1))\,dx$$
$$+\int_{\partial\Omega} S_M(u_1)|u_1|^{p(x)-2}u_1T_k(u_1-T_h(u_2))\,d\sigma$$
$$+\int_{\partial\Omega} S_M(u_2)|u_2|^{p(x)-2}u_2T_k(u_2-T_h(u_1))\,d\sigma$$
$$=\int_\Omega f\Big(S_M(u_1)T_k(u_1-T_h(u_2))+S_M(u_2)T_k(u_2-T_h(u_1))\Big)\,dx, \tag{2.32}$$

where $(S_M)$ is the sequence of functions in $\mathbb{S}$ defined in (2.14). While $M$ and $k$ are fixed, $h$ can be sent to infinity. Define the sets

$$E_1:=\{|u_1-u_2|\le k,\ |u_2|\le h\},\ E_2:=E_1\cap\{|u_1|\le h\},\ \text{and } E_3:=E_1\cap\{|u_1|>h\}.$$

We start with the first integral in (2.32). By (2.6), we have

$$\int_{\{|u_1-T_h(u_2)|\le k\}} S_M(u_1)a(x,\nabla u_1).\nabla(u_1-T_h(u_2))\,dx$$
$$=\int_{\{|u_1-T_h(u_2)|\le k,|u_2|\le h\}} S_M(u_1)a(x,\nabla u_1).\nabla(u_1-T_h(u_2))\,dx$$
$$+\int_{\{|u_1-T_h(u_2)|\le k,|u_2|>h\}} S_M(u_1)a(x,\nabla u_1).\nabla(u_1-T_h(u_2))\,dx$$
$$=\int_{\{|u_1-T_h(u_2)|\le k,|u_2|\le h\}} S_M(u_1)a(x,\nabla u_1).\nabla(u_1-u_2)\,dx$$
$$+\int_{\{|u_1-T_h(u_2)|\le k,|u_2|>h\}} S_M(u_1)a(x,\nabla u_1).\nabla u_1\,dx$$
$$\ge\int_{\{|u_1-T_h(u_2)|\le k,|u_2|\le h\}} S_M(u_1)a(x,\nabla u_1).\nabla(u_1-u_2)\,dx$$
$$=\int_{E_2} S_M(u_1)a(x,\nabla u_1).\nabla(u_1-u_2)\,dx+\int_{E_3} S_M(u_1)a(x,\nabla u_1).\nabla(u_1-u_2)\,dx$$
$$=\int_{E_2} S_M(u_1)a(x,\nabla u_1).\nabla(u_1-u_2)\,dx+\int_{E_3} S_M(u_1)a(x,\nabla u_1).\nabla u_1\,dx$$
$$-\int_{E_3} S_M(u_1)a(x,\nabla u_1).\nabla u_2\,dx$$

$$\geq \int_{E_2} S_M(u_1)a(x,\nabla u_1).\nabla(u_1-u_2)\,dx - \int_{E_3} S_M(u_1)a(x,\nabla u_1).\nabla u_2\,dx. \tag{2.33}$$

Using (2.7) and Hölder-type inequality, the last integral in (2.33) gives

$$\begin{aligned}&\left|\int_{E_3} S_M(u_1)a(x,\nabla u_1).\nabla u_2\,dx\right| \\ &\quad \leq C\sup_M \|S_M\|_{L^\infty}\left(\|\mathcal{M}\|_{p'(.)} + \left\||\nabla u_1|^{p(x)-1}\right\|_{L^{p'(.)}(\{h<|u_1|\leq h+k\})}\right) \\ &\quad \times \|\nabla u_2\|_{L^{p(.)}(\{h-k<|u_2|\leq h\})}\,. \end{aligned} \tag{2.34}$$

Now, we take $\phi = T_k(u_1 - T_h(u_1))$ as test function in (2.16) for $u_1$ and $S \in \mathbb{S}$ such that $S = S_{h+k+1}$. We get

$$\begin{aligned}&\int_\Omega S(u_1)a(x,\nabla u_1).\nabla T_k(u_1-T_h(u_1))\,dx \\ &\quad + \int_\Omega S'(u_1)a(x,\nabla u_1).(\nabla u_1)T_k(u_1-T_h(u_1))\,dx \\ &\quad + \int_\Omega b(u_1)S(u_1)T_k(u_1-T_h(u_1))\,dx \\ &\quad + \int_{\partial\Omega} S(u_1)|u_1|^{p(x)-2}u_1T_k(u_1-T_h(u_1))\,d\sigma \\ &= \int_\Omega fS(u_1)T_k(u_1-T_h(u_1))\,dx.\end{aligned}$$

Since the third and fourth integrals are non-negative, then one has

$$\begin{aligned}&\int_{\{h<|u_1|\leq h+k\}} a(x,\nabla u_1).\nabla u_1\,dx - k\int_{\{h+k<|u_1|\leq h+k+1\}} a(x,\nabla u_1).\nabla u_1\,dx \\ &\leq k\int_{\{|u_1|>h\}} |f|\,dx,\end{aligned}$$

and so

$$\begin{aligned}&\int_{\{h<|u_1|\leq h+k\}} a(x,\nabla u_1).\nabla u_1\,dx \\ &\quad \leq k\left(\int_{\{|u_1|>h\}} |f|\,dx + \int_{\{h+k<|u_1|\leq h+k+1\}} a(x,\nabla u_1).\nabla u_1\,dx\right).\end{aligned}$$

By using (2.8), we get

$$C_2 \int_{\{h<|u_1|\le h+k\}} |\nabla u_1|^{p(x)}\,dx$$
$$\le k\left(\int_{\{|u_1|>h\}} |f|\,dx + \int_{\{h+k<|u_1|\le h+k+1\}} a(x,\nabla u_1).\nabla u_1\,dx\right).$$

By (2.15) and since $meas(\{|u_1|>h\}) \to 0$ as $h\to\infty$, and since $f\in L^1(\Omega)$, we deduce that

$$\lim_{h\to\infty}\int_{\{h<|u_1|\le h+k\}} |\nabla u_1|^{p(x)}\,dx = 0, \text{ for any fixed number } k>0,$$

and so, by Lemma 2.2.2, we get $\lim_{h\to\infty}\left\| |\nabla u_1|^{p(x)-1}\right\|_{L^{p'(\cdot)}(\{h<|u_1|\le h+k\})} = 0.$

Similarly, taking $\phi = T_k(u_2 - T_h(u_2))$ as test function in (2.16) for $u_2$ with the same $S$ in $\mathbb{S}$, we get

$$\lim_{h\to\infty}\int_{\{h<|u_2|\le h+k\}} |\nabla u_2|^{p(x)}\,dx = 0, \text{ for any fixed number } k>0.$$

Hence,

$$\lim_{h\to\infty}\int_{\{h-k<|u_2|\le h\}} |\nabla u_2|^{p(x)}\,dx = \lim_{l\to\infty}\int_{\{l<|u_2|\le l+k\}} |\nabla u_2|^{p(x)}\,dx = 0,$$

for any fixed number $k>0$ with $l = h-k$.

So, by Lemma 2.2.2,

$$\|\nabla u_2\|_{L^{p(\cdot)}(\{h-k<|u_2|\le h\})} \to 0 \text{ as } h\to\infty, \text{ for any fixed number } k>0.$$

Therefore, from (2.33) and (2.34), we obtain

$$\int_{\{|u_1-T_h(u_2)|\le k\}} S_M(u_1)a(x,\nabla u_1).\nabla(u_1 - T_h(u_2))\,dx$$
$$\ge I_h + \int_{E_2} S_M(u_1)a(x,\nabla u_1).\nabla(u_1-u_2)\,dx, \tag{2.35}$$

where $I_h$ converges to zero as $h\to\infty$.

We may adopt the same procedure to treat the second term in (2.32) to obtain

$$\int_{\{|u_2-T_h(u_1)|\le k\}} S_M(u_2)a(x,\nabla u_2).\nabla(u_2 - T_h(u_1))\,dx$$

$$\geq J_h - \int_{E_2} S_M(u_2)a(x, \nabla u_2).\nabla(u_1 - u_2)\, dx, \tag{2.36}$$

where $J_h$ converges to zero as $h \to \infty$.

Now, for all $h, k > 0$, we set

$$K_h = \int_\Omega S_M(u_1)b(u_1)T_k(u_1 - T_h(u_2))\, dx + \int_\Omega S_M(u_2)b(u_2)T_k(u_2 - T_h(u_1))\, dx,$$

$$P_h = \int_{\partial\Omega} S_M(u_1)|u_1|^{p(x)-2}u_1 T_k(u_1 - T_h(u_2))d\sigma$$
$$+ \int_{\partial\Omega} S_M(u_2)|u_2|^{p(x)-2}u_2 T_k(u_2 - T_h(u_1))d\sigma,$$

$$R_h = \int_\Omega S'_M(u_1)a(x, \nabla u_1).(\nabla u_1)T_k(u_1 - T_h(u_2))\, dx$$
$$+ \int_\Omega S'_M(u_2)a(x, \nabla u_2).(\nabla u_2)T_k(u_2 - T_h(u_1))\, dx,$$

and

$$F_h = \int_\Omega f\Big(S_M(u_1)T_k(u_1 - T_h(u_2)) + S_M(u_2)T_k(u_2 - T_h(u_1))\Big)\, dx.$$

We have

$$S_M(u_1)b(u_1)T_k(u_1 - T_h(u_2)) \to S_M(u_1)b(u_1)T_k(u_1 - u_2) \text{ a.e. in } \Omega, \text{ as } h \to \infty,$$

and

$$|S_M(u_1)b(u_1)T_k(u_1 - T_h(u_2))| \leq k|b(u_1)| \in L^1(\Omega).$$

Then, by Lebesgue-dominated convergence theorem, we deduce that

$$\lim_{h\to\infty} \int_\Omega S_M(u_1)b(u_1)T_k(u_1 - T_h(u_2))\, dx = \int_\Omega S_M(u_1)b(u_1)T_k(u_1 - u_2)\, dx. \tag{2.37}$$

Similarly, we have

$$\lim_{h\to\infty} \int_\Omega S_M(u_2)b(u_2)T_k(u_2 - T_h(u_1))\, dx = \int_\Omega S_M(u_2)b(u_2)T_k(u_2 - u_1)\, dx. \tag{2.38}$$

Using (2.37) and (2.38), we get

$$\lim_{h\to\infty} K_h = \int_\Omega \big(S_M(u_1)b(u_1) - S_M(u_2)b(u_2)\big)T_k(u_1-u_2)\,dx. \tag{2.39}$$

By the same procedure as above, we use the Lebesgue-dominated convergence theorem to obtain

$$\lim_{h\to\infty} P_h = \int_{\partial\Omega} \big(S_M(u_1)|u_1|^{p(x)-2}u_1 - S_M(u_2)|u_2|^{p(x)-2}u_2\big)T_k(u_1-u_2)\,d\sigma, \tag{2.40}$$

$$\lim_{h\to\infty} R_h = \int_\Omega \big(S'_M(u_1)a(x,\nabla u_1).\nabla u_1 - S'_M(u_2)a(x,\nabla u_2).\nabla u_2\big)T_k(u_1-u_2)\,dx, \tag{2.41}$$

and

$$\lim_{h\to\infty} F_h = \int_\Omega f\big(S_M(u_1) - S_M(u_2)\big)T_k(u_1-u_2)\,dx. \tag{2.42}$$

Using (2.35), (2.36), (2.39)–(2.42), we get from (2.32) the following inequality as $h\to\infty$.

$$\begin{aligned}
&\int_{\{|u_1-u_2|\le k\}} \big(S_M(u_1)a(x,\nabla u_1) - S_M(u_2)a(x,\nabla u_2)\big).\nabla(u_1-u_2)\,dx\\
&\quad+\int_\Omega \big(S'_M(u_1)a(x,\nabla u_1).\nabla u_1 - S'_M(u_2)a(x,\nabla u_2).\nabla u_2\big)T_k(u_1-u_2)\,dx,\\
&\quad+\int_\Omega \big(S_M(u_1)b(u_1) - S_M(u_2)b(u_2)\big)T_k(u_1-u_2)\,dx\\
&\quad+\int_{\partial\Omega} \big(S_M(u_1)|u_1|^{p(x)-2}u_1 - S_M(u_2)|u_2|^{p(x)-2}u_2\big)T_k(u_1-u_2)\,d\sigma\\
&\le \int_\Omega f\big(S_M(u_1) - S_M(u_2)\big)T_k(u_1-u_2)\,dx.
\end{aligned} \tag{2.43}$$

Now, we fix $k>0$, and we pass to the limit in (2.43), as $M$ tends to infinity. The second term of the left-hand side of (2.43) is, in absolute value, smaller than

$$k\left(\int_{\{M-1\le|u_1|\le M\}} a(x,\nabla u_1).\nabla u_1 + \int_{\{M-1\le|u_2|\le M\}} a(x,\nabla u_2).\nabla u_2\right),$$

which converges to zero, as $M\to\infty$, thanks to relation (2.15) for $u_1$ and for $u_2$. Therefore, the second integral of (2.43) converges to zero as $M\to\infty$.

Since $S_M\to 1$ as $M\to\infty$, then $\big(S_M(u_1)a(x,\nabla u_1) - S_M(u_2)a(x,\nabla u_2)\big).\big(\nabla u_1 - \nabla u_2\big)$ converges a.e. to $\big(a(x,\nabla u_1) - a(x,\nabla u_2)\big).\big(\nabla u_1 - \nabla u_2\big)$, and moreover, thanks to (2.7) and to Hölder-type inequality, one has

$$\begin{aligned}&\big|\big(S_M(u_1)a(x,\nabla u_1)-S_M(u_2)a(x,\nabla u_2)\big).\big(\nabla u_1-\nabla u_2\big)\big|\\&\quad\le\big(|a(x,\nabla u_1)|+|a(x,\nabla u_2)|\big).\big(|\nabla u_1|+|\nabla u_2|\big)\in L^1(\Omega).\end{aligned}$$

Thus, by the Lebesgue-dominated convergence theorem, the first integral in (2.43) converges to $\displaystyle\int_{\{|u_1-u_2|\le k\}}(a(x,\nabla u_1)-a(x,\nabla u_2)).\nabla(u_1-u_2)\,dx.$

Similarly, one has the third and fourth integrals in (2.43) that converge, respectively, to $\displaystyle\int_\Omega(b(u_1)-b(u_2))T_k(u_1-u_2)\,dx$ and $\displaystyle\int_{\partial\Omega}\Big(|u_1|^{p(x)-2}u_1-|u_2|^{p(x)-2}u_2\Big)T_k(u_1-u_2)\,d\sigma$ by the dominated convergence theorem.

We next examine the right-hand side of (2.43). For all $k>0$,

$$f\big(S_M(u_1)-S_M(u_2)\big)T_k(u_1-u_2)\to 0 \text{ a.e. in } \Omega \text{ as } M\to\infty$$

and

$$|f\big(S_M(u_1)-S_M(u_2)\big)T_k(u_1-u_2)|\le 2k|f|\in L^1(\Omega).$$

The dominated convergence theorem allows us to write

$$\lim_{M\to\infty}\int_\Omega f\big(S_M(u_1)-S_M(u_2)\big)T_k(u_1-u_2)\,dx=0.$$

Thus, as $M\to\infty$, (2.43) gives

$$\begin{aligned}&\int_{\{|u_1-u_2|\le k\}}(a(x,\nabla u_1)-(u_2)a(x,\nabla u_2)).\nabla(u_1-u_2)\,dx\\&\quad+\int_\Omega(b(u_1)-b(u_2))T_k(u_1-u_2)\,dx\\&\quad+\int_{\partial\Omega}\Big(|u_1|^{p(x)-2}u_1-|u_2|^{p(x)-2}u_2\Big)T_k(u_1-u_2)\,d\sigma\le 0,\end{aligned}\tag{2.44}$$

for all $k>0$.

The functions $b$, $T_k$, and $t\longmapsto|t|^{p-2}t$ $(p>1)$ are non-decreasing and vanish at 0; then, by using (2.6), one has all the integrals in (2.44) that are non-negative. Therefore, for all $k>0$,

$$\int_{\{|u_1-u_2|<k\}}\big(a(x,\nabla u_1)-a(x,\nabla u_2)\big).\big(\nabla u_1-\nabla u_2\big)\,dx=0,\tag{2.45}$$

$$\int_\Omega(b(u_1)-b(u_2))T_k(u_1-u_2)\,dx=0,\tag{2.46}$$

and

$$\int_{\partial\Omega}\left(|u_1|^{p(x)-2}u_1 - |u_2|^{p(x)-2}u_2\right)T_k(u_1-u_2)\,d\sigma = 0. \tag{2.47}$$

From the strict monotonicity assumption (2.6), (2.45) gives

$$\nabla u_1 = \nabla u_2 \text{ a.e. on } \{|u_1-u_2|<k\}. \tag{2.48}$$

Because $k$ is arbitrary, as $k \to \infty$, we conclude that $\nabla u_1 = \nabla u_2$ a.e. on $\Omega$. Therefore,

$$u_1 - u_2 = c \text{ a.e. in } \Omega, \text{ where } c \text{ is a real constant.} \tag{2.49}$$

From (2.47), for all $k > 0$, there exists a subset $\Omega_k^{\partial} \subset \partial\Omega$ with $meas_{N-1}(\Omega_k^{\partial}) = 0$ such that for all $x \in \partial\Omega \setminus \Omega_k^{\partial}$,

$$\left(|u_1(x)|^{p(x)-2}u_1(x) - |u_2(x)|^{p(x)-2}u_2(x)\right)T_k\big(u_1(x)-u_2(x)\big) = 0. \tag{2.50}$$

Therefore,

$$\left(|u_1(x)|^{p(x)-2}u_1(x) - |u_2(x)|^{p(x)-2}u_2(x)\right)\big(u_1(x)-u_2(x)\big) = 0,$$
$$\forall x \in \partial\Omega \setminus \Big(\bigcup_{k\in\mathbb{N}^*}\Omega_k^{\partial}\Big). \tag{2.51}$$

But, since $p_- > 1$, then from relation

$$\left(|\xi|^{p(x)-2}\xi - |\eta|^{p(x)-2}\eta\right)(\xi-\eta) > 0 \text{ for all } \xi,\eta \in \mathbb{R},\ \xi \neq \eta \text{ (cf. [10])},$$

(2.51) gives

$$u_1 = u_2 \text{ a.e. on } \partial\Omega. \tag{2.52}$$

Now, for $k > 0$ fixed, one has, from (2.49), $T_k(u_1-u_2) = T_k(c) \in W^{1,1}(\Omega)$, and so, according to Lemma 2.2.5, one gets

$$\int_{\Omega}|T_k(u_1-u_2)|\,dx \le C_3\left(\int_{\{|u_1-u_2|\le k\}}|\nabla(u_1-u_2)|\,dx + \int_{\partial\Omega}|T_k(u_1-u_2)|\,d\sigma\right), \tag{2.53}$$

which gives, by using (2.48) and (2.52),

$$u_1 = u_2 \text{ a.e. in } \Omega. \tag{2.54}$$

□

## 2.4 Continuous Dependence of Renormalized Solution

Assume that (2.3)–(2.8) are verified, for all $n \in \mathbb{N}$, with the diffusion flux functions $a_n(.,.)$, the exponents $p_n : \overline{\Omega} \longrightarrow [p_-, p_+]$, and the non-negative functions $\mathcal{M}_n$ in $L^{p'_n(.)}(\Omega)$ such that the sequence $\left(\mathcal{M}_n^{p'_n(.)}\right)_{n\in\mathbb{N}}$ is equi-integrable, and with $C_1$, $C_2$, $p_+$, and $p_-$ independent of $n$.

According to theorems 2.3.1 and 2.3.2, there is a unique renormalized solution $u_n$ to problem $(Pb_n)$ under assumption that data $f_n$ are in $L^1(\Omega)$.

The purpose of this section is to prove that the sequence of solutions $(u_n)_{n\in\mathbb{N}}$ to problems $(Pb_n)$ converges to a function $u$ that is the solution of limit problem (2.1), when we have the following convergence assumptions:

$$\left|\begin{array}{l}\text{for all bounded subset K of } \mathbb{R}^N, \\ \sup_{\xi\in K} |a_n(.,\xi) - a(.,\xi)| \text{ converges to zero in measure on } \Omega,\end{array}\right. \tag{2.55}$$

where $a(x,\xi)$ verifies the assumptions (2.5)–(2.8) with the exponent $p$ verifying (2.3) such that

$$p_n \text{ converges to } p \text{ in measure on } \Omega. \tag{2.56}$$

Finally, assume that

$$f_n \text{ converges to } f \text{ weakly in } L^1(\Omega). \tag{2.57}$$

We further assume that the exponent $p$ verifies log-Hölder continuity assumption:

$$\exists c > 0, \forall x, y \in \overline{\Omega}, x \neq y, -(\log|x-y|)|p(x) - p(y)| \leq c. \tag{2.58}$$

*Remark 2.4.1* Note that several regularity results for Sobolev spaces with variable exponents can be obtained thanks to log-Hölder continuity condition (2.58); in particular, $C^\infty(\overline{\Omega})$ is dense in $W^{1,p(.)}(\Omega)$ (for more details, see [7]).

Now, through the theorem below, we establish a structural stability result for the renormalized solutions.

**Theorem 2.4.1** *Under the assumptions (2.55)–(2.57), let $(u_n)_{n\in\mathbb{N}}$ be the sequence of renormalized solutions of the problems $(Pb_n)$ associated to $a_n(.,.)$, $f_n$ and $p_n$.*

*Assume that the exponents $p$, $p_n$ verify log-Hölder continuity assumption (2.58).*

*Then, there exists a measurable function $u$ defined on $\Omega$ such that $u_n$ converges to $u$ a.e. in $\Omega$ and $\nabla u_n$ converges to $\nabla u$ a.e. in $\Omega$, as $n \to \infty$. The function $u$ is a renormalized solution of the problem (2.1) associated to $a(.,.)$, $f$ and $p$.* □

***Proof of Theorem** 2.4.1* We shall divide the proof into several steps. Throughout the proof, all extracted subsequences of a sequence will be still noted as this sequence and all constant independent of $n$ will be denoted by $C$.

**Lemma 2.4.1**

(*i*) *For all* $k > 0$, *the sequence* $\left(\|T_k(u_n)\|_{1,p_n(.)}\right)_{n\in\mathbb{N}}$ *is bounded.*
(*ii*) *The sequence of renormalized solutions* $(u_n)_{n\in\mathbb{N}}$ *of the problems* $(Pb_n)$ *verifies, for* $k > 0$ *large enough, the following estimates:*

$$meas(\{|u_n| > k\}) \leq \frac{C\,\|f_n\|_{L^1(\Omega)}}{\min\left(b(k), |b(-k)|\right)}, \tag{2.59}$$

$$\sup_n meas(\{|u_n| > k\}) \to 0, \text{ as } k \to \infty, \tag{2.60}$$

*and*

$$\lim_{k\to\infty}\sup_n \int_{\{k<|u_n|<k+1\}} |\nabla u_n|^{p_n(x)}\,dx = 0. \tag{2.61}$$

(*iii*) *There exists a measurable function* $u$ *on* $\Omega$ *such that, for all* $k > 0$, $T_k(u_n)$ *converges to* $T_k(u) \in W^{1,p(.)}(\Omega)$ *weakly in* $W^{1,p(.)}(\Omega)$. *Furthermore,* $u_n \to u$ *a.e. on* $\Omega$, *and, for all* $k > 0$, $\nabla T_k(u_n)$ *converges to a Young measure* $(\nu_x^k)_x$ *on* $\mathbb{R}^N$ *in the sense of the nonlinear weak-* convergence and*

$$\nabla T_k(u) = \int_{\mathbb{R}^N} \lambda d\nu_x^k(\lambda). \tag{2.62}$$

(*iv*) *There exists a function* $v \in L^{p(x)-1}(\partial\Omega)$ *such that for a.e.* $k > 0$,

$$T_k(v) = \tau(T_k(u)) \text{ a.e. on } \partial\Omega.$$

(*v*) *For all* $k > 0$,

$$\int_{\mathbb{R}^N\times\Omega} |\lambda|^{p(x)} d\nu_x^k(\lambda) dx < \infty \text{ et } T_k(u) \in W^{1,p(.)}(\Omega).$$

(*vi*) *One has*

$$\lim_{k\to\infty}\int_\Omega |\nabla(T_{k+1}(u) - T_k(u))|^{p(x)}\,dx = 0. \tag{2.63}$$

***Proof***

(*i*) In the renormalized formulation (2.16) of the problem $(Pb_n)$, we choose $S \in \mathbb{S}$ such that $S = S_{h+k}$, where $S_{h+k}$ is defined in (2.14) with $h, k > 0$, $h$ large enough. Also, since $T_k(u_n) \in W^{1,p_n(.)}(\Omega) \cap L^\infty(\Omega)$ because $u_n$ is a renormalized solution of the problem $(Pb_n)$, then we can take $\phi = T_k(u_n)$ as test function in the renormalized formulation (2.16) and view that the terms $\int_\Omega b(u_n)S(u_n)T_k(u_n)\,dx$ and $\int_{\partial\Omega} S(u_n)|u_n|^{p_n(x)-2}u_n T_k(u_n)\,d\sigma$ are non-negative, and we get

$$\int_\Omega a_n(x, \nabla T_k(u_n)).\nabla T_k(u_n)\,dx + \int_\Omega S'(u_n)a_n(x, \nabla u_n).(\nabla u_n)T_k(u_n)\,dx$$
$$\leq k\int_\Omega |f_n|\,dx.$$

While $k$ is fixed, $h$ can be sent to infinity. The second term of the left-hand side of this last inequality vanishes, as $h \to \infty$, due to (2.15). And, by using coercivity condition (2.8), we have

$$C_2\int_\Omega |\nabla T_k(u_n)|^{p_n(x)}\,dx \leq k\int_\Omega |f_n|\,dx.$$

Since the sequence $(f_n)_{n\in\mathbb{N}}$ converges weakly in $L^1(\Omega)$, then the right-hand side of this last inequality is uniformly bounded. So, we obtain

$$\int_\Omega |\nabla T_k(u_n)|^{p_n(x)}\,dx \leq Ck. \tag{2.64}$$

Moreover,

$$\int_\Omega |T_k(u_n)|^{p_n(x)}dx \leq \int_\Omega k^{p_n(x)}dx \leq \max\left(k^{p_+}, k^{p_-}\right) meas(\Omega). \tag{2.65}$$

From (2.64) and (2.65), we deduce that the sequence $\rho_{1,p_n(.)}(T_k(u_n))$ is uniformly bounded. By Lemma 2.2.3 and the fact that $p_n(.) \in [p_-, p_+]$, one has

$$\|T_k(u_n)\|_{1,p_n(.)} \leq \max\left(\rho_{1,p_n(.)}(T_k(u_n))^{1/p_-}, \rho_{1,p_n(.)}(T_k(u_n))^{1/p_+}\right).$$

We conclude that the sequence $\|T_k(u_n)\|_{1,p_n(.)}$ is uniformly bounded.

(*ii*) In the renormalized formulation (2.16) of problem $(Pb_n)$, we assume that $S \in \mathbb{S}$ is such that $S = S_k$, and we take $\phi = T_{\frac{1}{k}}(u_n)$ as test function, with $k > 0$ large enough. We obtain

$$\int_{\Omega} S(u_n)a_n(x, \nabla u_n).\nabla T_{\frac{1}{k}}(u_n)\,dx + \int_{\Omega} S'(u_n)a_n(x, \nabla u_n).(\nabla u_n)T_{\frac{1}{k}}(u_n)\,dx$$
$$+ \int_{\Omega} b(u_n)S(u_n)T_{\frac{1}{k}}(u_n)\,dx + \int_{\partial\Omega} S(u_n)|u_n|^{p_n(x)-2}u_n T_{\frac{1}{k}}(u_n)\,d\sigma$$
$$= \int_{\Omega} f_n S(u_n)T_{\frac{1}{k}}(u_n)\,dx,$$

which becomes

$$\int_{\Omega} a_n\left(x, \nabla T_{\frac{1}{k}}(u_n)\right).\nabla T_{\frac{1}{k}}(u_n)\,dx + \int_{\Omega} S'(u_n)a_n(x, \nabla u_n).(\nabla u_n)T_{\frac{1}{k}}(u_n)\,dx$$
$$+ \int_{\Omega} b(u_n)S(u_n)T_{\frac{1}{k}}(u_n)\,dx + \int_{\partial\Omega} S(u_n)|u_n|^{p_n(x)-2}u_n T_{\frac{1}{k}}(u_n)\,d\sigma$$
$$\le \frac{1}{k}\|f_n\|_{L^1(\Omega)}.$$

We deduce that

$$k\int_{\Omega} S'(u_n)a_n(x, \nabla u_n).(\nabla u_n)T_{\frac{1}{k}}(u_n)\,dx + k\int_{\Omega} b(u_n)S(u_n)kT_{\frac{1}{k}}(u_n)\,dx$$
$$\le \|f_n\|_{L^1(\Omega)}$$

and

$$k\int_{\Omega} S'(u_n)a_n(x, \nabla u_n).(\nabla u_n)T_{\frac{1}{k}}(u_n)\,dx$$
$$+k\int_{\partial\Omega} S(u_n)|u_n|^{p_n(x)-2}u_n T_{\frac{1}{k}}(u_n)\,d\sigma \le \|f_n\|_{L^1(\Omega)}.$$

The term $k\int_{\Omega} S'(u_n)a_n(x, \nabla u_n).(\nabla u_n)T_{\frac{1}{k}}(u_n)\,dx$ vanishes, as $k \to \infty$, due to (2.15). Also, $kT_{\frac{1}{k}}(u_n) \to sign(u_n)$ as $k \to \infty$. So, by using Fatou's lemma, we get, as $k \to \infty$,

$$\int_{\Omega} |b(u_n)|\,dx \le \|f_n\|_{L^1(\Omega)} \tag{2.66}$$

and

$$\int_{\partial\Omega} |u_n|^{p_n(x)-1}\,d\sigma \le \|f_n\|_{L^1(\Omega)}. \tag{2.67}$$

The inequality (2.66) becomes, for $k > 0$,

$$\int_{\{|u_n|>k\}} |b(u_n)|\, dx \leq \|f_n\|_{L^1(\Omega)}. \tag{2.68}$$

Therefore, since $|b(u_n)| \geq \min(b(k), |b(-k)|)$ on $\{|u_n| > k\}$, the relation (2.68) gives

$$\min(b(k), |b(-k)|) meas(\{|u_n| > k\}) \leq \|f_n\|_{L^1(\Omega)}$$

or again

$$meas(\{|u_n| > k\}) \leq \frac{\|f_n\|_{L^1(\Omega)}}{\min(b(k), |b(-k)|)}. \tag{2.69}$$

Being weakly convergent in $L^1(\Omega)$, the sequence $(f_n)_{n\in\mathbb{N}}$ is bounded, so the right-hand side of (2.69) tends to zero as $k \to \infty$, then $meas(\{|u_n| > k\})$ tends to zero as $k \to \infty$ uniformly in $n$, and (2.60) is proved.

For the proof of (2.61), let us take $\phi = T_{k+1}(u_n) - T_k(u_n)$ as test function and $S \in \mathbb{S}$ such that $S = S_{k+2}$ in the renormalized formulation (2.16). The function $T_{k+1}(u_n) - T_k(u_n)$ has a support contained in the set $\{|u_n| \geq k\}$ and is bounded by one. One deduces by (2.8)

$$\begin{aligned} C \int_{\{k<|u_n|<k+1\}} |\nabla u_n|^{p_n(x)}\, dx &+ \int_\Omega S'(u_n) a_n(x, \nabla u_n).(\nabla u_n)\phi\, dx \\ &\leq \int_{\{|u_n|\geq k\}} |f_n|\, dx. \end{aligned} \tag{2.70}$$

By the property (2.15) and by equi-integrability of $f_n$ and because of (2.60), for $k \to \infty$, one deduces, from (2.70), the estimate (2.61).

$(iii)$ From Lemma 2.4.1–(i), one gets

$$\begin{aligned} \|T_k(u_n)\|^{p_-}_{W^{1,p_-}(\Omega)} &= \int_\Omega |T_k(u_n)|^{p_-} dx + \int_\Omega |\nabla T_k(u_n)|^{p_-} dx \\ &\leq \int_\Omega \left(1 + |T_k(u_n)|^{p_n(x)}\right) dx + \int_\Omega \left(1 + |\nabla T_k(u_n)|^{p_n(x)}\right) dx \\ &\leq 2meas(\Omega) + \rho_{1,p_n(.)}(T_k(u_n)) \\ &\leq const(k). \end{aligned}$$

And so, the sequence $T_k(u_n)$ is uniformly bounded in $W^{1,p_-}(\Omega)$. Therefore, up to a subsequence, we can assume that the sequence $T_k(u_n)$ converges to a certain function $\sigma_k$ weakly in $W^{1,p_-}(\Omega)$, and by the compact imbedding theorem of $W^{1,p_-}(\Omega)$ in $L^{p_-}(\Omega)$, one has $T_k(u_n)$ that converges strongly to

$\sigma_k$ in $L^{p_-}(\Omega)$ and so a.e. in $\Omega$. Now, we have to prove that $\sigma_k = T_k(u)$ a.e. in $\Omega$. Let $s > 0$, and define the sets

$$E_n := \{|u_n| > k\}, \quad E_m := \{|u_m| > k\}, \text{ and } E_{n,m} := \{|T_k(u_n) - T_k(u_m)| > s\},$$

with $k > 0$. One has $\{|u_n - u_m| > s\} \subset E_n \cup E_m \cup E_{n,m}$, which gives

$$meas(\{|u_n - u_m| > s\}) \leq meas(E_n) + meas(E_m) + meas(E_{n,m}).$$

Let $\varepsilon > 0$. According to (2.60), we can choose $k = k(\varepsilon)$ to get

$$meas(E_n) \leq \frac{\varepsilon}{3} \text{ and } meas(E_m) \leq \frac{\varepsilon}{3}.$$

Since $T_k(u_n)$ converges strongly in $L^{p_-}(\Omega)$, then it is a Cauchy sequence in $L^{p_-}(\Omega)$. Hence, there exists $n_0 = n_0(\varepsilon, s) \in \mathbb{N}$ such that for all $n, m \geq n_0$,

$$meas(E_{n,m}) \leq \frac{1}{s^{p_-}} \int_\Omega |T_k(u_n) - T_k(u_m)|^{p_-} \, dx \leq \frac{\varepsilon}{3}.$$

So, we deduce that

$$meas(\{|u_n - u_m| > s\}) \leq \varepsilon, \text{ for all } n, m \geq n_0.$$

Finally, the sequence $(u_n)_{n\in\mathbb{N}}$ is a Cauchy sequence in measure. Hence, by extraction of subsequence, there exists a measurable function $u$ such that $u_n \to u$ a.e. in $\Omega$. Since $T_k$ is continuous, we have $T_k(u_n) \to T_k(u)$ a.e. in $\Omega$, and by the uniqueness of the limit, one has $\sigma_k = T_k(u)$ a.e. in $\Omega$ because $T_k(u_n) \to \sigma_k$ a.e. in $\Omega$.

Also, the weak convergence of $T_k(u_n)$ to $T_k(u)$ in $W^{1,p_-}(\Omega)$ leads to the weak convergence of $\nabla T_k(u_n)$ to $\nabla T_k(u)$ in $L^{p_-}(\Omega)$. Thanks to Theorem 2.2.1–$(i)$, $\nabla T_k(u_n)$ nonlinear weak-* converges to a Young measure $(\nu_x^k)_{x\in\Omega}$, and since its weak limit is $\nabla T_k(u)$, then $\nabla T_k(u)$ verifies the equality (2.62) according to (2.13).

$(iv)$ One has $T_k(u_n) \rightharpoonup T_k(u)$ in $W^{1,p_-}(\Omega)$ according to Lemma 2.4.1–$(iii)$, and since, for every $1 \leq p \leq \infty$, the trace operator

$$\tau : W^{1,p(.)}(\Omega) \longrightarrow L^{p(.)}(\partial\Omega), \; u \longmapsto \tau(u) = u_{|\partial\Omega},$$

is compact, $\tau(T_k(u_n))$ converges strongly to $\tau(T_k(u))$ in $L^{p_-}(\partial\Omega)$, and so, up to a subsequence, we can assume that $\tau(T_k(u_n))$ converges a.e. to $\tau(T_k(u))$ on $\partial\Omega$ for any $k > 0$, and so $u_n$ converges a.e. to $u$ on $\partial\Omega$. Then, since for a.e. $x \in \Omega$, $\big(T_k(u(x))\big)_k$ is monotone in $k$, we can define

$$v(x) := \lim_{k\to\infty} \tau(T_k(u(x))) \text{ a.e. on } \partial\Omega, \tag{2.71}$$

and one has $T_k(v) = \tau(T_k(u))$, for a.e. $k > 0$.

Further, from (2.67), we have

$$\int_{\partial\Omega} |u_n|^{p_n(x)-1} d\sigma \leq \|f_n\|_{L^1(\Omega)} \leq C.$$

So, by Fatou's lemma, one gets

$$\int_{\partial\Omega} |u|^{p(x)-1} d\sigma \leq C. \tag{2.72}$$

By (2.71) and (2.72), one has $v \in L^{p(x)-1}(\partial\Omega)$.

$(v)$ By assumption (2.56), $p_n \to p$ in measure on $\Omega$, and since $\nabla T_k(u_n) \rightharpoonup \nabla T_k(u)$ in $L^{p_-}(\Omega)$, then according to Theorem 2.2.1$-(ii)$, $(iii)$, for all $k \in \mathbb{N}$, the sequence $(p_n, \nabla T_k(u_n))_n$ converges to the Young measure $\delta_{p(x)} \otimes d\nu_x^k$ on $\mathbb{R} \times \mathbb{R}^N$.

Let us now consider the Carathéodory function

$$F_m : (x, (\lambda_0, \lambda)) \in \Omega \times (\mathbb{R} \times \mathbb{R}^N) \longmapsto |h_m(\lambda)|^{\lambda_0}, \; m \in \mathbb{N},$$

where $h_m$ is defined by (2.9). The sequence $(F_m(., (p_n(.), \nabla T_k(u_n))))_{n\in\mathbb{N}}$ is equi-integrable in $\Omega$ since it is uniformly bounded in $L^1(\Omega)$ according to (2.64). Then, we apply the nonlinear weak-* convergence property (2.12) to the function $F_m$ to get

$$\begin{aligned}
&\lim_{n\to\infty} \int_\Omega F_m(x, (p_n(x), \nabla T_k(u_n)(x)))\, dx \\
&= \int_\Omega \int_{\mathbb{R}\times\mathbb{R}^N} F_m(x, (\lambda_0, \lambda))\, d\delta_{p(x)}(\lambda_0) d\nu_x^k(\lambda)\, dx \\
&= \int_\Omega \int_{\mathbb{R}^N} F_m(x, (p(x), \lambda))\, d\nu_x^k(\lambda) dx \\
&= \int_{\Omega\times\mathbb{R}^N} |h_m(\lambda)|^{p(x)}\, d\nu_x^k(\lambda) dx.
\end{aligned}$$

Moreover,

$$\begin{aligned}
\lim_{n\to\infty} \int_\Omega F_m(x, (p_n(x), \nabla T_k(u_n)(x)))\, dx &= \lim_{n\to\infty} \int_\Omega |h_m(\nabla T_k(u_n))|^{p_n(x)}\, dx \\
&\leq \lim_{n\to\infty} \int_\Omega |\nabla T_k(u_n)|^{p_n(x)}\, dx \\
&\leq Ck \quad \text{according to (2.64).}
\end{aligned}$$

So,

$$\int_{\Omega\times\mathbb{R}^N} |h_m(\lambda)|^{p(x)}\, d\nu_x^k(\lambda)dx \leq Ck.$$

Since the sequence $(|h_m|)_{m\in\mathbb{N}}$ is increasing and $h_m(\lambda) \longrightarrow \lambda$ as $m \to \infty$, then by the monotone convergence theorem, we deduce that

$$\int_{\Omega\times\mathbb{R}^N} |\lambda|^{p(x)}\, d\nu_x^k(\lambda)dx \leq Ck.$$

By the formula (2.62) and Jensen inequality, one has

$$\begin{aligned}\int_\Omega |\nabla T_k(u)|^{p(x)}\, dx &= \int_\Omega \left| \int_{\mathbb{R}^N} \lambda d\nu_x^k(\lambda) \right|^{p(x)} dx \\ &\leq \int_{\Omega\times\mathbb{R}^N} |\lambda|^{p(x)} d\nu_x^k(\lambda)dx < Ck.\end{aligned}$$

Hence, we deduce that $\nabla T_k(u) \in L^{p(.)}(\Omega)$ and so $T_k(u) \in W^{1,p(.)}(\Omega)$.

$(vi)$ Up to subsequence, by $(iii)$, $T_{k+1}(u_n) - T_k(u_n)$ converges to $T_{k+1}(u) - T_k(u)$ a.e. on $\Omega$ and weakly in $W^{1,p_-}(\Omega)$. By arguing as in $(v)$, we get $\nabla(T_{k+1}(u) - T_k(u)) \in L^{p(.)}(\Omega)$, and its modular is upper bounded by

$$\begin{aligned}&\sup_n \int_\Omega |\nabla(T_{k+1}(u) - T_k(u))|^{p_n(x)}\, dx \\ &= \sup_n \int_{\{k<|u_n|<k+1\}} |\nabla u_n|^{p_n(x)}\, dx \to 0, \text{ as } k \to \infty,\end{aligned}$$

by (2.61). Thus, (2.63) follows. □

**Lemma 2.4.2**

$(i)$ *For all* $k > 0$*, the sequence* $(\mathcal{Y}_n^k)_{n\in\mathbb{N}}$, $\mathcal{Y}_n^k(x) := a_n(x, \nabla T_k(u_n(x)))$ *is equi-integrable on* $\Omega$*, and its weak limit* $\mathcal{Y}^k \in L^{p'(.)}(\Omega)$ *is such that*

$$\mathcal{Y}^k(x) := \int_{\mathbb{R}^N} a(x,\lambda)d\nu_x^k(\lambda),\ a.e. x \in \Omega. \tag{2.73}$$

$(ii)$ *For all* $\widehat{k} > k > 0$*, one has* $\mathcal{Y}^k = \mathcal{Y}^{\widehat{k}} \chi_{\{|u|<k\}}$.

***Proof***

$(i)$ We first show that the sequence $(\mathcal{Y}_n^k)_{n\in\mathbb{N}}$, $\mathcal{Y}_n^k := a_n(x, \nabla T_k(u_n))$ is equi-integrable in $\Omega$. The assumption (2.7) applied on $a_n(.,.)$ with exponent $p_n(x)$ implies, for all measurable subset $E \subset \Omega$,

$$\begin{aligned}\int_E |\mathcal{Y}_n^k|\,dx &\le C\int_E \left(1+\mathcal{M}_n+|\nabla T_k(u_n)|^{p_n(x)-1}\right)dx\\ &\le C\int_E (1+\mathcal{M}_n)\,dx+2C\left\||\nabla T_k(u_n)|^{p_n(x)-1}\right\|_{L^{p_n'(.)}}\|\chi_E\|_{L^{p_n(.)}}\\ &\le C\int_E (1+\mathcal{M}_n)\,dx+C'\max\left(\left(\rho_{p_n}(\chi_E)\right)^{1/p_+},\left(\rho_{p_n}(\chi_E)\right)^{1/p_-}\right)\\ &\le C\int_E (1+\mathcal{M}_n)\,dx+C'\max\left(meas(E)^{1/p_+},meas(E)^{1/p_-}\right)\end{aligned} \tag{2.74}$$

by using Hölder-type inequality and Lemma 2.2.2, where $2C\left\||\nabla T_k(u_n)|^{p_n(x)-1}\right\|_{L^{p_n'(.)}}$ is upper bounded by $C'$ by (2.64).

The whole right-hand side of (2.74) tends to zero when $meas(E)$ tends to zero because the sequence $(\mathcal{M}_n)_{n\in\mathbb{N}}$ is equi-integrable in $\Omega$. And so, the sequence $(\mathcal{Y}_n^k)_{n\in\mathbb{N}}$ is equi-integrable in $\Omega$. By Theorem 2.2.1–$(i)$, there exists a weak limit $\mathcal{Y}^k$ for the sequence $\mathcal{Y}_n^k$ in $L^1(\Omega)$.

In the following lines, we prove that the weak limit $\mathcal{Y}^k$ verifies the formula (2.73) and belongs to $L^{p'(.)}(\Omega)$.

We put the set

$$R_n := \{x\in\Omega;\ |p(x)-p_n(x)|<1/2\},$$

and we consider auxiliary functions $\tilde{\mathcal{Y}}_n^k := a\big(x,(\nabla T_k(u_n))\chi_{R_n}\big)$.

Let us show that the sequence $\left(\tilde{\mathcal{Y}}_n^k\right)_{n\in\mathbb{N}}$ is equi-integrable in $\Omega$. Indeed, we apply (2.7) with the exponent $p(.)$ on $a\,(.,.)$ to get

$$\int_E \tilde{\mathcal{Y}}_n^k\,dx \le C\int_E (1+\mathcal{M})\,dx + C\int_{E\cap R_n}|\nabla T_k(u_n)|^{p(x)-1}\,dx. \tag{2.75}$$

The first term of the right-hand side of (2.75) tends to zero when $meas(E)$ tends to zero. Also, for $x\in R_n$, one has $p(x)\le p_n(x)+1/2$, and, by using Hölder-type inequality, we have

$$\begin{aligned}&\int_{E\cap R_n}|\nabla T_k(u_n)|^{p(x)-1}\,dx\\ &\le \int_E \left(1+|\nabla T_k(u_n)|^{p_n(x)-1/2}\right)dx\\ &\le meas(E)+C\left\||\nabla T_k(u_n)|^{p_n(x)-1/2}\right\|_{L^{(2p_n(.))'}}\|\chi_E\|_{L^{2p_n(.)}}\\ &\le meas(E)+C\left\||\nabla T_k(u_n)|^{p_n(x)-1/2}\right\|_{L^{(2p_n(.))'}}\|\chi_E\|_{L^{2p_n(.)}}.\end{aligned} \tag{2.76}$$

But, by (2.64), one has

$$\rho_{(2p_n)'}\left(|\nabla T_k(u_n)|^{p_n(x)-1/2}\right) = \rho_{p_n}(\nabla T_k(u_n)) \leq Ck. \tag{2.77}$$

Also, by Proposition 2.2.3, one has

$$\begin{aligned} \|\chi_E\|_{L^{2p_n(.)}} &\leq \max\left(\left(\rho_{2p_n}(\chi_E)\right)^{1/(2p)_+}, \left(\rho_{2p_n}(\chi_E)\right)^{1/(2p)_-}\right) \\ &\leq \max\left((meas(E))^{1/(2p)_+}, (meas(E))^{1/(2p)_-}\right). \end{aligned} \tag{2.78}$$

From (2.76)–(2.78), the second term of the right-hand side of (2.75) is uniformly small for $meas(E)$ small, and the equi-integrability of $\left(\tilde{\mathcal{Y}}_n^k\right)_{n\in\mathbb{N}}$ follows.

Now, we assert that, by extraction of a subsequence, the sequence $\tilde{\mathcal{Y}}_n^k$ converges weakly to some function $\tilde{\mathcal{Y}}^k$ in $L^1(\Omega)$ as $n \to \infty$ thanks to Theorem 2.2.1–$(i)$.

It remains to prove that $\mathcal{Y}^k = \tilde{\mathcal{Y}}^k$. For that, it is sufficient to prove that $\mathcal{Y}_n^k - \tilde{\mathcal{Y}}_n^k$ converges strongly to zero in $L^1(\Omega)$.

Indeed, let $\varepsilon > 0$. By the Chebyshev inequality, one has

$$\begin{aligned} meas(\{|\nabla T_k(u_n)| > L\}) &\leq \left(\int_\Omega |\nabla T_k(u_n)|\, dx\right) \Big/ L \\ &\leq \int_\Omega \left(1 + |\nabla T_k(u_n)|^{p_n(x)}\right) dx \Big/ L \\ &\leq (meas(\Omega) + Ck)/L, \end{aligned}$$

by inequality (2.64).

It follows that $\sup_n (meas(\{|\nabla T_k(u_n)| > L\}) \to 0$ as $L \to \infty$. The sequence $\mathcal{Y}_n^k - \tilde{\mathcal{Y}}_n^k$ is equi-integrable in $\Omega$, so there exists $L_0 = L_0(\varepsilon)$ such that for $L > L_0$, one has

$$\int_{\{|\nabla T_k(u_n)|>L\}} |\mathcal{Y}_n^k - \tilde{\mathcal{Y}}_n^k|\, dx \leq \varepsilon/4, \text{ for all } n \in \mathbb{N}. \tag{2.79}$$

By the assumption (2.55), one has for all $\sigma > 0$

$$\lim_{n\to\infty} meas\left(\left\{x \in \Omega;\ \sup_{|\lambda|\leq L} |a_n(x,\lambda) - a(x,\lambda)| \geq \sigma\right\}\right) = 0.$$

Hence, by equi-integrability of $\mathcal{Y}_n^k - \tilde{\mathcal{Y}}_n^k$ on $\Omega$, there exists $n_0 = n_0(\sigma, L_0) \in \mathbb{N}$ such that for all $n > n_0$,

$$\int_{\left\{x\in\Omega;\ \sup\limits_{|\lambda|\le L} |a_n(x,\lambda) - a(x,\lambda)| \ge \sigma\right\}} |\mathcal{Y}_n^k - \tilde{\mathcal{Y}}_n^k|\, dx \le \varepsilon/4. \quad (2.80)$$

By the definition, one has $\tilde{\mathcal{Y}}_n^k = a(x, \nabla T_k(u_n))$ on the set $R_n$, and we consider the following set

$$R_n^{L,\sigma} := \left\{x \in R_n;\ \sup_{|\lambda|\le L} |a_n(x,\lambda) - a(x,\lambda)| < \sigma, |\nabla T_k(u_n)| \le L\right\}.$$

Since $|\nabla T_k(u_n)| \le L$ on $R_n^{L,\sigma}$, then one has

$$|a_n(x, \nabla T_k(u_n)) - a(x, \nabla T_k(u_n))| < \sigma \ \text{ on } \ R_n^{L,\sigma},$$

and so, for all $n$,

$$\int_{R_n^{L,\sigma}} |\mathcal{Y}_n^k - \tilde{\mathcal{Y}}_n^k|\, dx \le \sigma meas(\Omega) \le \varepsilon/4, \quad (2.81)$$

by taking $\sigma = \sigma(\varepsilon) < \varepsilon/(4meas(\Omega))$.

Also, by (2.79) and (2.80), we have

$$\int_{R_n\setminus R_n^{L,\sigma}} |\mathcal{Y}_n^k - \tilde{\mathcal{Y}}_n^k|\, dx \le \varepsilon/2, \text{ for all } n > n_0(\sigma(\varepsilon), L(\varepsilon)). \quad (2.82)$$

Since $p_n$ converges to $p$ in measure on $\Omega$, one has $meas(\Omega \setminus R_n) = meas\big(\{|p - p_n| \ge 1/2\}\big)$ that converges to zero as $n \to \infty$, and the equi-integrability of $\mathcal{Y}_n^k$ gives, for sufficiently large $n$,

$$\int_{\Omega\setminus R_n} |\mathcal{Y}_n^k - \tilde{\mathcal{Y}}_n^k|\, dx = \int_{\Omega\setminus R_n} |\mathcal{Y}_n^k|\, dx \le \varepsilon/4. \quad (2.83)$$

Now, by using (2.81), (2.82), and (2.83), we get, for $n > n_0(\sigma(\varepsilon), L(\varepsilon))$,

$$\int_\Omega |\mathcal{Y}_n^k - \tilde{\mathcal{Y}}_n^k|\, dx \le \varepsilon.$$

Hence, the sequence $\mathcal{Y}_n^k - \tilde{\mathcal{Y}}_n^k$ converges strongly to zero in $L^1(\Omega)$, as $n$ goes to infinity, and so, $\mathcal{Y}^k = \tilde{\mathcal{Y}}^k$.

Let us show the representation formula (2.73) for $\mathcal{Y}^k$. Since $meas(\Omega \setminus R_n) \to 0$ as $n \to \infty$, so, by the equi-integrability of $\nabla T_k(u_n)$ in $\Omega$, one has $\nabla T_k(u_n)(1 - \chi_{R_n})$, which converges to zero as $n \to \infty$. Therefore, the sequence $\nabla T_k(u_n)\chi_{R_n}$ converges to the same Young measure $\nu_x^k$ as the sequence $\nabla T_k(u_n)$. Now, fix $\psi \in \mathcal{D}(\Omega)$, and let us consider the Carathéodory function $a(.,.).\psi$. Since the sequence $\tilde{\mathcal{Y}}_n^k = a(x, \nabla T_k(u_n)\chi_{R_n})$ is equi-integrable in $\Omega$, then we can use the nonlinear weak-* convergence property (2.12) to get

$$\lim_{n\to\infty} \int_\Omega a(x, \nabla T_k(u_n)\chi_{R_n}).\psi \, dx = \int_{\Omega\times\mathbb{R}^N} a(x, \lambda).\psi \, d\nu_x^k(\lambda)dx. \quad (2.84)$$

Since $a(x, \nabla T_k(u_n)\chi_{R_n})$ converges weakly to $\tilde{\mathcal{Y}}^k$, (2.84) becomes

$$\int_\Omega \tilde{\mathcal{Y}}^k.\psi \, dx = \int_{\Omega\times\mathbb{R}^N} a(x, \lambda).\psi d\nu_x^k(\lambda)dx = \int_\Omega \left( \int_{\mathbb{R}^N} a(x, \lambda)d\nu_x^k(\lambda) \right).\psi dx,$$

which means that

$$\mathcal{Y}^k = \tilde{\mathcal{Y}}^k = \int_{\mathbb{R}^N} a(x, \lambda)d\nu_x^k(\lambda) \text{ in } \mathcal{D}'(\Omega) \text{ and so, a.e. on } \Omega.$$

Now, we end the proof with $\mathcal{Y}^k \in L^{p'(.)}(\Omega)$. One uses Jensen inequality, the assumption (2.7), and Lemma 2.4.1–($v$) to obtain

$$\begin{aligned}
\int_\Omega |\mathcal{Y}^k(x)|^{p'(x)}dx &= \int_\Omega \left| \int_{\mathbb{R}^N} a(x, \lambda)d\nu_x^k(\lambda) \right|^{p'(x)} dx \\
&\le \int_{\Omega\times\mathbb{R}^N} |a(x, \lambda)|^{p'(x)}d\nu_x^k(\lambda)dx \\
&\le \int_{\Omega\times\mathbb{R}^N} C(\mathcal{M}(x) + |\lambda|^{p(x)})d\nu_x^k(\lambda)dx < \infty.
\end{aligned}$$

($ii$) Since $\widehat{k} > k$, one has

$$T_k(u_n) \equiv T_k(T_{\widehat{k}}(u_n)),$$

and so

$$\nabla T_k(u_n) = \nabla T_{\widehat{k}}(u_n)\chi_{\{|T_{\widehat{k}}(u_n)|<k\}} = \nabla T_{\widehat{k}}(u_n)\chi_{\{|u_n|<k\}}.$$

Moreover, from assumption (2.5), one has $a_n(x, 0) = 0$ a.e. $x \in \Omega$. Hence,

$$a_n(x, \nabla T_{\widehat{k}}(u_n))\chi_{\{|u_n|<k\}} \equiv a_n(x, \nabla T_k(u_n)),$$

and the sequence $\left(h_n^k\right)_{n\in\mathbb{N}}$ converges weakly to $\mathcal{Y}^k$ in $L^1(\Omega)$, according to $(i)$.

Consider the sequence $\left(d_n^k\right)_{n\in\mathbb{N}}$ such that

$$d_n^k := g_n^k - h_n^k = a_n(x, \nabla T_{\widehat{k}}(u_n)) \left(\chi_{\{|u|<k\}} - \chi_{\{|u_n|<k\}}\right).$$

The function $\chi_{(-k,k)}(.)$ is continuous on the image of $\Omega$ by $u(.)$ for a.e. $k > 0$. Indeed, one has $meas\ (\{|u| = k\}) = 0$ for a.e. $k > 0$ by arguing as in the proof of Lemma 2.3.3. Therefore, since $u_n$ converges to $u$ a.e. in $\Omega$, then

$$\chi_{\{|u_n|<k\}} = \chi_{(-k,k)}(u_n) \to \chi_{(-k,k)}(u) = \chi_{\{|u|<k\}} \text{ a.e. in } \Omega \text{ as } n \to \infty.$$

So,

$$d_n^k \to 0 \text{ a.e. in } \Omega.$$

Moreover, by $(i)$, the sequence $\left(d_n^k\right)_{n\in\mathbb{N}}$ is equi-integrable in $\Omega$. Hence, by Vitali's theorem, the sequence $\left(d_n^k\right)_{n\in\mathbb{N}}$ converges strongly to zero in $L^1(\Omega)$. Therefore, $g_n^k = h_n^k + d_n^k$ tends to $\mathcal{Y}^k$ weakly in $L^1(\Omega)$. So, this ends the proof of $(ii)$. □

**Lemma 2.4.3**

$(i)$ *For all $k > 0$,*

$$\lim_{n\to\infty} \int_\Omega \mathcal{Y}_n^k.\nabla T_k(u_n)\, dx = \int_\Omega \mathcal{Y}^k.\nabla T_k(u)\, dx, \tag{2.85}$$

*and the "div-curl" inequality*

$$\int_{\Omega\times\mathbb{R}^N} (a(x,\lambda) - a(x, \nabla T_k(u))).(\lambda - \nabla T_k(u))d\nu_x^k(\lambda)dx \le 0 \tag{2.86}$$

*holds.*

$(ii)$ *For all $k > 0$,*

$$\mathcal{Y}^k(x) = a(x, \nabla T_k(u(x)))\ \textit{for a.e. } x \in \Omega, \tag{2.87}$$

*and $\nabla T_k(u_n)$ converges to $\nabla T_k(u)$ in measure in $\Omega$ as $n \to \infty$.*

***Proof***

$(i)$ Let $\psi \in C^\infty(\overline{\Omega})$. Since $p_n(.)$ is log-Hölder continuous, then $C^\infty(\overline{\Omega})$ is dense in $W^{1,p_n(.)}(\Omega)$. So, we can take $\psi$ as test function in the renormalized formulation (2.16) for $u_n$. We get

$$\left|\int_\Omega\Big(b(u_n)S(u_n)\psi + S(u_n)\mathcal{Y}_n^M.\nabla\psi - f_nS(u_n)\psi\Big)dx + \int_{\partial\Omega} S(u_n)|u_n|^{p_n(x)-2}u_n\psi d\sigma\right|$$
$$\leq \|\psi\|_{L^\infty}\int_\Omega |S'(u_n)|\mathcal{Y}_n^M.\nabla T_M(u_n)\,dx, \tag{2.88}$$

where $S \in \mathbb{S}$ with $suppS \subset [-M, M]$, $M > 0$.

We are going to pass to the limit in (2.88), as $n$ tends to infinity. By Lemma 2.4.1–$(iii)$, $u_n$ converges to $u$ a.e. in $\Omega$. By the continuity of $b$ and $S$, the term $b(u_n)S(u_n)$ converges a.e. in $\Omega$ to $b(u)S(u)$. Also, $|b(u_n)S(u_n)\psi| \leq \|S\|_{L^\infty}\max(b(M), |b(-M)|)|\psi| \in L^1(\Omega)$, and so, by the Lebesgue-dominated convergence theorem,

$$\int_\Omega b(u_n)S(u_n)\psi\,dx \longrightarrow \int_\Omega b(u)S(u)\psi\,dx, \quad \text{as } n \to \infty. \tag{2.89}$$

Since $p_n$ converges to $p$ in measure in $\Omega$ according to (2.56), then $p_n$ converges to $p$ a.e. in $\Omega$, up to a subsequence. Also, $u_n \longrightarrow u$ a.e. on $\Omega$. By the continuity of $S$, one has

$$S(u_n)|u_n|^{p_n(x)-2}u_n \to S(u)|u|^{p(x)-2}u \text{ a.e. on } \partial\Omega.$$

Moreover, since $suppS \subset [-M, M]$, then

$$\left|S(u_n)|u_n|^{p_n(x)-2}u_n\psi\right| \leq \max\left(M^{p_--1}, M^{p_+-1}\right)\|S\|_{L^\infty(\partial\Omega)}\,|\psi| \in L^1(\partial\Omega),$$

and so, the Lebesgue-dominated convergence theorem gives us

$$\int_{\partial\Omega} S(u_n)|u_n|^{p_n(x)-2}u_n\psi\,d\sigma \longrightarrow \int_{\partial\Omega} S(u)|u|^{p(x)-2}u\psi\,d\sigma, \quad \text{as } n \to \infty. \tag{2.90}$$

Let us prove now that

$$\int_\Omega f_nS(u_n)\psi\,dx \longrightarrow \int_\Omega fS(u)\psi\,dx, \quad \text{as } n \to \infty. \tag{2.91}$$

One has

$$\int_\Omega f_nS(u_n)\psi\,dx = \int_\Omega f_nS(u)\psi\,dx + \int_\Omega f_n(S(u_n) - S(u))\psi\,dx. \tag{2.92}$$

On the one hand, one has $\int_\Omega f_nS(u)\psi\,dx \longrightarrow \int_\Omega fS(u)\psi\,dx$ since $f_n \rightharpoonup f$ in $L^1(\Omega)$. On the other hand, one has, for $R > 0$,

$$
\begin{aligned}
\int_\Omega |f_n(S(u_n)-S(u))\psi|\,dx &= \int_{\{|f_n|>R\}} |f_n(S(u_n)-S(u))\psi|\,dx \\
&\quad + \int_{\{|f_n|\le R\}} |f_n(S(u_n)-S(u))\psi|\,dx \\
&\le 2\,\|\psi\|_{L^\infty}\,\|S\|_{L^\infty} \int_{\{|f_n|>R\}} |f_n|\,dx \\
&\quad + R\,\|\psi\|_{L^\infty} \int_\Omega |S(u_n)-S(u)|\,dx. \qquad (2.93)
\end{aligned}
$$

For $R > 0$ fixed, the second term of the right-hand side of the inequality (2.93) tends to zero as $n \to \infty$. Indeed, because of the continuity of $S$ and the compactness of $suppS$, $S(u_n)$ converges strongly to $S(u)$ in $L^1(\Omega)$ by the Lebesgue-dominated convergence theorem. By the Chebyshev inequality and since $f_n$ is bounded in $L^1(\Omega)$, one has

$$
\sup_n meas(\{|f_n| > R\}) \le \frac{\sup_n \|f_n\|_1}{R} \le \frac{C}{R} \longrightarrow 0 \text{ as } R \longrightarrow \infty.
$$

Since the sequence $f_n$ is equi-integrable on $\Omega$, then the first term in the right-hand side of (2.93) can be made as small as desired by the choice of $R$. Hence, the second term of the right-hand side of (2.92) tends to zero. And so, we deduce the convergence result (2.91).

To end the proof, let us prove that

$$
\int_\Omega S(u_n)\mathcal{Y}_n^M.\nabla\psi\,dx \to \int_\Omega S(u)\mathcal{Y}^M.\nabla\psi\,dx. \qquad (2.94)
$$

Indeed, for $R > 0$,

$$
\int_\Omega S(u_n)\mathcal{Y}_n^M.\nabla\psi\,dx = \int_{\{|\nabla\psi|<R\}} S(u_n)\mathcal{Y}_n^M.\nabla\psi\,dx + \int_{\{|\nabla\psi|\ge R\}} S(u_n)\mathcal{Y}_n^M.\nabla\psi\,dx. \qquad (2.95)
$$

For the first term of the right-hand side of (2.95), one has

$$
\begin{aligned}
\int_{\{|\nabla\psi|<R\}} S(u_n)\mathcal{Y}_n^M.\nabla\psi\,dx &= \int_{\{|\nabla\psi|<R\}} S(u)\mathcal{Y}_n^M.\nabla\psi\,dx \\
&\quad + \int_{\{|\nabla\psi|<R\}} (S(u_n)-S(u))\mathcal{Y}_n^M.\nabla\psi\,dx.
\end{aligned} \qquad (2.96)
$$

Since $\mathcal{Y}_n^M \rightharpoonup \mathcal{Y}^M$ in $L^{p'(.)}(\Omega)$ by Lemma 2.4.2–$(i)$, then the first term of the right-hand side of (2.96) tends to $\int_{\{|\nabla\psi|<R\}} S(u)\mathcal{Y}^M.\nabla\psi\,dx$ as $n \to \infty$.

For $\alpha > 0$ fixed, we can rewrite the second term of the right-hand side of (2.96) as follows:

$$\begin{aligned}
&\int_{\{|\nabla\psi|<R\}} |(S(u_n) - S(u))\mathcal{Y}_n^M.\nabla\psi|\,dx \\
&= \int_{\{|\nabla\psi|<R\}\cap\left\{|\mathcal{Y}_n^M|\leq\alpha\right\}} |(S(u_n) - S(u))\mathcal{Y}_n^M.\nabla\psi|\,dx \\
&+ \int_{\{|\nabla\psi|<R\}\cap\left\{|\mathcal{Y}_n^M|>\alpha\right\}} |(S(u_n) - S(u))\mathcal{Y}_n^M.\nabla\psi|\,dx \\
&\leq \alpha R \int_\Omega |S(u_n) - S(u)|\,dx \\
&+2R\,\|S\|_{L^\infty} \int_{\left\{|\mathcal{Y}_n^M|>\alpha\right\}} |\mathcal{Y}_n^M|\,dx.
\end{aligned} \tag{2.97}$$

The sequence $\mathcal{Y}_n^M$ is equi-integrable on $\Omega$ and is bounded in $L^1(\Omega)$ as it converges weakly in $L^1(\Omega)$, so using the same argument that leads to assert that the right-hand side of (2.93) tends to zero, as $n \to \infty$, in the inequality (2.97), then the second term of the right-hand side of (2.96) tends to zero as $n \to \infty$. Thus, the first term of the right-hand side of (2.95) converges to $\int_{\{|\nabla\psi|<R\}} S(u)\mathcal{Y}^M.\nabla\psi\,dx$ as $n \to \infty$.

For the second term of the right-hand side of (2.95), we note that, by Hölder-type inequality,

$$\left|\int_{\{|\nabla\psi|\geq R\}} \mathcal{Y}_n^M.(\nabla\psi S(u_n))\,dx\right| \leq C\,\|S\|_{L^\infty} \left\|\mathcal{Y}_n^M\right\|_{L^{p_n'(\cdot)}(\Omega)} \left\|\chi_{\{|\nabla\psi|\geq R\}}\nabla\psi\right\|_{L^{p_n(\cdot)}(\Omega)}. \tag{2.98}$$

One has $\left\|\mathcal{Y}_n^M\right\|_{L^{p_n'(\cdot)}(\Omega)} \leq C$ by Lemma 2.4.1. Also, since $\psi \in C^\infty(\overline{\Omega})$, then one has $\text{mes}(\{|\nabla\psi| \geq R\}) \to 0$ as $R \to \infty$. Hence,

$$\int_{\{|\nabla\psi|\geq R\}} |\nabla\psi|^{p_n(.)}\text{d}x \leq C\text{mes}(\{|\nabla\psi| \geq R\}) \to 0, \text{ as } R \to \infty,$$

where $C$ is constant that does not depend on $R$.

By Lemma 2.2.2–$(iii)$, $(iv)$, $\sup_n \left\|\chi_{\{|\nabla\psi|\geq R\}}\nabla\psi\right\|_{L^{p_n(\cdot)}(\Omega)}$ tends to zero as $R \to \infty$. So, the second term of right-hand side of (2.95) tends to zero as $R \to \infty$. Hence, as $n \to \infty$ and $R \to \infty$ in the equality (2.95), we deduce (2.94).

Thanks to convergences (2.89), (2.90), (2.91), and (2.94), we deduce, for $n$ large enough, that (2.88) gives

$$\left|\int_\Omega \Big(b(u)S(u)\psi + S(u)\mathcal{Y}^M.\nabla\psi - fS(u)\psi\Big)dx + \int_{\partial\Omega} S(u)|u|^{p(x)-2}u\psi\, d\sigma\right|$$
$$\le \|\psi\|_{L^\infty} \sup_n \int_\Omega |S'(u_n)|a_n(x,\nabla T_M(u_n)).\nabla T_M(u_n)\, dx. \tag{2.99}$$

Now, fix $k > 0$. By Lemma 2.4.1–($v$), one has $T_k(u) \in W^{1,p(.)}(\Omega) \cap L^\infty(\Omega)$. So, by the density of $C^\infty(\overline{\Omega})$ in $W^{1,p(.)}(\Omega)$, we can replace $\psi$ by $T_k(u)$ in (2.99).

Consider the sequence $(S_M)_M \subset \mathbb{S}$ such that:

- $S_M$ and $S'_M$ are uniformly bounded.
- $S_M = 1$ on $[-M+1, M-1]$, $suppS_M \subset [-M, M]$, for all $M \in \mathbb{N}^*$.
- The map $M \longmapsto b(z)S_M(z)$ is non-decreasing, for all $z \in \mathbb{R}$.

From now on, we replace $S$ by $S_M$ in (2.99).

According to Lemma 2.4.2–($ii$), for $M > k$, one has $\mathcal{Y}^k = \mathcal{Y}^M \chi_{\{|u|<k\}}$. Since $\nabla T_k(u) = 0$ outside $\{|u| < k\}$, then we can replace $\mathcal{Y}^M.\nabla T_k(u)$ by $\mathcal{Y}^k.\nabla T_k(u)$. Also, one has $suppS'_M \subset [-M, -M+1] \cup [M-1, M]$, and the sequence $S'_M$ is uniformly bounded, i.e., $\|S'_M\|_{L^\infty(\Omega)} \le C$, where $C$ is a positive constant independent of $M$. So, the term of the right-hand side of (2.99) is bounded by

$$C \sup_n \int_{\{M-1\le|u_n|\le M\}} a_n(x,\nabla T_M(u_n)).\nabla T_M(u_n)\, dx$$
$$\le C \sup_n \int_{\{M-1\le|u_n|\le M\}} \Big(\mathcal{M}_n|\nabla T_k(u_n)| + |\nabla T_M(u_n)|^{p_n(x)}\Big)\, dx$$
$$\le C \sup_n \left\|\mathcal{M}_n \chi_{\{|u_n|\ge M-1\}}\right\|_{L^{p'_n(.)}(\Omega)} \left\|\nabla T_k(u_n)\chi_{\{M-1\le|u_n|\le M\}}\right\|_{L^{p_n(.)}(\Omega)}$$
$$+C \sup_n \int_{\{M-1\le|u_n|\le M\}} |\nabla T_M(u_n)|^{p_n(x)}\, dx. \tag{2.100}$$

Thanks to Lemma 2.2.2, (2.60), (2.61), and the fact that $\mathcal{M}_n$ is equi-integrable, one has the term of the right-hand side of (2.99) that tends to zero when $M \to \infty$.

By the monotone convergence theorem, since $b(u)S_M(u)$ is non-decreasing and converges a.e. in $\Omega$ to $b(u)$, then $b(u)S_M(u)\psi$ converges strongly to $b(u)\psi$ in $L^1(\Omega)$. Moreover, by the Lebesgue-dominated convergence theorem, the terms $S_M(u)\mathcal{Y}^k.\nabla\psi$, $fS_M(u)\psi$, and $S_M(u)|u|^{p(x)-2}u\psi$ converge, respectively, strongly to $\mathcal{Y}^k.\nabla\psi$, to $f\psi$ in $L^1(\Omega)$ and to $|u|^{p(x)-2}u\psi$ in $L^1(\partial\Omega)$. Hence, the inequality (2.99) becomes, with $\psi$ replaced by $T_k(u)$, as $M \to \infty$,

$$\int_\Omega \big(b(u)T_k(u) + \mathcal{Y}^k.\nabla T_k(u) - fT_k(u)\big)\mathrm{d}x + \int_{\partial\Omega} S(u)|u|^{p(x)-2}uT_k(u)\mathrm{d}\sigma = 0. \tag{2.101}$$

Now, we consider the renormalized formulation (2.16) for $u_n$ where we take $T_k(u_n)$ as test function and $S \in \mathbb{S}$ with $S = S_h$,

$$\int_\Omega \Big( S_h(u_n)\mathcal{Y}^k_n.\nabla T_k(u_n) + S'_h(u_n)a_n(x,\nabla u_n).(\nabla u_n)T_k(u_n)$$

$$+b(u_n)S_h(u_n)T_k(u_n)\Big)\,dx$$

$$+\int_{\partial\Omega} S_h(u_n)|u_n|^{p(x)-2}u_n T_k(u_n)\,d\sigma = \int_\Omega f_n S_h(u_n)T_k(u_n)\,dx. \quad (2.102)$$

We are going to pass to the limit in (2.102), as $h \to \infty$.

We use the property (2.15) to pass to the limit, as $h \to \infty$, in the term containing the factor $S'_h(u_n)$, and since $S_h$ is monotone in $h$, we use monotone convergence theorem to pass to the limit in the terms containing the factor $S_h(u_n)$. As $h \to \infty$ in (2.102), we get then

$$\int_\Omega \Big(\mathcal{Y}^k_n.\nabla T_k(u_n) + b(u_n)T_k(u_n)\Big)\,dx + \int_{\partial\Omega} |u_n|^{p(x)-2}u_n T_k(u_n)\,d\sigma$$

$$= \int_\Omega f_n T_k(u_n)\,dx. \quad (2.103)$$

Since $u_n$ converges to $u$ a.e. in $\Omega$, and also because $f_n \rightharpoonup f$ in $L^1(\Omega)$ and $\|T_k\| < \infty$, arguing as in (2.92) and (2.93), we have

$$\int_\Omega f_n T_k(u_n)\,dx = \int_\Omega f_n T_k(u)\,dx + \int_\Omega f_n\,(T_k(u_n) - T_k(u))\,dx$$

$$\to \int_\Omega f T_k(u)\,dx, \text{ as } n \to \infty.$$

In the sequel, since $b(u_n)T_k(u_n) \geq 0$ and $|u_n|^{p(x)-2}u_n T_k(u_n) \geq 0$, by Fatou's lemma, one deduces

$$\int_\Omega \big(b(u)T_k(u) - fT_k(u)\big)\,dx + \int_{\partial\Omega} \big(|u|^{p(x)-2}uT_k(u)\big)\,d\sigma$$

$$\leq \liminf_{n\to\infty}\left(\int_\Omega \big(b(u_n)T_k(u_n) - f_nT_k(u_n)\big)\,dx + \int_{\partial\Omega} |u_n|^{p(x)-2}u_nT_k(u_n)\,d\sigma\right).$$

And so, from the inequality above and by using (2.103) and (2.101), we get (2.85).

Now, let us go to the proof of the "div-curl" inequality (2.86). Thanks to Lemma 2.2.1, we know that the sequence

$$\left(a_n(x, h_m(\nabla T_k(u_n))).h_m(\nabla T_k(u_n))\right)_{m>0}$$

is upper bounded by $\mathcal{Y}_n^k.\nabla T_k(u_n)$ because it converges while growing to $\mathcal{Y}_n^k.\nabla T_k(u_n)$, as $m \to \infty$. So, one has, by (2.85),

$$\int_\Omega \mathcal{Y}^k.\nabla T_k(u)dx \geq \liminf_{n\to\infty} \int_\Omega a_n(x, h_m(\nabla T_k(u_n))).h_m(\nabla T_k(u_n))dx, \text{ for all } m > 0.$$

Since $\int_\Omega \lambda d\nu_x^k(\lambda)$ and $\int_\Omega a(x,\lambda)d\nu_x^k(\lambda)$ are, respectively, the weak limits of $\nabla T_k(u_n)$ and $a_n(x, \nabla T_k(u_n))$, then using the nonlinear weak-* convergence property (2.12), we get

$$\begin{aligned}&\lim_{n\to\infty} \int_\Omega a_n(x, h_m(\nabla T_k(u_n))).h_m(\nabla T_k(u_n))\, dx\\ &\quad= \int_{\Omega\times\mathbb{R}^N} a(x, h_m(\lambda)).h_m(\lambda)d\nu_x^k(\lambda)dx\end{aligned}$$

and so

$$\int_\Omega \mathcal{Y}^k.\nabla T_k(u)\, dx \geq \int_{\Omega\times\mathbb{R}^N} a(x, h_m(\lambda)).h_m(\lambda)\, d\nu_x^k(\lambda)dx.$$

Now, thanks to Lemma 2.2.1, we can apply the monotone convergence theorem on the sequence $(a(x, h_m(\lambda)).h_m(\lambda))_m$ to deduce that, as $m \to \infty$,

$$\int_\Omega \mathcal{Y}^k.\nabla T_k(u)\, dx \geq \int_{\Omega\times\mathbb{R}^N} a(x,\lambda).\lambda\, d\nu_x^k(\lambda)dx. \tag{2.104}$$

Now using the representation formulas (2.62) and (2.73), and the fact that $\nu_x^k(\lambda)$ is a probability measure on $\mathbb{R}^N$ for a.e. $x \in \Omega$, we find

$$\begin{aligned}&\int_{\Omega\times\mathbb{R}^N} \big(a(x,\lambda) - a(x, \nabla T_k(u))\big).\big(\lambda - \nabla T_k(u)\big)d\nu_x^k(\lambda)dx\\ &= \int_{\Omega\times\mathbb{R}^N} a(x,\lambda).\lambda d\nu_x^k(\lambda)dx - \int_\Omega \left(\int_{\mathbb{R}^N} a(x,\lambda)d\nu_x^k(\lambda)\right) \nabla T_k(u)dx\\ &\quad - \int_\Omega a(x, \nabla T_k(u)) \left(\int_{\mathbb{R}^N} \lambda d\nu_x^k(\lambda)\right) dx\\ &\quad + \int_\Omega (a(x, \nabla T_k(u)).\nabla T_k(u)) \left(\int_{\mathbb{R}^N} d\nu_x^k(\lambda)\right) dx\end{aligned}$$

$$= \int_{\Omega\times\mathbb{R}^N} a(x,\lambda).\lambda d\nu_x^k(\lambda)dx - \int_\Omega \left(\int_{\mathbb{R}^N} a(x,\lambda)d\nu_x^k(\lambda)\right)\left(\int_{\mathbb{R}^N} \lambda d\nu_x^k(\lambda)\right)dx$$

$$= \int_{\Omega\times\mathbb{R}^N} a(x,\lambda).\lambda d\nu_x^k(\lambda)dx - \int_\Omega \mathcal{Y}^k.\nabla T_k(u)\,dx.$$

From (2.104), we deduce (2.86).

$(ii)$ We prove (2.87), i.e., $\mathcal{Y}^k = a(x, \nabla T_k(u))$ a.e. in $\Omega$.

Thanks to the "div-curl" inequality (2.86) and the strict monotonicity assumption (2.6) on $a(x,.)$, one has

$$\Big(a(x,\lambda) - a(x,\nabla T_k(u))\Big).\Big(\lambda - \nabla T_k(u)\Big)d\nu_x^k(\lambda) = 0 \text{ for a.e. } x \in \Omega,$$

and subsequently for a.e. $x \in \Omega$, $\lambda = \nabla T_k(u)$ w.r.t. the measure $\nu_x^k$ on $\mathbb{R}^N$. Since, by the representation formula (2.62), $\nabla T_k(u) = \int_\Omega \lambda d\nu_x^k(\lambda)$, then the measure $\nu_x^k$ reduces to the Dirac measure $\delta_{\nabla T_k(u)}$. Now, from the representation formula (2.73), we can deduce (2.87). Indeed, one has

$$\mathcal{Y}^k(x) = \int_{\mathbb{R}^N} a(x,\lambda)d\nu_x^k(\lambda) = \int_{\mathbb{R}^N} a(x,\lambda)d\delta_{\nabla T_k(u(x))}(\lambda) = a(x, \nabla T_k(u(x))).$$

Moreover, the sequence $\nabla T_k(u_n)$ generates the Young measure $\nu_x^k = \delta_{\nabla T_k(u)}$ a.e. on $\Omega$. So, from Theorem 2.2.1–$(ii)$, $\nabla T_k(u_n)$ converges to $\nabla T_k(u)$ in measure on $\Omega$ as $n \to \infty$. □

**Lemma 2.4.4** *For a.e. $k > 0$, $a_n(x, \nabla T_k(u_n)).\nabla T_k(u_n)$ converges to $a(x, \nabla T_k(u)).\nabla T_k(u)$ strongly in $L^1(\Omega)$.* □

***Proof*** By Lemma 2.4.3–(ii) and (2.55), up to a subsequence, we have $a_n(x, \nabla T_k(u_n)).\nabla T_k(u_n)$ that converges to $a(x, \nabla T_k(u)).\nabla T_k(u)$ a.e. in $\Omega$. Since $a_n(x, \nabla T_k(u_n)).\nabla T_k(u_n) \geq 0$, by Fatou's lemma, one has

$$\int_\Omega a(x,\nabla T_k(u)).\nabla T_k(u)\,dx \leq \liminf_{n\to\infty} \int_\Omega a_n(x,\nabla T_k(u_n)).\nabla T_k(u_n)\,dx$$

and so, by (2.85), we have

$$\liminf_{n\to\infty} \int_\Omega a_n(x,\nabla T_k(u_n)).\nabla T_k(u_n)\,dx = \int_\Omega a(x,\nabla T_k(u)).\nabla T_k(u)\,dx.$$

Thus, by the Scheffé's theorem (see [24]), one has $a_n(x, \nabla T_k(u_n)).\nabla T_k(u_n)$ that converges to $a(x, \nabla T_k(u)).\nabla T_k(u)$ strongly in $L^1(\Omega)$, up to subsequence. □

**Lemma 2.4.5** *$u$ is a renormalized solution of (2.1).* □

***Proof*** The first two requirements in Definition 2.3.1 are satisfied by Lemma 2.4.1–(iv), (v).

Now, we prove that $b(u) \in L^1(\Omega)$. Indeed, from (2.101), one has

$$\int_\Omega b(u)T_k(u)\,dx \le \int_\Omega f T_k(u)\,dx$$

or

$$\int_\Omega b(u)\frac{1}{k}T_k(u)\,dx \le \|f\|_{L^1(\Omega)},$$

which becomes, for $k \to 0$,

$$\int_\Omega |b(u)|\,dx \le \|f\|_{L^1(\Omega)},$$

by Fatou's lemma, since $\frac{1}{k}T_k(u) \to sign(u)$ as $k \to 0$. Hence, $b(u) \in L^1(\Omega)$.

Now, we prove (2.15) with the diffusion flux $a(.,.)$. By (2.7) and Hölder-type inequality, we get

$$\begin{aligned}\int_{\{k<|u|<k+1\}} a(x,\nabla u).\nabla u\,dx &\le C\int_{\{k<|u|<k+1\}}\Big(\mathcal{M}|\nabla u| + |\nabla u|^{p(x)}\Big)\,dx\\ &\le C\big\|\mathcal{M}\chi_{\{|u|>k\}}\big\|_{L^{p'(.)}(\Omega)}\big\|(\nabla u)\chi_{\{k<|u|<k+1\}}\big\|_{L^{p(.)}(\Omega)}\\ &\quad + C\int_{\{k<|u|<k+1\}}|\nabla u|^{p(x)}\,dx. \end{aligned} \tag{2.105}$$

Thus, (2.15) follows from (2.63).

It remains to prove (2.16) for $u$. Because $C^\infty(\overline{\Omega})$ is dense in $W^{1,p(.)}(\Omega)$ and in $W^{1,p_n(.)}(\Omega)$ since $p$ and $p_n$ verify (2.58), we can take test functions in $C^\infty(\overline{\Omega})$. So, let $\psi \in C^\infty(\overline{\Omega})$ a test function for the renormalized formulation (2.16) for $u_n$. One has

$$\begin{aligned}\int_\Omega \Big(S(u_n)a_n(x,\nabla u_n).\nabla\psi + S'(u_n)a_n(x,\nabla u_n).\nabla u_n\psi + b(u_n)S(u_n)\psi\Big)\,dx\\ + \int_{\partial\Omega}|u_n|^{p(x)-2}u_n S(u_n)\psi\,d\sigma = \int_\Omega f_n S(u_n)\psi\,dx,\end{aligned} \tag{2.106}$$

where $S \in \mathbb{S}$ with $suppS \subset [-M, M]$.

As $n \to \infty$ in (2.106), reasoning as above to pass from (2.88) to (2.99), we get the different limits given in (2.89)–(2.91) and (2.94). So, we should direct especially our attention to the term

$$\int_\Omega S'(u_n)a_n(x,\nabla u_n).(\nabla u_n)\psi\,dx = \int_\Omega S'(u_n)\mathcal{Y}_n^M.(\nabla T_M(u_n))\psi\,dx.$$

The sequence $S'(u_n)$ is uniformly bounded and converges to $S'(u)$ a.e. in $\Omega$. Thanks to Lemma 2.4.4 and by using Lebesgue generalized convergence theorem, this term converges to

$$\int_\Omega S'(u)\mathcal{Y}^M.\nabla T_M(u)\psi\,dx = \int_\Omega S'(u)a(x,\nabla u).\nabla u\psi\,dx.$$

We deduce the renormalized formulation (2.16) for $u$ with test functions in $C^\infty(\overline{\Omega})$ and by density with test functions in $W^{1,p(.)}(\Omega)\cap L^\infty(\Omega)$, which end the proof of Theorem 2.4.1. □

## References


1. B. Andreianov, M. Bendahmane, S. Ouaro, Structural stability for variable exponent elliptic problems. I. The $p(x)$-Laplacian kind problems. Nonlinear Anal. **73**, 2–24 (2010)
2. F. Andreu, N. Igbida, J.M. Mazón, J. Toledo, $L^1$ existence and uniqueness results for quasi-linear elliptic equations with nonlinear boundary conditions. Ann. Inst. H. Poincaré. Anal. Non Linéaire **24**(1), 61–89 (2007)
3. F. Andreu, J.M. Mazón, S. Segura De Léon, J. Toledo, Quasi-linear elliptic and parabolic equations in $L^1$ with nonlinear boundary conditions. Adv. Math. Sci. Appl. **7**(1), 183–213 (1997)
4. P. Bénilan, L. Boccardo, T. Gallouët, R. Gariepy, M. Pierre, J.L. Vazquez, An $L^1$-theory of existence and uniqueness of solutions of nonlinear elliptic equations, Ann. Sc. Norm. Super. Pisa Cl. Sci. (4) **22**(2), 241–273 (1995)
5. Y. Chen, S. Levine, M. Rao, Variable exponent, linear growth functionals in image restoration. SIAM J. Appl. Math. **66**(4), 1383–1406 (2006)
6. L. Diening, Theoretical and numerical results for electrorheological fluids, Ph.D. thesis, University of Freiburg, Germany, 2002
7. L. Diening, Riesz potential and Sobolev embeddings on generalized Lebesgue and Sobolev spaces $L^{p(.)}$ *and* $W^{k,p(.)}$. Math. Nachr. **268**, 31–43 (2004)
8. L. Diening, P. Harjulehto, P. Hästö, M. Růžička, in *Lebesgue and Sobolev Spaces with Variable Exponents*. Lecture Notes in Mathematics, vol. 2017 (Springer Verlag, Heidelberg, 2011)
9. R. Eymard, Th. Gallouët, R. Herbin, Finite volume methods, in *Handbook of Numerical Analysis*, vol. VII (North-Holland, 2000), pp. 713–1020
10. X. Fan, Q. Zhang, Existence of solutions for p(x)-Laplacian Dirichlet problem. Nonlinear Anal. **52**, 1843–1852 (2003)
11. X. Fan, D. Zhao, On the spaces $L^{p(x)}(\Omega)$ and $W^{m,p(x)}(\Omega)$. J. Math. Anal. Appl. **263**, 424–446 (2001)
12. P. Gwiazda, A. Swierczewska, On non-Newtonian fluids with a property of rapid thickening under different stimulus. Math. Models Methods Appl. Sci. **18**(7), 1073–1092 (2008)
13. P. Harjulehto, P. Hästö, V. Latvala, Minimizers of the variable exponent, non-uniformly convex Dirichlet energy. J. Math. Pures Appl. (9) **89**(2), 174–197 (2008)
14. N. Hungerbühler, Quasi-linear parabolic systems in divergence form with weak monotonicity. Duke Math. J. **107**(3), 497–520 (2001)
15. O. Kovacik, J. Rakosnik, On spaces $L^{p(x)}$ and $W^{1,p(x)}$. Czech. Math. J. **41**, 592–618 (1991)

16. S. Ouaro, A. Tchousso, Well posedness result for a nonlinear elliptic problem involving variable exponent and Robin type boundary condition. Afr. Diaspora J. Math. **11**(2), 36–64 (2011)
17. A. Prignet, Conditions aux limites non homogènes pour des problèmes elliptiques avec second membre mesure. Ann. Fac. Sci. Toulouse Math. **6**(2), 297–318 (1997)
18. K.R. Rajagopal, M. Růžička; Mathematical modelling of electrorheological materials. Contin. Mech. Thermodyn. **13**, 59–78 (2001)
19. M. Růžička, in *Electrorheological Fluids: Modeling and Mathematical Theory*. Lecture Notes in Mathematics, vol. 1748 (Springer, Berlin, 2000)
20. M. Růžička; Modeling, mathematical and numerical analysis of electrorheological fluids. Appl. Math. **49**(6), 565–609 (2004)
21. L. Wang, Y. Fan, W. Ge, Existence and multiplicity of solutions for a Neumann problem involving the $p(x)$-Laplace operator. Nonlinear Anal. **71**, 4259–4270 (2009)
22. J. Yao, Solutions for Neumann boundary value problems involving $p(x)-$Laplace operators. Nonlinear Anal. **68**, 1271–1283 (2008)
23. V.V. Zhikov, Averaging of functionals of the calculus of variations and elasticity theory. Izv. Akad. Nauk SSSR Ser. Mat. **50**(4), 675–710 (1986) (in Russian); Engl. transl. in Mah. Ussr Izvestiya **29**, 33–66 (1987)
24. V.V. Zhikov, On the Lavrentiev effect. Dokl. Akad. Nauk **345**(1), 10–14 (1995) (in Russian)

# Chapter 3
# Weak Solutions of Anti-periodic Discrete Nonlinear Problems

**Rodrigue Sanou, Idrissa Ibrango, Blaise Koné, and Aboudramane Guiro**

**Abstract** We consider the existence of weak solutions for discrete nonlinear problems. The proof of the main result is based on a minimization method.

**Keywords** Discrete nonlinear problems · Minimization method · Anti-periodic

## 3.1 Introduction

In this chapter, we investigate the existence of weak solution for the following anisotropic nonlinear discrete anti-periodic boundary problem:

$$\begin{cases} -\Delta\big[\alpha(k-1)a(k-1, \Delta u(k-1))\big] = f(k, u(k)), \; k \in \mathbb{N}[1, N], \\ \\ u(0) = -u(N); \quad u(1) = -u(N+1), \end{cases} \tag{3.1}$$

R. Sanou
Laboratoire de Mathématiques et Informatique (LAMI), UFR, Sciences et Techniques, Université Thomas SANKARA, Ouagadougou, Burkina Faso

I. Ibrango
Laboratoire de Mathématiques, d'Informatique et Applications (LaMIA), Institut Universitaire de Technologie, Université Nazi BONI, Bobo Dioulasso, Burkina Faso

B. Koné (✉)
Laboratoire de Mathématiques et Informatique (LAMI), Institut Burkinabé des Arts et Métiers, Université Joseph KI-Zerbo, Ouagadougou, Burkina Faso

A. Guiro
Laboratoire de Mathématiques, d'Informatique et Applications (LaMIA), UFR, Sciences Exactes et Appliquées, Université Nazi BONI, Bobo Dioulasso, Burkina Faso



T. Diagana et al. (eds.), *Partial Differential Equations and Applications*,
Springer Proceedings in Mathematics & Statistics 420,
https://doi.org/10.1007/978-3-031-27661-3_3

where $\Delta u(k) = u(k+1) - u(k)$ is the forward difference operator, $\mathbb{N}[1, N] = \{1, \ldots, N\}$, and $a, \alpha, f$ are functions to be defined later and where $N$ is a fixed integer $\geq 3$.

The theory of difference equations occupies now a central position in applicable analysis. We just refer to the recent results of Agarwal et al.[1], Yu and Guo [21], Koné and Ouaro [12], Guiro et al.[9], Cai and Yu [4], Zhang and Liu [22], Mihăilescu et al.[17], Candito and D'Agui [5], Cabada et al.[3], Jiang and Zhou [10], and the references therein. In [22], the authors studied the following problem:

$$\begin{aligned} &\Delta^2 y(k-1) + \lambda f((y(k)) = 0, \ k \in \mathbb{N}[1, T], \\ &y(0) = y(T+1) = 0, \end{aligned} \tag{3.2}$$

where $\lambda > 0$ is a parameter, $\Delta^2 y(k) = \Delta(\Delta y(k))$, and $f : [0, +\infty) \longrightarrow \mathbb{R}$ is a continuous function satisfying the condition

$$f(0) = -a, \ \textit{where a is positive constant}. \tag{3.3}$$

The problem (3.2) is referred as the "semipositone" problem in the literature, which was introduced by Castro and Shivaji [6]. Semipositone problems arise in bulking of mechanical systems, design of suspension bridges, chemical reactions, astrophysics, combustion, and management of natural resources.

The studies regarding problems such as (3.1) or (3.2) can be placed at interface of certain mathematical fields such as nonlinear partial differential equations and numerical analysis. On the other hand, they are strongly motivated by their applicability in mathematical physics as mentioned above.

In [10], Jiang and Zhou studied the following problem:

$$\begin{aligned} &\Delta^2 y(k-1) = f(k, u(k)) \ \ k \in \mathbb{N}[1, T], \\ &u(0) = \Delta u(T) = 0, \end{aligned} \tag{3.4}$$

where $T$ is a fixed positive integer and $f : \mathbb{N}[1, T] \times \mathbb{R} \longrightarrow \mathbb{R}$ is a continuous function.

Jiang and Zhou proved the existence of nontrivial solutions for (3.4) by using strongly monotone operator principle and critical point theory.

In [11], it is considered a discrete variant of the variable exponent anisotropic problem

$$
\begin{aligned}
&-\sum_{k=1}^{N} \frac{\partial}{\partial x_i} a_i\left(x, \frac{\partial u}{\partial x_i}\right) = f(x) \quad in \quad \Omega \\
&u = 0 \;\; on \quad \Gamma_1 \\
&\tfrac{\partial u}{\partial \eta} = 0 \;\; on \quad \Gamma_2,
\end{aligned}
\tag{3.5}
$$

where $\Omega \subset \mathbb{R}^N$ (N $\geq$ 3) is a bounded domain with smooth boundary $\Gamma_1 \cup \Gamma_2 = \partial\Omega$, $f \in L^{\infty}(\Omega)$, and $p_i$ is continuous on $\overline{\Omega}$ such that $1 < p_i(x) < N$ for all $x \in \overline{\Omega}$ and all $i \in \mathbb{N}[1, N]$, where $p_i^- := ess \inf_{x\in\Omega} p_i(x)$ and $\sum_{i=1}^{N}\left(\frac{1}{p_i^-}\right) > 1$.

The first equation of (3.5) was recently analysed by Koné et al.[13] and Ouaro [18] and generalized to a Radon measure data by Koné et al. [14] for a homogeneous Dirichlet boundary condition ($u = 0$ on $\partial\Omega$). The study (3.5) will be done in a forthcoming work. Problem such as (3.5) has been intensively studied in the last decades since they can model various phenomena arising from the study of elastic mechanics (see [20, 23]), electrorheological fluids (see [8, 19, 20]), and image restoration (see [7]). In [7], Chen et al. studied a functional with variable exponent $1 \leq p(x) \leq 2$ that provides a model for image denoising, enhancement, and restoration. Their paper created another interest for the study of problems with variable exponent.

Note that Mihăilescu et al. (see [15, 16]) were the first authors who study anisotropic elliptic problems with variable exponent. In general, the interested reader can find more information about difference equation in [1, 3, 5, 9, 11, 12].

Our goal in this chapter is to use a minimization method in order to establish some existence results of solutions of (3.1). The idea of the proof is to transfer the problem of the existence of solutions for (3.1) into the problem of existence of a minimizer for some associated energy functional. This method was successfully used by Bonanno et al.[2] for the study of an eigenvalue nonhomogeneous Neumann problem, where, under an appropriate oscillating behaviour of the nonlinear term, they proved the existence of a determined open interval of positive parameters for which the problem considered admits infinitely many weak solutions that strongly converge to zero, in an appropriate Orlicz–Sobolev space. Let us point out that, to our best knowledge, discrete problems such as (3.1) involving anisotropic exponents have been discussed for the first time by Mihăilescu et al. [17], in a second time by Koné and Ouaro [12], and in a third time by Guiro et al.[9]. In [17], the authors proved by using critical point theory the existence of a continuous spectrum of eigenvalues for the problem

$$
\begin{aligned}
-\Delta\left(|\Delta u(k-1)|^{p(k-1)-2}\Delta u(k-1)\right) &= \lambda |u(k)|^{q(k)-2} u(k), \; k \in \mathbb{N}[1, T], \\
u(0) = u(T+1) &= 0,
\end{aligned}
\tag{3.6}
$$

where $T \geq 2$ is the positive integer, and the functions $p : \mathbb{N}[0, T] \longrightarrow [2, +\infty)$ and $q : \mathbb{N}[1, T] \longrightarrow [2, +\infty)$ are bounded, while $\lambda$ is positive constant.

In [12], Koné and Ouaro proved, by using minimization method, existence and uniqueness of weak solutions for the following problem:

$$\begin{cases} -\Delta\left(a\left(k-1, \Delta u\left(k-1\right)\right)\right) = f\left(k\right), \quad k \in \mathbb{N}[1, T] \\ u\left(0\right) = u\left(T+1\right) = 0, \end{cases} \tag{3.7}$$

where $T \geq 2$ is a positive integer. The function $(a\left(k-1, \Delta u\left(k-1\right)\right))$ that appears in the left-hand side of problem (3.1) is more general than the one that appears in (3.6).

In [9], Guiro et al. studied the following two-point boundary-value problems:

$$\begin{cases} -\Delta\left(a\left(k-1, \Delta u\left(k-1\right)\right)\right) + |u\left(k\right)|^{p(k)} u\left(k\right) = f\left(k\right), \quad k \in \mathbb{N}[1, T] \\ \Delta u\left(0\right) = \Delta u\left(T\right) = 0. \end{cases} \tag{3.8}$$

The function $(a\left(k-1, \Delta u\left(k-1\right)\right))$ has the same properties as in [12], but the boundary conditions are different. For this reason, Guiro et al. defined a new norm in the Hilbert space considered in order to get, by using minimization methods, existence of unique weak solution (which is also a classical solution since the Hilbert space associated is of finite dimension). Indeed, they used the following norm:

$$\|u\| = \left(\sum_{k=1}^{T+1} |\Delta u\left(k-1\right)|^2 + \sum_{k=1}^{T} |u\left(k\right)|^2\right)^{\frac{1}{2}}, \tag{3.9}$$

which is associated to the Hilbert space

$$W = \{\, v : \mathbb{N}[0, T+1] \longrightarrow \mathbb{R};\ \textit{such that}\ \ \Delta v\left(0\right) = \Delta v\left(T\right) = 0\}. \tag{3.10}$$

In order to get the coercivity of the energy functional, the authors of [9] assumed the following on the exponent:

$$p : \mathbb{N}[0, N] \longrightarrow (2, +\infty). \tag{3.11}$$

In this chapter, we assume that the exponent $p : \mathbb{N}[0, N] \longrightarrow [2, +\infty)$.

The remaining part of this chapter is organized as follows. Section 3.2 is devoted to mathematical preliminaries. The main existence and uniqueness result is stated and proved in Sect. 3.3. In Sect. 3.4, we have the extension of the problem (3.1).

## 3.2 Mathematical Background

By a solution to problem (3.1), we mean such a function $u : \mathbb{N}[0, N+1] \longrightarrow \mathbb{R}$ that satisfies the given equation on $\mathbb{N}[1, N]$ and the boundary conditions. In the $N$-dimensional Hilbert space,

$$X = \{u : \mathbb{N}[0, N+1] \longrightarrow \mathbb{R} : u(0) = -u(N); \quad u(1) = -u(N+1)\},$$

with the inner product

$$\langle x, y\rangle = \sum_{k=1}^{N+1} \Delta x(k-1)\Delta y(k-1), \qquad \forall\, x, y \in X,$$

we consider the following norm:

$$\|x\| = \left(\sum_{k=1}^{N+1} |\Delta x(k-1)|^2\right)^{\frac{1}{2}}. \tag{3.12}$$

Let

$$p,\ r : \mathbb{N}[0, N] \longrightarrow [2, +\infty) \tag{3.13}$$

and denoted by

$$p^- = \min_{k\in\mathbb{N}[0,N]} p(k), \quad p^+ = \max_{k\in\mathbb{N}[0,N]} p(k), \quad r^- = \min_{k\in\mathbb{N}[0,N]} r(k), \qquad \text{and}$$

$$r^+ = \max_{k\in\mathbb{N}[0,N]} r(k).$$

For the data $\alpha$, $a$, and $f$, we assume what follows:

$$(H_1). \begin{cases} a(k,.) : \mathbb{R} \to \mathbb{R}, \ \ k \in \mathbb{N}[0, N] \text{ is continuous and there exists} \\ \quad A(.,.) : \mathbb{N}[0, N] \times \mathbb{R} \to \mathbb{R} \\ \text{which satisfies } a(k,\xi) = \frac{\partial}{\partial \xi} A(k,\xi) \text{ and } A(k, 0) = 0, \ \text{ for all } k \in \mathbb{N}[0, N]. \end{cases}$$

$(H_2)$. For all $k \in \mathbb{N}[0, N]$ and $\xi \neq \eta$,

$$(a(k,\xi) - a(k,\eta)).(\xi - \eta) > 0. \tag{3.14}$$

$(H_3)$. For any $k \in \mathbb{N}[0, N]$, $\xi \in \mathbb{R}$, we have

$$A(k,\xi) \geq \frac{1}{p(k)}|\xi|^{p(k)}. \tag{3.15}$$

$(H_4)$. For each $k \in \mathbb{N}[0, N]$, the function $f(k, .) : \mathbb{R} \longrightarrow \mathbb{R}$ is continuous, and there exists a constant $C_1 > 0$ such that

$$|f(k, \xi)| \le C_1\big(1 + |\xi|^{r(k)-1}\big). \tag{3.16}$$

We denote

$$F(k, \xi) = \int_0^{\xi} f(k, s)ds \text{ for } (k, \xi) \in \mathbb{N}[0, N] \times \mathbb{R}, \tag{3.17}$$

and we deduce that there exists a constant $C_2 > 0$ such that

$$|F(k, \xi)| \le C_2\big(1 + |\xi|^{r(k)}\big). \tag{3.18}$$

$(H_5)$. The function $\alpha : \mathbb{N}[0, N] \longrightarrow (0, +\infty)$ is such that for all $k \in \mathbb{N}[0, N]$,

$$0 < \alpha^- = \min_{k\in\mathbb{N}[0,N]} (\alpha(k)) \le \alpha(k) \le \alpha^+ = \max_{k\in\mathbb{N}[0,N]} (\alpha(k)) < +\infty. \tag{3.19}$$

$(H_6)$. For each $k \in \mathbb{N}[0, N]$, $r(k) < p^-$.

*Example 3.2.1* There are many functions satisfying both $(H_1) - (H_5)$. Let us mention the following:

- $A(k, \xi) = \dfrac{1}{p(k)}\Big(\big(1 + |\xi|^2\big)^{p(k)/2} - 1\Big)$, where $a(k, \xi) = \big(1 + |\xi|^2\big)^{(p(k)-2)/2}\xi$, $\forall\, k \in \mathbb{N}[0, N], \xi \in \mathbb{R}$.
- $f(k, \xi) = 1 + |\xi|^{r(k)-1}$, $\forall\, k \in \mathbb{N}[0, N]$, and $\xi \in \mathbb{R}$.
- $\alpha(k) = 1$, $\forall\, k \in \mathbb{N}[0, N]$.

□

Moreover, we may consider $X$ with the following norm:

$$|x|_m = \left(\sum_{k=1}^{N} |x(k)|^m\right)^{\frac{1}{m}}, \quad \forall\, x \in X \quad \text{and} \quad m \ge 2. \tag{3.20}$$

We have the following inequalities (see [4]):

$$N^{(2-m)/(2m)}|x|_2 \le |x|_m \le N^{1/m}|x|_2, \quad \forall\, x \in X \quad \text{and} \quad m \ge 2. \tag{3.21}$$

We need the following auxiliary results throughout our paper (see [17]):

**Lemma 3.2.1**

1. *There exist two positive constant* $C_3,\ C_4$ *such that*

$$\sum_{k=1}^{N+1} |\Delta x(k-1)|^{p(k-1)} \geq C_3 \|x\|^{p^-} - C_4 \tag{3.22}$$

*for all* $x \in X$ *with* $\|x\| > 1$.

2. *For any* $m \geq 2$, *there exists a positive constant* $c_m$ *such that*

$$\sum_{k=1}^{N} |x(k)|^m \leq c_m \sum_{k=1}^{N+1} |\Delta x(k-1)|^m, \quad \forall x \in X. \tag{3.23}$$

## 3.3 Existence of Weak Solutions

In this section, we study the existence of weak solution of problem (3.1).

**Definition 3.3.1** A weak solution of problem (3.1) is $u \in X$ such that:

$$\sum_{k=1}^{T+1} \alpha(k-1) a(k-1, \Delta u(k-1)) \Delta v(k-1) = \sum_{k=1}^{T} f(k, u(k)) v(k) \tag{3.24}$$

for all $v \in X$. □

Note that since $X$ is a finite-dimensional space, the weak solutions coincide with the classical solution of the problem (3.1).

**Theorem 3.3.1** *Assume that* $(H_1) - (H_6)$ *hold. Then, there exists a weak solution of the problem (3.1).*

We define the energy functional $J : X \longrightarrow \mathbb{R}$ by

$$J(u) = \sum_{k=1}^{N+1} \alpha(k-1) A\big(k-1, \Delta u(k-1)\big) - \sum_{k=1}^{N} F\big(k, u(k)\big). \tag{3.25}$$

**Lemma 3.3.1** *The functional* $J$ *is well-defined on* $X$ *and is of class* $C^1\big(X, \mathbb{R}\big)$ *with the derivative given by*

$$\langle J'(u), v\rangle = \sum_{k=1}^{N+1} \alpha(k-1)a(k-1, \Delta u(k-1))\Delta v(k-1) - \sum_{k=1}^{N} f\big(k, u(k)\big)v(k), \tag{3.26}$$

*for all* $u, v \in X$. □

***Proof*** Let $I(u) = \sum_{k=1}^{N+1} \alpha(k-1)A\big(k-1, \Delta u(k-1)\big)$ and $\Lambda(u) = \sum_{k=1}^{N} F\big(k, u(k)\big)$.
As in [9], Lemma 3.4, we can prove that the functional $I$ derivative is given by

$$\langle I'(u), v\rangle = \sum_{k=1}^{N+1} \alpha(k-1)a\big(k-1, \Delta u(k-1)\big)\Delta v(k-1). \tag{3.27}$$

On the other hand, for all $u, v \in X$, we have

$$\langle \Lambda'(u), v\rangle = \sum_{k=1}^{N} f(k, u(k)).$$

The functional $J$ is clearly of class $C^1$. □

**Proposition 3.3.1** *The functional $J$ is coercive and bounded from below.* □

Indeed, according to (3.18) and (3.15) , we have

$$\begin{aligned}
J(u) &= \sum_{k=1}^{N+1} \alpha(k-1)A\big(k-1, \Delta u(k-1)\big) - \sum_{k=1}^{N} F\big(k, u(k)\big) \\
&\geq \alpha^- \sum_{k=1}^{N+1} A\big(k-1, \Delta u(k-1)\big) - C_2 \sum_{k=1}^{N} |u(k)|^{r(k)} - C_2 \\
&\geq \alpha^- \sum_{k=1}^{N+1} \frac{1}{p(k-1)} \big|\Delta u(k-1)\big|^{p(k-1)} - C_2 \sum_{k=1}^{N} |u(k)|^{r(k)} - C_2 \\
&\geq \frac{\alpha^-}{p^+} \sum_{k=1}^{N+1} \big|\Delta u(k-1)\big|^{p(k-1)} - C_2 \sum_{k=1}^{N} |u(k)|^{r(k)} - C_2.
\end{aligned}$$

To prove the coercivity of $J$, we may assume that $||u|| > 1$, and we deduce from (3.22) that

$$J(u) \geq \frac{C_3\ \alpha^-}{p^+} ||u||^{p^-} - C_4 - C_2 \sum_{k=1}^{N} |u(k)|^{r(k)} - C_2$$

$$\geq \frac{C_3\,\alpha^-}{p^+}||u||^{p^-} - C_4 - C_2\sum_{k=1}^{N}|u(k)|^{r^-} - C_2\sum_{k=1}^{N}|u(k)|^{r^+} - C_2.$$

Using (3.23), we see that

$$J(u) \geq \frac{C_3\,\alpha^-}{p^+}||u||^{p^-} - C_4 - C_2(C_{r^-})\sum_{k=1}^{N}|\Delta u(k)|^{r^-} - C_2(C_{r^+})\sum_{k=1}^{N}|\Delta u(k)|^{r^+} - C_2.$$

By using (3.21), there exist positive constants $K_1$ and $K_2$ such that

$$J(u) \geq \frac{C_3\,\alpha^-}{p^+}||u||^{p^-} - C_4 - K_1||u||^{r^-} - K_2||u||^{r^+} - C_2. \tag{3.28}$$

Since $p^- > r^+$, $J$ is coercive.

Besides, for $||u|| \leq 1$, we see from the Weierstrass theorem that $J$ is bounded from below there. Recall that $X$ is finite-dimensional. Then it follows that summarizing $J$ is bounded from below. □

***Proof of Theorem** 3.3.1* Since $J$ is continuous, bounded from below, and coercive on $X$, using the relation between critical points of $J$ and problem (3.1), we deduce that $J$ has a minimizer that is a weak solution of problem (3.1).

## 3.4 An Extension

In this section, we show that the existence result obtained for (3.1) can be extended to a more general discrete boundary-value problem of the form

$$\begin{cases} -\Delta\big[\alpha(k-1)a(k-1,\Delta u(k-1))\big] + |u|^{q(k)-2}u(k) = f(k,u(k)), \; k \in \mathbb{N}[1,N], \\ u(0) = -u(N); \quad u(1) = -u(N+1), \end{cases} \tag{3.29}$$

with $q : \mathbb{N}[1,N] \longrightarrow (1,+\infty)$.

A function $u \in X$ is a solution of problem (3.29) if for any $v \in X$,

$$\sum_{k=1}^{N+1}\alpha(k-1)a(k-1,\Delta u(k-1))\Delta v(k-1) + \sum_{k=1}^{N}|u(k)|^{q(k)-2}u(k)v(k)$$
$$-\sum_{k=1}^{N} f\big(k,u(k)\big)v(k) = 0. \tag{3.30}$$

**Theorem 3.4.1** *Under asymptotes* $(H_1) - (H_6)$*, there exists a weak solution* $u \in X$ *of problem (3.29).*

***Proof*** For $u \in X$, we defined the energy functional $J$ by

$$J(u) = \sum_{k=1}^{N+1} \alpha(k-1)A\big(k-1, \Delta u(k-1)\big) + \sum_{k=1}^{N} \frac{1}{q(k)}|u|^{q(k)} - \sum_{k=1}^{N} F\big(k, u(k)\big).$$

The functional $J$ is well-defined, continuous, and of class $C^1(X, \mathbb{R})$ with a derivative given by

$$\langle J'(u), v\rangle = \sum_{k=1}^{N+1} \alpha(k-1)a(k-1, \Delta u(k-1))\Delta v(k-1) + \sum_{k=1}^{N} |u|^{q(k)-2}u(k)v(k)$$
$$- \sum_{k=1}^{N} f\big(k, u(k)\big)v(k),$$

for all $u, v \in X$.

Since

$$\sum_{k=1}^{N} \frac{1}{q(k)}|u|^{q(k)} \geq 0,$$

we have

$$J(u) \geq \sum_{k=1}^{N+1} \alpha(k-1)A\big(k-1, \Delta u(k-1)\big) - \sum_{k=1}^{N} F\big(k, u(k)\big), \tag{3.31}$$

and, according to Proposition 3.3.1, that arguments applied above also work.

# References


1. R.P. Agarwal, K. Perera, D. O'Regan, Multiple positive solutions of singular and non-singular discrete problems via variational methods. Nonlinear Anal. **58**, 69–73 (2004)
2. G. Bonanno, G. Molica Bisci, V. Radulescu, Arbitrarily small weak solutions for nonlinear eigenvalue problem in Orlicz-Sobolev spaces. Monatshefte fur Mathematik **165**(3–4), 305–318 (2012)
3. A. Cabada, A. Iannizzoto, S. Tersian, Multiple solutions for discrete boundary value problems. J. Math. Anal. Appl. **356**, 418–428 (2009)
4. X. Cai, J. Yu, Existence theorems for second-order discrete boundary value problems. J. Math. Anal. Appl. **320**, 649–661 (2006)

5. P. Candito, G. D'Agui, Three solutions for a discrete nonlinear Neumann problem involving p-Laplacian. Adv. Differ. Equ. **11**, 862016 (2010)
6. A. Castro, R. Shivaji, Nonnegative solutions for a class of radically symmetric nonpositone problems. Proce. Am. Math. Soc. **106**, 735–740 (1989)
7. Y. Chen, S. Levine, M. Rao, Variable exponent, linear growth functionals in image restoration. SIAM J. Appl. Math. **66**(4), 1383–1406 (2006)
8. L. Diening, Theoretical and numerical results for electrorheological fluids. Ph.D. Thesis. University of Freiburg, 2002
9. A. Guiro, I. Nyanquini, S. Ouaro, On the solvability of discrete nonlinear Neumann problems involving the $p(x)$-Laplacian. Adv. Differ. Equ. **32** (2011)
10. L. Jiang, Z. Zhou, Existence of nontrivial solutions for discrete nonlinear two point boundary value problems. Appl. Math. Comput. **180**(1), 318–329 (2006)
11. B. Koné, S. Ouaro, On the solvability of discrete nonlinear two-point boundary value problems. Int. J. Math. Math. Sci. 2012, 927607 (2012)
12. B. Koné, S. Ouaro, Weak solutions for anisotropic discrete boundary value problems. J. Differ. Equ. Appl. **16**(2), 1–11 (2010)
13. B. Koné, S. Ouaro, S. Traoré, Weak solutions for anisotropic nonlinear elliptic equations with variable exponent. Electron. J. Differ. Equ. **144**, 1–11 (2009)
14. B. Koné, S. Ouaro, S. Soma, Weak solutions for anisotropic nonlinear elliptic problem with variable exponent and measure data. Int. J. Evol. Equ. **5**(3), 1–23 (2011)
15. M. Mihăilescu, P. Pucci, V. Raduslescu, Nonhomogeneous boundary value problems in anisotropic Sobolev spaces, C. R. Math. **345**(10), 561–566 (2007)
16. M. Mihăilescu, P. Pucci, V. Radulescu, Eigenvalue problems for anisotropic quasilinear elliptic equations with variables exponent. J. Math. Anal. Appl. **340**(1), 687–698 (2008)
17. M. Mihailescu, V. Radulescu, S. Tersian, Eigenvalue problems for anisotropic discrete boundary value problems. J. Differ. Equ. Appl. **15**, 557–567 (2009)
18. S. Ouaro, Well-posedness results for anisotropic nonlinear elliptic equations with variable exponent and $L^1$-data. Cubo. A Math. J. **12**(1), 133–148 (2010)
19. K.R. Rajagopal, M. Ruzicka, Mathematical modeling of electrorheological materials. Continuum Mech. Thermodyn. **13**, 59–78 (2001)
20. M. Ruzicka, in *Electrorheological Fluids: Modeling and Mathematical Theory*. Lecture Notes in Mathematics, vol. 1748 (Springer, Berlin, 2000)
21. J. Yu, Z. Guo, On boundary value problems for a discrete generalized Emden-Fowler equation. J. Math. Anal. Appl. **231**, 18–31 (2006)
22. G. Zhang, S. Liu, On a class of semipositone discrete boundary value problem. J. Math. Anal. Appl. **325**, 175–182 (2007)
23. V. Zhikov, Averaging of functionals in the calculus of variations and elasticity. Math. USSR-Izvestiya **29**, 33–66 (1987)

# Chapter 4
# Boundary Feedback Controller over a Bluff Body for Prescribed Drag and Lift Coefficients

**Evrad M. D. Ngom, Abdou Sène, and Daniel Y. Le Roux**

**Abstract** This chapter presents an improved boundary feedback controller for the two- and three-dimensional Navier–Stokes equations, in a bounded domain $\Omega$, *for prescribed drag and lift coefficients*. In order to determine the feedback control law, we consider an extended system coupling the equations governing the Navier–Stokes problem with an equation satisfied by the control on the bluff body, which is a part of the domain boundary. By using the Faedo–Galerkin method and a priori estimation techniques, a boundary control is built. This control law ensures the controllability of the discrete system. Then, a compactness result then allows us to pass to the limit in the non-linear system satisfied by the approximated solutions.

**Keywords** Navier–Stokes system · Boundary feedback stabilization · Bluff body · Drag and lift coefficients

## 4.1 Introduction

Flow over a bluff body is a common occurrence associated with fluid flowing over an obstacle or with the movement of a natural or artificial body. Evident examples are the flows past an airplane, a submarine, and wind blowing past a bridge or a high-rise building. This chapter presents an improved boundary feedback control for the two- and three-dimensional Navier–Stokes equations around a bluff body. Let $\Omega$ be

E. M. D. Ngom (✉)
CEA-MITIC, Université Gaston Berger, Dakar, Senegal
e-mail: evrad-diokel.ngom@ugb.edu.sn

A. Sène
CEA-MITIC, Université Virtuelle du Sénégal, Dakar-Fann, Senegal
e-mail: abdou.sene@uvs.edu.sn

D. Y. Le Roux
Université de Lyon, CNRS, Université Lyon 1, Institut Camille Jordan, Villeurbanne, France
e-mail: dleroux@math.univ-lyon1.fr



T. Diagana et al. (eds.), *Partial Differential Equations and Applications*,
Springer Proceedings in Mathematics & Statistics 420,
https://doi.org/10.1007/978-3-031-27661-3_4

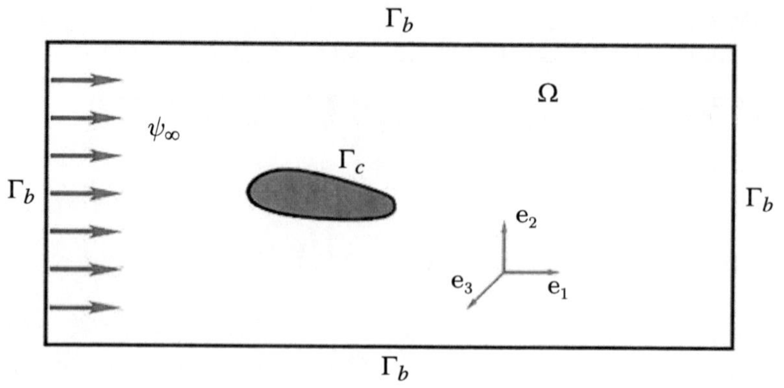


**Fig. 4.1** Description of the domain $\Omega$ and of the two connected components $\Gamma_b$ and $\Gamma_c$

a bounded and connected domain in $\mathbb{R}^d$ $(d = 2, 3)$, with a boundary $\Gamma$ of class $C^2$, and composed of two connected components $\Gamma_b$ and $\Gamma_c$ such that $\Gamma = \Gamma_b \cup \Gamma_c$. Such a boundary decomposition is schematized in Fig. 4.1. In particular, the boundary $\Gamma_c$ represents the contour of the bluff body, and it is the part of $\Gamma$ where a Dirichlet boundary control in feedback form has to be determined.

For $\mathbf{e}_i = (\delta_{1i}, \delta_{2i}, \delta_{3i})$, $i = 1, \ldots, d$, with $\delta_{ij}$, the Kronecker symbol $\Gamma_c$ is chosen such that

$$\int_{\Gamma_c} \mathbf{e}_i \cdot \mathbf{n}\, d\zeta = 0, \tag{4.1}$$

where $\mathbf{n}$ denotes the unit outer normal vector to $\Gamma$.

For example, condition (4.1) holds when $\Gamma_c$ is a sphere with centre $(0, 0, 0)$ and radius $r$. Indeed, in that case, $\Gamma_c$ is the locus of all points $\mathbf{X} = (x, y, z)^t$ such that $f(\mathbf{X}) = \|\mathbf{X}\|^2 - r^2 = 0$, which lead to

$$\mathbf{n} = -\frac{\nabla f(\mathbf{X})}{\|\nabla f(\mathbf{X})\|} = -\frac{\mathbf{X}}{\|\mathbf{X}\|},$$

and hence, (4.1) is obtained. Condition (4.1) also holds in the case where $f(x, y, z)$ is the contour of a circular cylinder. More generally, when $f(x, y, z)$ represents the boundary $\Gamma_c$, condition (4.1) is satisfied if $\nabla f(x, y, z)$ is odd with respect to each variable $x, y, z$, supplemented with specific symmetries for $f(x, y, z)$ and $\Gamma_c$.

Let $T > 0$ be a fixed real number, $Q = [0, T[\times\Omega$, $\Sigma_b = [0, T[\times\Gamma_b$, $\Sigma_c = [0, T[\times\Gamma_c$, and $\mathbf{V}^{1/2}(\widetilde{\Gamma})$, $\widetilde{\Gamma} \subset \Gamma$, is defined as the space of trace functions whose extension by zero over $\Gamma$ belongs to $\mathbf{H}^{1/2}(\Gamma)$. We consider the perturbed trajectory $(\mathbf{u}, \pi)$, solution of the non-stationary Navier–Stokes model

$$
\begin{cases}
\dfrac{\partial \mathbf{u}}{\partial t} - \nu \Delta \mathbf{u} + (\mathbf{u} \cdot \nabla)\mathbf{u} + \nabla \pi = \mathbf{f} & \text{in } Q, \\
\nabla \cdot \mathbf{u} = 0 & \text{in } Q, \\
\mathbf{u} = \boldsymbol{\psi}_\infty(\mathbf{x}) & \text{on } \Sigma_b, \\
\mathbf{u} = \mathbf{v}_c(t, \mathbf{x}) & \text{on } \Sigma_c, \\
\mathbf{u}(t = 0, \mathbf{x}) = \mathbf{u}_0(\mathbf{x}) & \text{in } \Omega,
\end{cases}
\tag{4.2}
$$

where $\mathbf{u}$ and $\pi$ are the velocity field and the pressure, respectively, $\nu$ is the kinematic viscosity, and $\mathbf{f} \in \mathbf{H}^{-1}(\Omega)$ represents body forces acting on the fluid. Further, $\mathbf{u}_0(\mathbf{x})$ is the initial condition, and $\mathbf{v}_c(t, \mathbf{x})$ represents the control input on $\Sigma_c$, while the specified Dirichlet boundary condition $\boldsymbol{\psi}_\infty$ is such that

$$
\boldsymbol{\psi}_\infty \in \mathbf{V}^{1/2}(\Gamma_b) \quad \text{and} \quad \int_{\Gamma_b} \boldsymbol{\psi}_\infty \cdot \mathbf{n}\, d\zeta = 0. \tag{4.3}
$$

The different regimes of the flow are given by the values of the Reynolds number $\mathcal{R}_e = \frac{\overline{\boldsymbol{\psi}}_\infty D}{\nu}$, with $D$ and $\overline{\boldsymbol{\psi}}_\infty$ being the characteristic dimension (e.g., the size of $\Gamma_c$) and the characteristic velocity, respectively.

For low Reynolds numbers, due to the highly viscous body, the force exerted on the body is mainly attributed to skin friction. However, when the Reynolds number $\mathcal{R}_e$ exceeds a certain critical value, small perturbations destabilize the solution of the system (4.2) and yield a periodic solution $(\mathbf{u}, \pi)$ represented by the well-known von Kármán vortex street. In fluid dynamics, a von Kármán vortex street is a repeating pattern of swirling vortices caused by the unsteady separation of flow of a fluid around blunt bodies. This vortex shedding is responsible for such phenomena as the "singing" of suspended telephone or power lines, and the vibration of a car antenna at certain speeds that may lead to structural failure or reduction in performance. Further, vortex shedding occurs over a wide range of Reynolds numbers, causing significant increases in the mean drag and lift fluctuations. Therefore, the effective control of vortex shedding is important in engineering applications.

Recall that in fluid dynamics, the drag coefficient, denoted by $C_x$, is a dimensionless quantity that is used to quantify the drag or resistance of an object in a fluid environment, such as air or water. A low drag coefficient indicates the object will have less aerodynamic or hydrodynamic drag. The lateral lift coefficient and the vertical lift coefficient denoted by $C_y$ and $C_z$, respectively, are dimensionless coefficients that relate the lift generated by a lifting body to the density of the fluid around the body. It is common to show, for a particular airfoil section, the relationship between section lift coefficient and angle of attack. It is also useful to show the relationship between section lift coefficient and drag coefficient.

The coefficients $C_x$, $C_y$, and $C_z$, which are always associated with a particular surface area $S$, are defined [2, 15, 34] as

$$C_x(t) = \frac{2\,F_1(\mathbf{u}, \pi)}{\rho \overline{\psi}_\infty^2 S}, \quad C_y(t) = \frac{2\,F_2(\mathbf{u}, \pi)}{\rho \overline{\psi}_\infty^2 S}, \quad C_z(t) = \frac{2\,F_3(\mathbf{u}, \pi)}{\rho \overline{\psi}_\infty^2 S}, \tag{4.4}$$

where the fluid density $\rho$ is taken to $\rho = 1$ in the present paper and

$$F_i(\mathbf{u}, \pi) = -\int_{\Gamma_c} [\nu \nabla \mathbf{u} \cdot \mathbf{n} - \pi \mathbf{n}] \cdot \mathbf{e}_i \, d\zeta, \quad i = 1, \ldots, d. \tag{4.5}$$

The control of the unsteady viscous flow past bluff bodies has been studied by a number of authors, e.g., [3, 8, 13, 16, 22] for the passive control, [1, 4, 5, 12, 23, 24, 37] for the active open-loop control, and [2, 10, 21, 25, 26, 31] for active closed-loop control, also called a feedback control. Feedback control methods are an attractive choice over passive and active open-loop controls in that the control input is continuously modified according to the response of the flow system. For more examples of control over a bluff body, one can refer to the review work of H. Choi et al. [14].

In the above-mentioned papers, the authors aim at decreasing the mean drag coefficient, suppressing the vortex shedding, narrowing the wake width, and/or stabilizing the system around a given steady-state flow. In particular, the reduction of the drag coefficient remains a difficult and challenging issue, and an important question arises: what is the lowest possible drag achievable from control in the case of bluff bodies? For example, by employing a high-frequency rotation of the circular cylinder, Tokumaru and Dimotakis [37] experimentally obtained approximately 80% drag reduction at $R_e = 15{,}000$. A significant drag reduction is also obtained by Amitay et al. [1], Glezer and Amitay [19], for high Reynolds numbers ranging from 31,000 to 131,000, by applying a high-frequency forcing from a synthetic jet to flow over a circular cylinder.

Apart from experimental and numerical simulations studies, a number of theoretical works have focussed about the stabilization around a prescribed equilibrium state, e.g., [6, 7, 17, 29, 30, 32, 33]. In most of these theoretical stabilization results, and thanks to the employed control laws, the authors aim to suppress the vortex shedding and narrow the wake width. Further, in [29] (in finite dimension) and in [32] (in infinite dimension), the stabilization result is obtained via enough small initial perturbations. However, if the above-mentioned studies aim to find an equilibrium state, such an equilibrium state is not reached by prescribing the drag coefficient $C_x$ and the lift coefficients $C_y$ and $C_z$.

This is why the present paper aims to present a theoretical study regarding the feedback control over a bluff body for prescribed drag and lift coefficients (which can be as small as desired). To our knowledge, such a study has not been conducted previously, and it is the main objective of the present paper.

For prescribed time functions $\widetilde{\lambda}_i(t)$, $i = 1, \ldots, d$, we need to find a feedback control $\mathbf{v}_c = \mathcal{M}(\mathbf{u})$, where $\mathcal{M}$ is the feedback law, such that $F_i$ in (4.5) satisfies

$$F_i(\mathbf{u}, \pi) = \widetilde{\lambda}_i(t), \; i = 1, \ldots, d. \tag{4.6}$$

To this end, the boundary control $\mathbf{v}_c$ in (4.2) is written on the form

$$\mathbf{v}_c(t,\mathbf{x}) = \sum_{i=1}^{d} \alpha_i(t)\,\mathbf{e}_i(\mathbf{x}) \quad \text{on } \Sigma_c, \tag{4.7}$$

where the quantities $\alpha_i$, $i = 1,\ldots,d$, are a priori unknown and have to be determined in the feedback form. In order to determine $\alpha_i$, leading to the determination of the boundary control $\mathbf{v}_c$, we consider the trajectory $(\boldsymbol{\psi}, q) \in \mathbf{H}^1(\Omega) \times L_0^2(\Omega)$ solution of the stationary Navier–Stokes model [18]:

$$\begin{cases} -\nu\Delta\boldsymbol{\psi} + (\boldsymbol{\psi}\cdot\nabla)\boldsymbol{\psi} + \nabla q = \mathbf{f}, & \text{in } \Omega, \\ \nabla\cdot\boldsymbol{\psi} = 0 & \text{in } \Omega, \\ \boldsymbol{\psi} = 0 & \text{on } \Gamma_c, \\ \boldsymbol{\psi} = \boldsymbol{\psi}_\infty & \text{on } \Gamma_b, \end{cases} \tag{4.8}$$

and we substitute $(\mathbf{u}, \pi)$ by $(\mathbf{v} + \boldsymbol{\psi}, p + q)$ in (4.2) and (4.6). Consequently, we get this extended system that is considered in the following:

$$\begin{cases} (a)\ \dfrac{\partial\mathbf{v}}{\partial t} - \nu\Delta\mathbf{v} + (\mathbf{v}\cdot\nabla)\boldsymbol{\psi} + (\boldsymbol{\psi}\cdot\nabla)\mathbf{v} + (\mathbf{v}\cdot\nabla)\mathbf{v} + \nabla p = 0 & \text{in } Q, \\ (b)\ \nabla\cdot\mathbf{v} = 0 & \text{in } Q, \\ (c)\ \mathbf{v} = 0 & \text{on } \Sigma_b, \\ (d)\ \mathbf{v} = \sum_{i=1}^{d} \alpha_i(t)\,\mathbf{e}_i(\mathbf{x}) & \text{on } \Sigma_c, \\ (e)\ \mathbf{v}(t=0,\mathbf{x}) = \mathbf{v}_0(\mathbf{x}) = \mathbf{u}_0(\mathbf{x}) - \boldsymbol{\psi}(\mathbf{x}) & \text{in } \Omega, \\ (f)\ \langle(-\nu\nabla\mathbf{v} + Ip)\cdot\mathbf{n}, \mathbf{e}_i\rangle_{\mathbf{H}^{-\frac{1}{2}}(\Gamma_c),\mathbf{H}^{\frac{1}{2}}(\Gamma_c)} = \lambda_i,\ i = 1,\ldots,d, & \end{cases} \tag{4.9}$$

where $\lambda_i(t) = -F_i(\boldsymbol{\psi}, q) - \widetilde{\lambda}_i(t)$, $i = 1,\ldots,d$. As in [29, 30] where the authors stabilize the two- and three-dimensional Navier–Stokes problem around a given stationary state, system (4.9) is solved via a Galerkin procedure. Such a procedure consists in building a sequence of approximated solutions using an adequate Galerkin basis.

This chapter is organized as follows. In Sect. 4.2, the notations and mathematical preliminaries are given. In Sect. 4.3, the existence of at least one solution of the non-linear extended system (4.9) is established by applying the Galerkin method.

## 4.2 Notation and Preliminaries

### 4.2.1 Function Spaces

The usual function spaces $L^2(\Omega)$, $H^1(\Omega)$, $H_0^1(\Omega)$ are used, and we let $\mathbf{L}^2(\Omega) = (L^2(\Omega))^d$, $\mathbf{H}^1(\Omega) = (H^1(\Omega))^d$, $\mathbf{H}_0^1(\Omega) = (H_0^1(\Omega))^d$. Further, we denote by $\|\cdot\| = \|\cdot\|_{\mathbf{L}^2(\Omega)}$ the norm in $\mathbf{L}^2(\Omega)$. Finally, if $\mathbf{u} \in \mathbf{L}^2(\Omega)$ is such that $\nabla \cdot \mathbf{u} \in L^2(\Omega)$, the normal trace of $\mathbf{u}$ in $\mathbf{H}^{-\frac{1}{2}}(\Gamma)$ is $\mathbf{u} \cdot \mathbf{n}$.

A few spaces are now introduced:

$$\mathbf{V}^1(\Omega) = \{\mathbf{u} \in \mathbf{H}^1(\Omega) \ : \ \nabla \cdot \mathbf{u} = 0 \text{ in } \Omega\}, \tag{4.10}$$

$$\mathbf{V}_0^1(\Omega) = \{\mathbf{u} \in \mathbf{H}_0^1(\Omega) \ : \ \nabla \cdot \mathbf{u} = 0 \text{ in } \Omega\}, \tag{4.11}$$

$$\mathbf{V}(\Omega) = \{\mathbf{u} \in \mathbf{V}^1(\Omega), \ \mathbf{u} = 0 \text{ on } \Gamma_b, \int_{\Gamma_c} \mathbf{u} \cdot \mathbf{n}\, d\varsigma = 0\}, \tag{4.12}$$

$$\mathbf{H}(\Omega) = \{\mathbf{u} \in \mathbf{L}^2(\Omega) \ : \ \nabla \cdot \mathbf{u} = 0 \text{ in } \Omega, \ \mathbf{u} \cdot \mathbf{n} = 0 \text{ on } \Gamma_b\}. \tag{4.13}$$

$\mathbf{H}(\Omega)$ is a Hilbert space endowed with $\mathbf{L}^2$-norm, and $\mathbf{V}(\Omega)$ is Hilbert space endowed with $\mathbf{H}^1$-norm. Denoting by $\mathbf{V}^{-1}(\Omega) = (\mathbf{V}_0^1(\Omega))'$ the dual space of $\mathbf{V}_0^1(\Omega)$ and considering $\mathbf{H}(\Omega)$ identified with its own dual, we have $\mathbf{V}(\Omega) \subset \mathbf{H}(\Omega) \subset \mathbf{V}^{-1}(\Omega)$ algebraically and topologically with compact injections.

Finally, the solution $\mathbf{v}$ of (4.9) is searched in the space

$$\mathbf{W}(\Omega) = \left\{\mathbf{v} \in \mathbf{V}(\Omega), \exists\, \boldsymbol{\alpha} = (\alpha_1, \cdots, \alpha_d) \text{ such that } \mathbf{v} = \sum_{i=1}^{d} \alpha_i \mathbf{e}_i \text{ on } \Gamma_c\right\}, \tag{4.14}$$

where the orthonormal basis $\mathbf{e}_i$ of $\mathbb{R}^3$ is such that $\mathbf{e}_i \in \mathbf{V}^{1/2}(\Gamma_c)$, $i = 1, \ldots, d$.

### 4.2.2 Linear Forms and a Few Inequalities

In order to define a weak form of the Navier–Stokes equations, we introduce the continuous bilinear form

$$a(\mathbf{v}_1, \mathbf{v}_2) = \int_\Omega \nabla \mathbf{v}_1 : \nabla \mathbf{v}_2 \, d\mathbf{x}, \quad \forall \mathbf{v}_1, \mathbf{v}_2 \in \mathbf{H}^1(\Omega),$$

and the trilinear form

$$b(\mathbf{v}_1,\mathbf{v}_2,\mathbf{v}_3)=\int_\Omega(\mathbf{v}_1\cdot\nabla)\mathbf{v}_2\cdot\mathbf{v}_3\,d\mathbf{x},\quad \forall\mathbf{v}_1,\mathbf{v}_2,\mathbf{v}_3\in\mathbf{H}^1(\Omega).$$

Thanks to Hölder inequality, we obtain

$$|b(\mathbf{v}_1,\mathbf{v}_2,\mathbf{v}_3)|\le\|\mathbf{v}_1\|_{\mathbf{L}^3(\Omega)}\|\nabla\mathbf{v}_2\|\|\mathbf{v}_3\|_{\mathbf{L}^6(\Omega)}.$$

Further, due to the generalized Sobolev's inequality, there exists a positive constant $C$ such that

$$\|\mathbf{v}_1\|_{\mathbf{L}^3(\Omega)}\le C\|\mathbf{v}_1\|^{\frac12}\|\nabla\mathbf{v}_1\|^{\frac12}\ \text{ and }\ \|\mathbf{v}_3\|_{\mathbf{L}^6(\Omega)}\le C\|\nabla\mathbf{v}_3\|,\quad \text{for } d=2,3,$$

and hence,

$$|b(\mathbf{v}_1,\mathbf{v}_2,\mathbf{v}_3)|\le C\|\mathbf{v}_1\|^{\frac12}\|\nabla\mathbf{v}_1\|^{\frac12}\|\nabla\mathbf{v}_2\|\|\nabla\mathbf{v}_3\|. \tag{4.15}$$

By using Hölder inequality, we obtain

$$|b(\mathbf{v},\mathbf{u},\mathbf{v})|\le\|\mathbf{v}\|^2_{\mathbf{L}^4(\Omega)}\|\nabla\mathbf{u}\|,\quad \forall\mathbf{u},\mathbf{v}\in\mathbf{H}^1(\Omega),$$

and hence thanks to [11, Remark III.2.17], we deduce

$$|b(\mathbf{v},\mathbf{u},\mathbf{v})|\le C\|\mathbf{v}\|^{2-\frac{d}{2}}\|\nabla\mathbf{v}\|^{\frac{d}{2}}\|\nabla\mathbf{u}\|,\quad \forall\mathbf{u},\mathbf{v}\in\mathbf{H}^1(\Omega). \tag{4.16}$$

By employing integration by parts, the following property holds true

$$b(\mathbf{u},\mathbf{v},\mathbf{v})=\frac12\int_{\Gamma_c}|\mathbf{v}|^2(\mathbf{u}\cdot\mathbf{n})\,d\zeta,\quad \forall\mathbf{u}\in\mathbf{V}^1(\Omega)\text{ and }\forall\mathbf{v}\in\mathbf{V}(\Omega). \tag{4.17}$$

For all $\mathbf{v}=\sum_{i=1}^d\alpha_i\mathbf{e}_i$ and $\widetilde{\mathbf{v}}=\sum_{i=1}^d\widetilde{\alpha}_i\mathbf{e}_i$ on $\Gamma_c$, we have

$$\mathbf{v}\cdot\widetilde{\mathbf{v}}=\sum_{i=1}^d\alpha_i\widetilde{\alpha}_i\ \text{ on }\Gamma_c\quad\text{and}\quad \mathbf{v}\cdot\mathbf{n}=\sum_{i=j}^d\alpha_j(\mathbf{e}_j\cdot\mathbf{n})\ \text{ on }\Gamma_c. \tag{4.18}$$

From the trace theorem and the Poincaré inequality, we obtain $\|\mathbf{v}\|_{\mathbf{L}^2(\Gamma)}\le C\|\nabla\mathbf{v}\|$, $\forall\mathbf{v}\in\mathbf{W}(\Omega)$, and hence,

$$\|\mathbf{v}\|_{\mathbf{L}^2(\Gamma_c)}=\sqrt{\sum_{i=1}^d\alpha_i^2}\le C\|\nabla\mathbf{v}\|. \tag{4.19}$$

In the next section, the variational formulation of the control problem (4.9) is given.

## 4.3 Existence Result

### 4.3.1 The Variational Formulation

We consider the variational formulation for the extended system (4.9).

**Definition 4.3.1** Let $T > 0$ be an arbitrary real number, $\lambda_i(t)$ in $L^2(0,T)$, $i = 1,\dots,d$ and $\mathbf{v}_0 \in \mathbf{H}(\Omega)$; we shall say that $\mathbf{v}$ is a weak solution of (4.9) on $[0,T)$ if:

- $\mathbf{v} \in L^\infty(0,T;\mathbf{H}(\Omega)) \cap L^2(0,T;\mathbf{V}(\Omega))$.
- $\exists \boldsymbol{\alpha} = (\alpha_1,\dots,\alpha_d) \in \mathbf{L}^2(0,T)$ such that $\mathbf{v} = \sum_{i=1}^{d} \alpha_i\, \mathbf{e}_i$ on $\Gamma_c$,

$$\begin{cases} (a) \quad \dfrac{d}{dt}\displaystyle\int_\Omega \mathbf{v}\cdot\widetilde{\mathbf{v}}\,d\mathbf{x} + \nu a(\mathbf{v},\widetilde{\mathbf{v}}) + b(\mathbf{v},\boldsymbol{\psi},\widetilde{\mathbf{v}}) \\ \qquad\qquad + b(\boldsymbol{\psi},\mathbf{v},\widetilde{\mathbf{v}}) + b(\mathbf{v},\mathbf{v},\widetilde{\mathbf{v}}) = \sum_{i=1}^{d} \widetilde{\alpha}_i\lambda_i, \\ (b) \quad \left(\displaystyle\int_\Omega \mathbf{v}\cdot\widetilde{\mathbf{v}}\,d\mathbf{x}\right)(0) = \displaystyle\int_\Omega \mathbf{v}_0\cdot\widetilde{\mathbf{v}}\,d\mathbf{x}, \end{cases} \tag{4.20}$$

$\forall \widetilde{\mathbf{v}} \in \mathbf{W}(\Omega)$ with $\widetilde{\mathbf{v}} = \sum_{i=1}^{d} \widetilde{\alpha}_i\, \mathbf{e}_i$ on $\Gamma_c$. □

Note that the initial condition $(4.20)_b$ makes sense because for any solution $\mathbf{v}$ of $(4.20)_a$, function $t \to \int_\Omega \mathbf{v}(t)\cdot\widetilde{\mathbf{v}}\,d\mathbf{x}$ is continuous (see [11] Corollary II.4.2).

We now first establish the a priori estimates for the extended system (4.9).

### 4.3.2 A Priori Estimates

Taking $\widetilde{\mathbf{v}} = \mathbf{v}$ in $(4.20)_a$ leads to

$$\frac{1}{2}\frac{d}{dt}\|\mathbf{v}\|^2 + \nu\|\nabla\mathbf{v}\|^2 + b(\mathbf{v},\mathbf{v},\mathbf{v}) + b(\boldsymbol{\psi},\mathbf{v},\mathbf{v}) + b(\mathbf{v},\boldsymbol{\psi},\mathbf{v}) = \sum_{i=1}^{d}\alpha_i\lambda_i. \tag{4.21}$$

First, let us estimate the terms of $b(\cdot,\cdot,\cdot)$ in (4.21). Using (4.17) yields

$$b(\mathbf{v},\mathbf{v},\mathbf{v}) = \frac{1}{2}\int_{\Gamma_c} |\mathbf{v}|^2(\mathbf{v}\cdot\mathbf{n})\,d\zeta, \quad \forall \mathbf{v} \in \mathbf{W}(\Omega). \tag{4.22}$$

Using (4.1) and (4.18) in (4.22), we obtain

$$b(\mathbf{v},\mathbf{v},\mathbf{v}) = \frac{1}{2}\left(\sum_{i=1}^{d}\alpha_i^2\right)\sum_{j=1}^{d}\alpha_j\int_{\Gamma_c}\mathbf{e}_j\cdot\mathbf{n}\,d\zeta = 0. \tag{4.23}$$

Second, from (4.17), we have

$$b(\boldsymbol{\psi},\mathbf{v},\mathbf{v}) = \frac{1}{2}\int_{\Gamma_c}|\mathbf{v}|^2(\boldsymbol{\psi}\cdot\mathbf{n})\,d\zeta, \quad \forall\mathbf{v}\in\mathbf{W}(\Omega),$$

and since $\boldsymbol{\psi} = 0$ on $\Gamma_c$, we deduce

$$b(\boldsymbol{\psi},\mathbf{v},\mathbf{v}) = 0. \tag{4.24}$$

Finally, using (4.15) and Young's inequality leads to

$$\begin{aligned}
|b(\mathbf{v},\boldsymbol{\psi},\mathbf{v})| &\le C\|\mathbf{v}\|^{\frac{1}{2}}\|\nabla\mathbf{v}\|^{\frac{1}{2}}\|\nabla\boldsymbol{\psi}\|\|\nabla\mathbf{v}\| \\
&\le \frac{\epsilon_1}{2}\|\nabla\mathbf{v}\|^2 + \frac{C^2}{2\epsilon_1}\|\nabla\boldsymbol{\psi}\|^2\|\mathbf{v}\|\|\nabla\mathbf{v}\| \\
&\le \frac{\epsilon_1+\epsilon_2}{2}\|\nabla\mathbf{v}\|^2 + \frac{1}{2\epsilon_2}\left(\frac{C^4}{4\epsilon_1^2}\|\nabla\boldsymbol{\psi}\|^4\right)\|\mathbf{v}\|^2,
\end{aligned}$$

and taking $\epsilon_1 = \epsilon_2 = \frac{\nu}{4}$, we obtain

$$|b(\mathbf{v},\boldsymbol{\psi},\mathbf{v})| \le \frac{\nu}{4}\|\nabla\mathbf{v}\|^2 + \left(\frac{8C^4}{\nu^3}\|\nabla\boldsymbol{\psi}\|^4\right)\|\mathbf{v}\|^2. \tag{4.25}$$

We now estimate the term in the right-hand side of (4.21). Using (4.19), we obtain $|\alpha_i| \le C\|\nabla\mathbf{v}\|$, and hence,

$$\sum_{i=1}^{d}\alpha_i\lambda_i \le C\,\|\nabla\mathbf{v}\|\left(\sum_{i=1}^{d}|\lambda_i|\right) \le \frac{\nu}{4}\|\nabla\mathbf{v}\|^2 + M_\lambda(t), \tag{4.26}$$

where $M_\lambda(t) = \frac{1}{\nu}\left(C\sum_{i=1}^{d}|\lambda_i(t)|\right)^2$. Using (4.23)–(4.26) in (4.21), the following inequality holds

$$\frac{1}{2}\frac{d}{dt}\|\mathbf{v}\|^2 + \frac{\nu}{2}\|\nabla\mathbf{v}\|^2 \le \left(\frac{8C^4}{\nu^3}\|\nabla\boldsymbol{\psi}\|^4\right)\|\mathbf{v}\|^2 + M_\lambda(t). \tag{4.27}$$

Consequently, thanks to Gronwall lemma, we deduce from (4.27) the following estimation:

$$\sup_{t\leq T} \|\mathbf{v}(t)\|^2 + \nu \int_0^T \|\nabla \mathbf{v}(t)\|^2 dt \leq C_\lambda(T), \tag{4.28}$$

where $C_\lambda(T)$ depends on $T$, $M_\lambda$, $\|\nabla \boldsymbol{\psi}\|$, and $\|\mathbf{v}_0\|$ .

Let us estimate $\dfrac{d\mathbf{v}}{dt}$. By using integration by parts and the technics used in (4.23)–(4.24), we show that

$$b(\mathbf{v}, \boldsymbol{\psi}, \widetilde{\mathbf{v}}) = -b(\mathbf{v}, \widetilde{\mathbf{v}}, \boldsymbol{\psi}),$$

$$b(\mathbf{v}, \mathbf{v}, \widetilde{\mathbf{v}}) = -b(\mathbf{v}, \widetilde{\mathbf{v}}, \mathbf{v}).$$

Moreover, by employing (4.15) and (4.16), we obtain

$$|b(\boldsymbol{\psi}, \mathbf{v}, \widetilde{\mathbf{v}})| \leq C\|\boldsymbol{\psi}\|^{\frac{1}{2}} \|\nabla \boldsymbol{\psi}\|^{\frac{1}{2}} \|\nabla \mathbf{v}\| \|\nabla \widetilde{\mathbf{v}}\|,$$

$$|b(\mathbf{v}, \widetilde{\mathbf{v}}, \boldsymbol{\psi})| \leq C\|\boldsymbol{\psi}\|^{\frac{1}{2}} \|\nabla \boldsymbol{\psi}\|^{\frac{1}{2}} \|\nabla \mathbf{v}\| \|\nabla \widetilde{\mathbf{v}}\|,$$

$$|b(\mathbf{v}, \widetilde{\mathbf{v}}, \mathbf{v})| \leq C\|\mathbf{v}\|^{2-\frac{d}{2}} \|\nabla \mathbf{v}\|^{\frac{d}{2}} \|\nabla \widetilde{\mathbf{v}}\|;$$

hence, from (4.20), by taking $\widetilde{\alpha}_i = 0$, yielding $\widetilde{\mathbf{v}} \in \mathbf{V}_0^1(\Omega)$, we deduce

$$\left\| \frac{d\mathbf{v}}{dt} \right\|_{\mathbf{V}^{-1}(\Omega)} \leq \nu\|\nabla \mathbf{v}\| + C\|\boldsymbol{\psi}\|^{\frac{1}{2}} \|\nabla \boldsymbol{\psi}\|^{\frac{1}{2}} \|\nabla \mathbf{v}\| + C\|\mathbf{v}\|^{2-\frac{d}{2}} \|\nabla \mathbf{v}\|^{\frac{d}{2}} := \mathcal{G}(t),$$

where $\mathcal{G}(t)$ is bounded in $L^{\frac{4}{d}}(]0, T[)$ according to estimate (4.28). Therefore,

$$\left\| \frac{d\mathbf{v}}{dt} \right\|_{L^{\frac{4}{d}}(]0,T[;\mathbf{V}^{-1}(\Omega))} \leq \left( \int_0^T \mathcal{G}^{\frac{4}{d}}(t)\, dt \right)^{\frac{d}{4}} \leq C_\lambda(T). \tag{4.29}$$

**Theorem 4.3.1** *Assume that* (4.1) *is satisfied. For an arbitrary function* $\lambda_i$ *in* $L^2(0, T)$, $i = 1, \ldots, d$, *and an arbitrary initial data* $\mathbf{v}_0$ *in* $\mathbf{H}(\Omega)$, *there exists a solution* $\mathbf{v}$ *in the sense of Definition 4.3.1 and a distribution* $p$ *on* $Q$ *such that (4.9) holds. Moreover,* $\frac{d\mathbf{v}}{dt}$ *belongs to* $L^{\frac{d}{4}}(]0, T[; \mathbf{V}^{-1}(\Omega))$. □

***Proof*** In the first step, a Galerkin basis is built for the space $\mathbf{W}(\Omega)$ defined in (4.14), while in the second step we prove the existence of a weak solution $\mathbf{v}$. Finally, we prove the existence of the pressure.

### 4.3.3 A Galerkin Basis for the Space $\mathbf{W}(\Omega)$

For $i = 1, \ldots, d$, we consider the following Stokes problem:

$$\begin{cases} -\Delta \mathbf{w}_i + \nabla q_i = 0, & \text{in } \Omega, \\ \nabla \cdot \mathbf{w}_i = 0 & \text{in } \Omega, \\ \mathbf{w}_i = 0 & \text{on } \Gamma_b, \\ \mathbf{w}_i = \mathbf{e}_i & \text{on } \Gamma_c. \end{cases} \tag{4.30}$$

From condition (4.1), $\int_{\Gamma_c} \mathbf{e}_i \cdot \mathbf{n} \, d\zeta = 0$. Thus, system (4.30) admits a unique solution $(\mathbf{w}_i, q_i)$ belonging to $\mathbf{H}^1(\Omega) \times L_0^2(\Omega)$ (see [11, Theorem IV.6.5]). Moreover, for all $\mathbf{z} \in \mathbf{V}_0^1(\Omega)$ defined in (4.11) and for all $\alpha_i \in \mathbb{R}$, we have $\mathbf{v} = \mathbf{z} + \sum_{i=1}^d \alpha_i \mathbf{w}_i \in \mathbf{W}(\Omega)$, where $\mathbf{w}_i$ satisfies (4.30). Indeed, we have $\mathbf{z}, \mathbf{w}_i \in \mathbf{V}(\Omega)$, and since $\mathbf{z} = 0$ on $\Gamma_c$, we obtain $\mathbf{v} = \sum_{i=1}^d \alpha_i \mathbf{w}_i$ on $\Gamma_c$. When $(\mathbf{z}_n)_{n \in \mathbb{N}}$ defines a countable orthonormal basis of $\mathbf{V}_0^1(\Omega)$, since $\mathbf{w}_i = \mathbf{e}_i$ on $\Gamma_c$, the sequence $\mathbf{w}_1, \ldots, \mathbf{w}_d, \mathbf{z}_1, \mathbf{z}_2, \mathbf{z}_3, \ldots$, is then linearly independent. Consequently, $\mathbf{W}(\Omega)$ can be rewritten as

$$\mathbf{W}(\Omega) = \text{span}(\mathbf{w}_i)_{\{1 \le i \le d\}} \oplus \text{span}(\mathbf{z}_n)_{\{n \in \mathbb{N}^*\}}, \tag{4.31}$$

and $\mathbf{v}$ is expressed as

$$\mathbf{v} = \mathbf{z} + \sum_{i=1}^{d} \alpha_i \mathbf{w}_i, \quad \text{with} \quad \mathbf{z} = \sum_{i=1}^{\infty} \theta_i \mathbf{z}_i.$$

### 4.3.4 Existence of Weak Solution

The proof of the existence follows a standard procedure [30]. In a first step, a sequence of approximate solutions using a Galerkin method is built. A compactness result allows us to pass to the limit in the system satisfied by the approximated solutions.

#### 4.3.4.1 The Galerkin Method

Let $m \in \mathbb{N}^*$; we define the space

$$W_m = \text{span}(\mathbf{w})_{\{1 \le i \le d\}} \oplus \text{span}(\mathbf{z}_i)_{\{1 \le i \le m\}},$$

and we express $\mathbf{v}_m \in W_m$ as

$$\mathbf{v}_m = \sum_{i=1}^{d+m} \alpha_{im} \boldsymbol{\varphi}_i,$$

where $\boldsymbol{\varphi}_i = \mathbf{w}_i$, for $i = 1, \ldots, d$, and $\boldsymbol{\varphi}_i = \mathbf{z}_{i-d}$, for $i = d+1, d+2, \ldots, d+m$. We consider the following finite-dimensional problem:

$$\begin{cases} (a) \quad \dfrac{d}{dt} \displaystyle\int_\Omega \mathbf{v}_m \cdot \boldsymbol{\varphi}_j \, d\mathbf{x} + \nu a(\mathbf{v}_m, \boldsymbol{\varphi}_j) + b(\mathbf{v}_m, \boldsymbol{\psi}, \boldsymbol{\varphi}_j) + b(\boldsymbol{\psi}, \mathbf{v}_m, \boldsymbol{\varphi}_j) \\ \qquad\qquad + b(\mathbf{v}_m, \mathbf{v}_m, \boldsymbol{\varphi}_j) = \displaystyle\sum_{i=1}^{d} \delta_{ij} \lambda_i, \\ (b) \quad \displaystyle\int_\Omega \mathbf{v}_m(0) \cdot \boldsymbol{\varphi}_j \, d\mathbf{x} = \int_\Omega \mathbf{v}_0 \cdot \boldsymbol{\varphi}_j \, d\mathbf{x}, \quad \text{for } j = 1, 2, \ldots, d+m. \end{cases} \tag{4.32}$$

**Lemma 4.3.1** *The discrete problem (4.32) has a unique solution* $\mathbf{v}_m$ *belonging to* $C^1(0, T_m; W_m)$. *Moreover, the solution satisfies*

$$\|\mathbf{v}_m\|_{L^\infty(0,T;\mathbf{L}^2(\Omega))} + \|\mathbf{v}_m\|_{L^2(0,T;\mathbf{H}^1(\Omega))} \leq C_\lambda(T), \tag{4.33}$$

$$\left\| \frac{d\mathbf{v}_m}{dt} \right\|_{L^{\frac{4}{d}}(]0,T[;\mathbf{V}^{-1}(\Omega))} \leq C_\lambda(T), \tag{4.34}$$

*where* $C_\lambda(T)$ *is a positive constant independent of* $m$. □

***Proof*** Classical results of non-linear ODEs lead to the existence of the greatest $T_m$ in $(0, T)$ such that the discrete problem (4.32) has a unique solution $\mathbf{v}_m \in C^1(0, T_m; W_m)$. Indeed, the resulting mass matrix defined as $M_{ij} = \int_\Omega \boldsymbol{\varphi}_i \cdot \boldsymbol{\varphi}_j \, d\mathbf{x}$ $(1 \leq i, j \leq d+m)$ is nonsingular. In order to show that $T_m$ is independent of $m$, it is sufficient to verify the boundedness of the $L^2$-norm of $\mathbf{v}_m$ independently of $m$. Following the same procedure as for the derivation of the a priori estimates (4.28) and (4.29) yields (4.33) and (4.34). If $T_m < T$, then $\|\mathbf{v}_m\|$ should tend to $+\infty$ as $t \to T_m$ because of the explosion criteria. However, this does not happen since $\|\mathbf{v}_m\|$ is bounded independently of $m$ in (4.33), and therefore $T_m = T$. □

For a subsequence of $\mathbf{v}_m$ (still denoted by $\mathbf{v}_m$), the estimates in (4.33) and (4.34) yield the following weak convergences as $m$ tends to $\infty$ :

$$\begin{cases} \mathbf{v}_m \rightharpoonup \mathbf{v} \text{ weakly in } L^2(]0, T[; \mathbf{V}(\Omega)), \\ \mathbf{v}_m \rightharpoonup \mathbf{v} \text{ weakly* in } L^\infty(]0, T[; \mathbf{H}(\Omega)), \\ \dfrac{d\mathbf{v}_m}{dt} \rightharpoonup \dfrac{d\mathbf{v}}{dt} \text{ weakly in } L^{\frac{4}{d}}(]0, T[; \mathbf{V}^{-1}(\Omega)). \end{cases} \tag{4.35}$$

Nevertheless, the convergences in (4.35) are not sufficient to pass to the limit in the weak formulation (4.32) because of the presence of the convection term. Consequently, in order to utilize the compactness theory on the sequence of approximated solution $\mathbf{v}_m$, we apply the Aubin theorem [27, Théorème 5.1, page 58] with $p_0 = 2$, $p_1 = \frac{4}{d}$ and $B_0 = \mathbf{V}(\Omega)$, $B_1 = \mathbf{V}^{-1}(\Omega)$ and $B = \mathbf{H}(\Omega)$. Note that $B_0 \subset B \subset B_1$, and the imbedding from $B_0$ to $B$ is compact. Wet set

$$U = \{\mathbf{v}, \mathbf{v} \in L^2(]0, T[; \mathbf{V}(\Omega)), \mathbf{v} \in L^{\frac{4}{d}}(]0, T[; \mathbf{V}^{-1}(\Omega))\},$$

and equipped with the norm $\|\mathbf{v}\|_{L^2(]0,T[;\mathbf{V}(\Omega))} + \|\mathbf{v}\|_{L^{\frac{4}{d}}(]0,T[;\mathbf{V}^{-1}(\Omega))}$, $U$ is a Banach space. Then, by applying the Aubin compacity theorem, we prove that the imbedding $U \subset L^2(]0, T[; \mathbf{H}(\Omega))$ is compact; hence, we obtain the following strong convergence (at least for a subsequence of $\mathbf{v}_m$ still denoted by $\mathbf{v}_m$)

$$\mathbf{v}_m \to \mathbf{v} \text{ strongly in } L^2(0, T; \mathbf{L}^2(\Omega)). \tag{4.36}$$

Using the above strong convergence result and (4.35) enables us to pass to the limit in the weak formulation, obtained from (4.32) after multiplication by $\varphi \in \mathcal{D}([0, T))$ and integration by parts with respect to time. Hence, for all $\widetilde{\mathbf{v}}_j = \widetilde{\alpha}_j \boldsymbol{\varphi}_j$, $j = 1, 2, \ldots, d + m$, passing to the limit yields

$$\begin{aligned}
&-\int_0^T \int_\Omega \mathbf{v} \cdot \widetilde{\mathbf{v}}_j \varphi'(t)\, d\mathbf{x}\, dt + \int_\Omega \mathbf{v}_0 \widetilde{\mathbf{v}}_j \varphi(0)\, d\mathbf{x} + \nu \int_0^T \int_\Omega \nabla \mathbf{v} : \nabla \widetilde{\mathbf{v}}_j \varphi(t)\, d\mathbf{x}\, dt \\
&\quad + \int_0^T \int_\Omega (\mathbf{v} \cdot \nabla \mathbf{v}) \cdot \widetilde{\mathbf{v}}_j \varphi(t)\, d\mathbf{x}\, dt + \int_0^T \int_\Omega (\mathbf{v} \cdot \nabla \boldsymbol{\psi}) \cdot \widetilde{\mathbf{v}}_j \varphi(t)\, d\mathbf{x}\, dt \\
&\quad + \int_0^T \int_\Omega (\boldsymbol{\psi} \cdot \nabla \mathbf{v}) \cdot \widetilde{\mathbf{v}}_j \varphi(t)\, d\mathbf{x}\, dt = \int_0^T \widetilde{\alpha}_j \delta_{jk} \lambda_k(t) \varphi(t)\, dt.
\end{aligned} \tag{4.37}$$

By linearity, Eq. (4.37) holds for all $\widetilde{\mathbf{v}}$ combination of finite $\widetilde{\mathbf{v}}_j$ and, by density, for any element of $\mathbf{W}(\Omega)$. □

Now we can retrieve the controlled problem (4.9).

### *4.3.5 Retrieving the Controlled Problem*

First, we prove the existence of the pressure.

**Lemma 4.3.2** *There exists $p \in \mathcal{D}'(]0, T[; L^2(\Omega))$ such that $(\mathbf{v}, p)$ satisfies $(4.9)_a$ in the distribution sense.* □

***Proof*** By choosing $\varphi \in \mathcal{D}(0,T)$ in (4.37), $\forall \widetilde{\mathbf{v}} = \widetilde{\mathbf{z}} + \widetilde{\alpha}_j \mathbf{w}_j \in \mathbf{W}(\Omega),\ j = 1, \dots, d$, and $\widetilde{\mathbf{z}} \in \mathbf{V}_0^1(\Omega)$, we obtain

$$\begin{aligned}
&\int_0^T \int_\Omega \frac{\partial \mathbf{v}}{\partial t} \cdot \widetilde{\mathbf{v}} \varphi(t)\, d\mathbf{x}\, dt + \nu \int_0^T \int_\Omega \nabla \mathbf{v} : \nabla \widetilde{\mathbf{v}} \varphi(t)\, d\mathbf{x}\, dt \\
&\quad + \int_0^T \int_\Omega (\mathbf{v} \cdot \nabla \mathbf{v}) \cdot \widetilde{\mathbf{v}} \varphi(t)\, d\mathbf{x}\, dt + \int_0^T \int_\Omega (\mathbf{v} \cdot \nabla \boldsymbol{\psi}) \cdot \widetilde{\mathbf{v}} \varphi(t)\, d\mathbf{x}\, dt \\
&\qquad + \int_0^T \int_\Omega (\boldsymbol{\psi} \cdot \nabla \mathbf{v}) \cdot \widetilde{\mathbf{v}} \varphi(t)\, d\mathbf{x}\, dt = \int_0^T \widetilde{\alpha}_j \lambda_j(t)\, \varphi(t) dt;
\end{aligned} \tag{4.38}$$

hence,

$$\begin{aligned}
&\int_\Omega \frac{\partial \mathbf{v}}{\partial t} \cdot \widetilde{\mathbf{v}}\, d\mathbf{x} + \nu \int_\Omega \nabla \mathbf{v} : \nabla \widetilde{\mathbf{v}}\, d\mathbf{x} + \int_\Omega (\mathbf{v} \cdot \nabla \mathbf{v}) \cdot \widetilde{\mathbf{v}}\, d\mathbf{x} + \int_\Omega (\mathbf{v} \cdot \nabla \boldsymbol{\psi}) \cdot \widetilde{\mathbf{v}}\, d\mathbf{x} \\
&\quad + \int_\Omega (\boldsymbol{\psi} \cdot \nabla \mathbf{v}) \cdot \widetilde{\mathbf{v}}\, d\mathbf{x} = \widetilde{\alpha}_j \lambda_j(t) \quad \text{in } \mathcal{D}'(0,T).
\end{aligned} \tag{4.39}$$

Further, taking $\widetilde{\alpha}_j = 0,\ j = 1, \dots, d$, yielding $\widetilde{\mathbf{v}} \in \mathbf{V}_0^1(\Omega)$, we deduce

$$\begin{aligned}
&\int_\Omega \frac{\partial \mathbf{v}}{\partial t} \cdot \widetilde{\mathbf{v}}\, d\mathbf{x} + \nu \int_\Omega \nabla \mathbf{v} : \nabla \widetilde{\mathbf{v}}\, d\mathbf{x} + \int_\Omega (\mathbf{v} \cdot \nabla \mathbf{v}) \cdot \widetilde{\mathbf{v}}\, d\mathbf{x} \\
&\quad + \int_\Omega (\mathbf{v} \cdot \nabla \boldsymbol{\psi}) \cdot \widetilde{\mathbf{v}}\, d\mathbf{x} + \int_\Omega (\boldsymbol{\psi} \cdot \nabla \mathbf{v}) \cdot \widetilde{\mathbf{v}}\, d\mathbf{x} = 0 \quad \text{in } \mathcal{D}'(0,T).
\end{aligned} \tag{4.40}$$

Then, for $\mathbf{f}$ defined as

$$\mathbf{f} = \frac{\partial \mathbf{v}}{\partial t} - \nu \Delta \mathbf{v} + (\mathbf{v} \cdot \nabla) \boldsymbol{\psi} + (\boldsymbol{\psi} \cdot \nabla) \mathbf{v} + (\mathbf{v} \cdot \nabla) \mathbf{v},$$

using (4.40) leads to $\mathbf{f} \in \mathcal{D}'(]0,T[;\ \mathbf{H}^{-1}(\Omega))$ and $\langle \mathbf{f}, \widetilde{\mathbf{v}} \rangle_{\mathbf{H}^{-1}(\Omega), \mathbf{H}_0^1(\Omega)} = 0$, for all $\widetilde{\mathbf{v}}$ in $\mathbf{V}_0^1(\Omega)$. Hence, due to de Rham's theorem[36], there exists $p \in \mathcal{D}'(]0,T[;\ L^2(\Omega))$ such that $\mathbf{f} = -\nabla p$. □

Next, we prove that $(\mathbf{v}, p)$ satisfies $(4.9)_f$. By writing $(4.9)_a$ in the form

$$\frac{\partial \mathbf{v}}{\partial t} + \nabla \cdot (-\nu \nabla \mathbf{v} + Ip) + (\mathbf{v} \cdot \nabla) \boldsymbol{\psi} + (\boldsymbol{\psi} \cdot \nabla) \mathbf{v} + (\mathbf{v} \cdot \nabla) \mathbf{v} = 0 \text{ in } Q$$

and using [36, Chap I, Theorem 1.2], we obtain

$$\int_\Omega \frac{\partial \mathbf{v}}{\partial t} \cdot \widetilde{\mathbf{v}}\, d\mathbf{x} + \int_\Omega (\nu \nabla \mathbf{v} - Ip) : \nabla \widetilde{\mathbf{v}}\, d\mathbf{x} + \langle (-\nu \nabla \mathbf{v} + Ip) \cdot \mathbf{n}, \widetilde{\mathbf{v}} \rangle_{\mathbf{H}^{-\frac{1}{2}}(\Gamma), \mathbf{H}^{\frac{1}{2}}(\Gamma)}$$

$$+\int_{\Omega}(\mathbf{v}\cdot\nabla\mathbf{v})\cdot\widetilde{\mathbf{v}}\,d\mathbf{x}+\int_{\Omega}(\mathbf{v}\cdot\nabla\boldsymbol{\psi})\cdot\widetilde{\mathbf{v}}\,d\mathbf{x}+\int_{\Omega}(\boldsymbol{\psi}\cdot\nabla\mathbf{v})\cdot\widetilde{\mathbf{v}}\,d\mathbf{x}=0, \qquad (4.41)$$

for all $\widetilde{\mathbf{v}}$ in $\mathbf{W}(\Omega)$. By comparing (4.39) and (4.41), we retrieve $(4.9)_f$, namely

$$\langle(-\nu\nabla\mathbf{v}+Ip)\cdot\mathbf{n},\mathbf{e}_i\rangle_{\mathbf{H}^{-\frac{1}{2}}(\Gamma_c),\mathbf{H}^{\frac{1}{2}}(\Gamma_c)}=\lambda_i.$$

Finally, it remains to verify that the initial condition $(4.9)_e$ belongs to $\mathbf{W}'(\Omega)$. In this purpose, we multiply $(4.9)_a$ by $\widetilde{\mathbf{v}}\varphi$, with $\varphi(T)=0$, and integrate with respect to time and space

$$-\int_0^T\int_{\Omega}\mathbf{v}\cdot\widetilde{\mathbf{v}}\varphi'(t)\,d\mathbf{x}\,dt+\int_{\Omega}\mathbf{v}(0)\widetilde{\mathbf{v}}\varphi(0)\,d\mathbf{x}+\nu\int_0^T\int_{\Omega}\nabla\mathbf{v}:\nabla\widetilde{\mathbf{v}}\varphi(t)\,d\mathbf{x}\,dt$$
$$+\int_0^T\int_{\Omega}(\mathbf{v}\cdot\nabla\mathbf{v})\cdot\widetilde{\mathbf{v}}\varphi(t)\,d\mathbf{x}\,dt+\int_0^T\int_{\Omega}(\mathbf{v}\cdot\nabla\boldsymbol{\psi})\cdot\widetilde{\mathbf{v}}\varphi(t)\,d\mathbf{x}\,dt$$
$$+\int_0^T\int_{\Omega}(\boldsymbol{\psi}\cdot\nabla\mathbf{v})\cdot\widetilde{\mathbf{v}}\varphi(t)\,d\mathbf{x}\,dt=\int_0^T\sum_{i=1}^{d}\widetilde{\alpha}_i\lambda_i(t)\,\varphi(t)dt. \qquad (4.42)$$

By comparing (4.37) and (4.42), we obtain $\int_{\Omega}(\mathbf{v}(0)-\mathbf{v}_0)\cdot\widetilde{\mathbf{v}}\varphi(0)\,d\mathbf{x}=0$, and choosing $\varphi$ such that $\varphi(0)=1$ yields

$$\int_{\Omega}(\mathbf{v}(0)-\mathbf{v}_0)\cdot\widetilde{\mathbf{v}}\,d\mathbf{x}=0 \quad \forall\widetilde{\mathbf{v}}\in\mathbf{W}(\Omega);$$

hence, $\mathbf{v}(0)=\mathbf{v}_0$ in $\mathbf{W}'(\Omega)$. We conclude that $\mathbf{v}$ is the solution of (4.9).

## 4.4 Concluding Remarks

In this chapter, the control of the two- and three-dimensional Navier–Stokes equations in a bounded domain is studied *around prescribed drag and lift coefficients*, using a boundary feedback control. In order to determine a feedback law, an extended system coupling the Navier–Stokes equations with an equation satisfied by the control on the domain boundary is considered. We first assume that on the bluff body $\Sigma_c$ (a part of the domain boundary), the trace of the fluid velocity $\mathbf{v}_c$ is a linear combination of a given velocity field represented by $\mathbf{e}_i=(\delta_{1i},\ldots,\delta_{di})^T$, $i=1,\ldots,d$, and the proportionality coefficient $\alpha_i$, such that $\mathbf{v}_c=\sum_{i=1}^{d}\alpha_i\mathbf{e}_i$. The quantity $\alpha_i$ is an unknown of the problem and is written in a feedback form. By using the Galerkin method, $\alpha_i$ is determined such that the Dirichlet boundary control $\mathbf{v}_c$ is satisfied on $\Sigma_c$, and the controlling boundary control is built. Finally, we show that the feedback control (4.6) provides control of the Navier–Stokes problem *around given drag and lift coefficients*.

## References

1. M. Amitay, B.L. Smith, A. Glezer, Aerodynamic flow control using synthetic jet technology, in 36th AIAA Aerospace Sciences Meeting and Exhibit, Reno, Nevada, AIAA Pap. No. 98-0208 (1998)
2. L. Amodei, J.-M. Buchot, A stabilization algorithm of the Navier-Stokes equations based on algebraic Bernoulli equation. Numer. Linear Algebra Appl. **19**, 700–727 (2012)
3. E. Anderson, A. Szewczyk, Effects of a splitter plate on the near wake of a circular cylinder in 2 and 3-dimensional flow configurations. Exp. Fluids **23**, 161–174 (1997)
4. D. Arcas, L. Redekopp, Aspects of wake vortex control through base blowing/suction. Phys. Fluids **16**, 452–456 (2004)
5. G. Artana, R. Sosa, E. Moreau, G. Touchard, Control of the near-wake flow around a circular cylinder with electro-hydrodynamic actuators. Exp. Fluids **35**, 580–588 (2003)
6. M. Badra, T. Takahashi, Stabilization of parabolic nonlinear systems with finite-dimensional feedback or dynamical controllers: Application to the Navier-Stokes system. SIAM J. Control Optim. **49**, 420–463 (2011)
7. V. Barbu, Stabilization of Navier-Stokes equations by oblique boundary feedback controllers. SIAM J. Control Optim. **50**, 2288–2307 (2012)
8. P.W. Bearman, J.K. Harvey, Control of circular cylinder flow by the use of dimples. AIAA J. **31**, 1753–1756 (1993)
9. M. Bercovier, O. Pironneau, V. Sastri, Finite elements and characteristics for some parabolic-hyperbolic problems. Appl. Math. Model **7**, 89–96 (1983)
10. M. Bergmann, L. Cordier, J. Brancher, Optimal rotary control of the cylinder wake using proper orthogonal decomposition reduced-order model. Phys. Fluids **17**, 097101:1-21 (2005)
11. F. Boyer, P. Fabrie, *Mathematical Tools for the Study of the Incompressible Navier-Stokes Equations and Related Models*, vol. 183 (Springer, Berlin, 2013)
12. J. Carberry, J. Sheridan, D. Rockwell, Controlled oscillations of a cylinder: a new wake state. J. Fluids Struct. **17**, 337–343 (2003)
13. J. Choi, W. Jeon, H. Choi, Mechanism of drag reduction by dimples on a sphere. Phys. Fluids **18**, 041702: 1–4 (2006)
14. H. Choi, W.P. Jeon, J. Kim, Control of flow over a bluff body. Annu. Rev. Fluid Mech. **40**, 113–139 (2008)
15. L.J. Clancy, *Aerodynamics* (Pitman Publishing, London, 1975)
16. C. Dalton, Y. Xu, J.C. Owen, The suppression of lift on a circular cylinder due to vortex shedding at moderate Reynolds numbers. J. Fluids Struct. **15**, 617–62 (2001)
17. A.V. Fursikov, Stabilization for the 3D Navier-Stokes system by feedback boundary control. Partial differential equations and applications. Discrete and Cont. Dyn. Syst. **10**, 289–314 (2004)
18. G.P. Galdi, *An Introduction to the Mathematical Theory of the Navier-Stokes Equations, Vol. II, Nonlinear Steady Problems*. Springer Tracts in Natural Philosophy, vol. 39 (Springer, New York, 1994)
19. A. Glezer, M. Amitay, Synthetic jets. Annu. Rev. Fluid Mech. **34**, 503–529 (2002)
20. C. Grandmont, B. Maury, A. Soualah, Multiscale modelling of the respiratory tract: A theoretical framework. ESAIM: Proc. **23**, 10–29 (2008)
21. J-W. He, R. Glowinski, R. Metcalfe, A. Nordlander, J. Periaux, Active control and drag optimization for flow past a circular cylinder. 1: oscillatory cylinder rotation. J. Comput. Phys. **163**, 83–117 (2000)
22. J-Y. Hwang, K-S. Yang, S-H. Sun, Reduction of flow-induced forces on a circular cylinder using a detached splitter plate. Phys. Fluids **15**, 2433–2436 (2003)
23. J. Kim, H. Choi, Distributed forcing of flow over a circular cylinder. Phys. Fluids **17**, Article ID 033103 (2005)
24. E. Konstantinidis, S. Balabani, M. Yianneskis, The timing of vortex shedding in a cylinder wake imposed by periodic inflow perturbations. J. Fluid Mech. **543**, 45–55 (2005)

25. F. Li, N. Aubry, Feedback control of a flow past a cylinder via transverse motion. Phys. Fluids **15**, 2163–2176 ( 2003)
26. Z. Li, I. Navon, M. Hussaini, F. Le Dimet, Optimal control of cylinder wakes via suction and blowing. Comput. Fluids **32**, 149–171 (2003)
27. J.L. Lions, *Quelques méthodes de résolution des problèmes aux limites non linéaires*, (Dunod, Malakoff, 2002)
28. J. Moran, *An Introduction to Theoretical and Computational Aerodynamics* (Dover, Illinois, 2003)
29. E.M.D. Ngom, A. Sene, D.Y. LeRoux, Global stabilization of the Navier-stokes equations around an unstable equilibrium state with a boundary feedback controller. Evol. Eq. Control Theory **4**(1), 89–106 (2015)
30. E.M.D. Ngom, A. Sène, D.Y. Le Roux, Boundary stabilization of the Navier-Stokes equations with feedback controller via a Galerkin method. Evol. Equ. Control Theory **3**, 147–166 (2014)
31. S.S. Ravindran, Optimal boundary feedback flow stabilization by model reduction. Comput. Methods Appl. Mech. Engrg. **196**, 2555–2569 (2007)
32. S.S. Ravindran, Stabilisation of Navier-Stokes equations by boundary feedback. Int. J. Numer. Anal. Model. **4**, 608–624 (2007)
33. J.-P. Raymond, L. Thevenet, Boundary feedback stabilization of the two-dimensional Navier-Stokes equations with finite-dimensional controllers. Discrete Contin. Dynam. Syst. **27**, 1159–1187 (2010)
34. M. Schäfer, S. Turek, Benchmark computations of laminar flow around a cylinder, in *Flow Simulation with High-Performance Computers II*. Notes on Numerical Fluid Mechanics, vol. 1 (Vieweg+Teubner Verlag, Wiesbaden, 1996), pp. 547–566
35. A. Sene, B.A. Wane, D.Y. Le Roux, Control of irrigation channels with variable bathymetry and time dependent stabilization rate. C. R. Acad. Sci. Paris Ser. I **346**, 1119–1122 (2008)
36. R. Temam, *Navier-Stokes Equations, Theory and Numerical Analysis* (American Mathematical Society, Providence, 2001)
37. P.T. Tokumaru, P.E. Dimotakis, Rotary oscillatory control of a cylinder wake. J. Fluid Mech. **224**, 77–90 (1991)

# Chapter 5
# Discrete Potential Boundary-Value Problems of Kirchhoff Type

**Abdoul Aziz Kalifa Dianda and Stanislas Ouaro**

**Abstract** In this chapter, we prove the existence of solutions for some discrete nonlinear difference equations of Kirchhoff type subjected to a potential boundary-value condition. We make use of a variational technique that relies on Szulkin's critical point theory, which ensures the existence of solutions by ground state and mountain pass methods.

**Keywords** Szulkin critical point theory · Ground states method · Mountain pass theorem · Potentiel boundary-value condition

## 5.1 Introduction

This chapter is concerned with the existence of solutions to problems of the form

$$\begin{cases} -M(A(k-1,\Delta u(k-1)))\Delta(a(k-1,\Delta u(k-1))) = f(k,u(k)); \, k \in \mathbb{Z}[1,T] \\ \\ (a(0,\Delta u(0)), -a(T,\Delta u(T)) \in \partial j(u(0); u(T+1)), \end{cases} \tag{5.1}$$

where $T \geq 2$ is a positive integer and $\Delta u(k) = u(k+1) - u(k)$ is the forward difference operator. Here and hereafter, we denote by $\mathbb{Z}[a,b]$ the discrete interval $\{a, a+1, a+2, \ldots, b\}$, where $a$ and $b$ are integers with $a < b$. $f : \mathbb{Z}[1,T] \times \mathbb{R} \to \mathbb{R}$ is a continuous and monotone function with respect to the second variable, $j : \mathbb{R} \times \mathbb{R} \to (-\infty, \infty)$ is convex, proper (i.e., $D(j) := \{z \in \mathbb{R} \times \mathbb{R} : j(z) < +\infty\} \neq \emptyset$),

A. A. K. Dianda · S. Ouaro (✉)
Département de Mathématiques, Université Joseph KI-ZERBO, Unité de Formation et de Recherche en Sciences Exactes et Appliquées, Ouagadougou, Burkina Faso



T. Diagana et al. (eds.), *Partial Differential Equations and Applications*, Springer Proceedings in Mathematics & Statistics 420,
https://doi.org/10.1007/978-3-031-27661-3_5

and lower semicontinuous (in short l.s.c.), and $\partial j$ denotes the subdifferential of $j$. Recall that, for $z \in \mathbb{R} \times \mathbb{R}$, the set $\partial j$ is defined by

$$\partial j(z) = \{\zeta \in \mathbb{R} \times \mathbb{R} : j(\xi) - j(z) \geq (\zeta \mid \xi - z), \text{ for all } \xi \in \mathbb{R} \times \mathbb{R}\}, \tag{5.2}$$

where $(. \mid .)$ stands for the usual inner product in $\mathbb{R} \times \mathbb{R}$.

We also consider the function space $X = \{v : \{0, 1, \ldots, T+1\} \to \mathbb{R}\}$ with the inner product

$$(u, v) = \sum_{k=1}^{T+1} \Delta u(k-1)\Delta v(k-1), \text{ for all } u, v \in X.$$

We assume that

$$a(k, .) : \mathbb{R} \to \mathbb{R} \text{ is continuous for all } k \in \{0, 1, \ldots, T\}, \tag{5.3}$$

and there exists a mapping $\mathrm{A} : \mathbb{Z}[1, T] \times \mathbb{R} \to \mathbb{R}$ that satisfies

$$A(k, 0) = 0, \text{ for all } k \in \mathbb{Z}[0, T]. \tag{5.4}$$

$$a(k, \xi) = \frac{\partial}{\partial \xi} A(k, \xi), \text{ for all } k \in \mathbb{Z}[0, T]. \tag{5.5}$$

We also assume that:

- There exists a constant $C_1$ such that

$$|a(k, \xi)| \leq C_1(1 + |\xi|^{p(k)-1}), \text{ for all } k \in \mathbb{Z}[0, T]. \tag{5.6}$$

- The following holds true.

$$(a(k, \xi) - a(k, \eta)).(\xi - \eta) > 0, \text{ for all } k \in \mathbb{Z}[0, T] \text{ and } \xi, \eta \in \mathbb{R} \text{ such that } \xi \neq \eta. \tag{5.7}$$

- The following holds true.

$$|\xi|^{p(k)} \leq a(k, \xi)\xi \leq p(k)A(k, \xi), \text{ for all } k \in \mathbb{Z}[0, T] \text{ and } \xi \in \mathbb{R}. \tag{5.8}$$

We also assume that

$$p : \mathbb{Z}[0, T] \to (1, \infty), \tag{5.9}$$

$M : (0, \infty) \to (0, \infty)$ is continuous and nondecreasing, and there exist positive reals $B_1$, $B_2$ with $B_1 \leq B_2$ and $\alpha \geq 1$ such that

$$B_1 t^{\alpha-1} \leq M(t) \leq B_2 t^{\alpha-1} \text{ for } t \geq t^* > 0. \tag{5.10}$$

As examples of functions satisfying assumptions (5.3)–(5.10), we can give the following:

- $M(A(k,\xi)) = M\left(\frac{1}{p(k)}|\xi|^{p(k)}\right) = 1$, where $M(t) = 1$ and $a(k,\xi) = |\xi|^{p(k)-2}\xi$, for $k \in \mathbb{Z}[0,T]$ and $\xi \in \mathbb{R}$.
- $M(A(k,\xi)) = b + \frac{c}{p(k)}\left[\left(1+|\xi|^2\right)^{p(k)/2} - 1\right]$, where $M(t) = b + ct$ and $a(k,\xi) = \left(1+|\xi|^2\right)^{(p(k)-2)/2}\xi$, for all $k \in \mathbb{Z}[0,T]$ and $\xi \in \mathbb{R}$.

If we take $M(t) = 1$, (5.1) is reduced to a problem studied by Kyelem et al. in [16].

In [16], the authors proved the existence of solutions for discrete potential boundary-value problem, by using variational techniques that rely on Szulkin's critical point theory and ensure the existence of solutions by ground state and mountain pass methods.

Problem (5.1) has its origin in the theory of nonlinear vibration and can be seen as a generalization of the problem studied in [16]. The equations of the type (5.1) were firstly proposed by Kirchhoff in 1876 (see [13]). After that, several physicists also considered such equation for their researches in the theory of nonlinear vibrations. The first study that deals with anisotropic discrete boundary-value problems of p(.) Kirchhoff type difference equation was done by Yucedag (see [23]).

In [15], Koné et *al.* studied the problem

$$\begin{cases} -M(A(k-1,\Delta u(k-1)))\Delta(a(k-1,\Delta u(k-1))) = f(k); \quad k \in \mathbb{Z}[1,T] \\ \\ u(0) = \Delta u(T) = 0, \end{cases} \tag{5.11}$$

where $T \geq 2$ is a positive integer and $\Delta u(k) = u(k+1) - u(k)$ is the forward difference operator. They proved the existence of weak solutions to a family of discrete boundary-value problems whose right-hand side belongs to a discrete Hilbert space.

It is usually seen that nonlinear multivalued boundary condition includes particular cases of classical boundary conditions; these are obtained by appropriate choices of $j$ (see, e.g., Ch.2 in [11]). For other choices of $j$ yielding various boundary conditions, we refer the reader to Gasinski and Papageorgiou [8] and Jebelean and Serban [12].

The study of boundary-value problems with discrete Laplacian using variational approaches was developed in the last years. Most of the papers deal with classical boundary conditions such as Dirichlet boundary conditions (see, e.g., Agarwal et al. [1], Cabada et al. [4]), Neumann boundary conditions (see, e.g., Candito and D'Agui [5], Tian and Ge [22]), and periodic boundary conditions (see, e.g., He and Chen [10], Jebelean and Serban [12]).

Recently, boundary-value problems with discrete Laplacian subjected to Dirichlet, Neumann, or Periodic boundary conditions were studied by Bereanu et al. [2], Galewski and Glab [6, 7], Guiro et al. [9], Koné and Ouaro [14], Mashiyev et al. [17], and Mihailescu et al. [18, 19].

In [2], the authors are concerned with the existence of solutions of the following periodic and Neumann boundary $p(.)$-Laplacian problem:

$$\begin{cases} -\Delta_{p(k-1)}x(k-1) = f(k, x(k)) \text{ for } k \in \mathbb{Z}[1, T], \\ x(0) - x(T+1) = 0 = \Delta x(0) - \Delta x(T+1), \end{cases} \tag{5.12}$$

and the following Neumann boundary $p(.)$-Laplacian problem:

$$\begin{cases} -\Delta_{p(k-1)}x(k-1) = f(k, x(k)) \text{ for } k \in \mathbb{Z}[1, T], \\ \Delta x(0) = 0 = \Delta x(T+1). \end{cases} \tag{5.13}$$

Bereanu, Jebelean, and Serban obtained in [2] the existence results of solutions to problems (5.12) and (5.13) in appropriate discrete spaces using variational methods and some applications of lower and upper solution theorems for both considered cases. In [3], Bereanu et al. made use of variational approach to obtain ground state and mountain pass solutions for the following problem:

$$\begin{cases} -\Delta_{p(k-1)}\Big(u(k-1)\Big) = f(k, u(k)) \text{ for } k \in \mathbb{Z}[1, T], \\ (h_{p(0)}(\Delta u(0)), -h_{p(T)}(\Delta u(T)) \in \partial j(u(0), u(T+1)), \end{cases}$$

where $\Delta u(k) = u(k+1) - u(k)$ is the forward difference operator and $\Delta_{p(\cdot)}$ is a discrete $p(\cdot)$-Laplacian operator that is

$$-\Delta_{p(k-1)}(u(k-1)) := \Delta(h_{p(k-1)}(\Delta u(k-1))),$$

with $h_{p(k)} : \mathbb{R} \to \mathbb{R}$ defined by $h_{p(k)}(u(k)) = |u(k)|^{p(k)-2}u(k)$.

In this chapter, we prove the existence of solutions to problem (5.1) by using ground state method and mountain pass technique. This chapter is organized as follows: Sect. 5.2 is devoted to mathematical preliminary, Sect. 5.3 deals with the existence of solutions to problem (5.1) using ground state method. In Sect. 5.4, we deal with the existence of non-trivial solutions to problem (5.1) by using mountain pass technique.

## 5.2 Preliminary

Our approach for the boundary-value problem (5.1) relies on the critical point theory developed by Szulkin [21].

We consider the following norm:

$$\|u\|_{p(\cdot)} := \inf\left\{\lambda > 0 : \sum_{k=1}^{T} \frac{1}{p(k)}\left|\frac{u(k)}{\lambda}\right|^{p(k)} \leq 1\right\},$$

and we introduce the following function:

$$p : \mathbb{Z}[1, T] \longrightarrow (1, \infty).$$

Let us denote

$$p^- := \min_{k\in\mathbb{Z}[0,T]} p(k) \text{ and } p^+ := \max_{k\in\mathbb{Z}[0,T]} p(k).$$

Let $\varphi : X \longrightarrow \mathbb{R}$ be defined by

$$\varphi(u) = \widehat{M}\left(\sum_{k=1}^{T+1} A\left(k-1, \Delta u(k-1)\right)\right), \text{ for all } u \in X, \tag{5.14}$$

where $\widehat{M}(t) = \int_0^t M(s)ds$.

Using the functional $j$, we introduce the functional $J : X \longrightarrow (-\infty; \infty)$ given by

$$J(u) = j(u(0); u(T+1)), \text{ for all } u \in X. \tag{5.15}$$

Note that, as $j$ is proper, convex, and l.s.c., the same properties hold true for $J$.

Let us set

$$\psi = \varphi + J. \tag{5.16}$$

Let us also define

$$F(k, t) = \int_0^t f(k, \tau)d\tau, \text{ for all } k \in \mathbb{Z}[1, T], \text{ for all } t \in \mathbb{R}$$

and

$$\Phi(u) = -\sum_{k=1}^{T} F(k, u(k)), \ \forall u \in X. \tag{5.17}$$

The energy functional associated to problem (5.1) is given by

$$I = \Phi + \psi, \tag{5.18}$$

with $\psi$ given by (5.16) and $\Phi$ given by (5.17).

**Lemma 5.2.1** *Let $u \in X$ and $p^+ < \infty$; then $\|u\|_{p(.)}$ is equivalent to the Luxemburg norm defined by*

$$\|u\|_e := \inf\left\{\lambda > 0; \sum_{k=1}^{T}\left|\frac{u(k)}{\lambda}\right|^{p(k)} \leq 1\right\}.$$

***Proof*** We have

$$\sum_{k=1}^{T}\frac{1}{p(k)}\left|\frac{u(k)}{\lambda}\right|^{p(k)} \geq \frac{1}{p^+}\sum_{k=1}^{T}\left|\frac{u(k)}{\lambda}\right|^{p(k)};$$

thus,

$$\|u\|_{p(.)} \geq \lambda_1\|u\|_e,$$

$$\sum_{k=1}^{T}\frac{1}{p(k)}\left|\frac{u(k)}{\lambda}\right|^{p(k)} \leq \frac{1}{p^-}\sum_{k=1}^{T}\left|\frac{u(k)}{\lambda}\right|^{p(k)};$$

therefore,

$$\|u\|_{p(.)} \leq \lambda_2\|u\|_e.$$

We infer that

$$\lambda_1\|u\|_e \leq \|u\|_{p(.)} \leq \lambda_2\|u\|_e.$$

□

Now, let us present some basic properties of the general critical point theory.

Let $I : X \to \mathbb{R} \cup \{\infty\}$ be the energy functional associated to problem (5.1) given by

$$(\mathbf{H}) : I = \Phi + \psi,$$

with $\Phi : X \to \mathbb{R}$ a $C^1(X, \mathbb{R})$ function and $\psi : X \to \mathbb{R} \cup \{\infty\}$ a convex, lower semicontinuous, and proper function.

**Definition 5.2.1** An element $u \in X$ satisfying (**H**) is called a critical point of the functional $I : X \to \mathbb{R} \cup \{\infty\}$ if $\langle \Phi'(u), v - u\rangle + \psi(v) - \psi(u) \geq 0,$ for all $v \in X$.

**Definition 5.2.2** The functional $I : X \to \mathbb{R} \bigcup\{\infty\}$ satisfying (**H**) is said to satisfy the Palais–Smale (in short (PS)) condition in the sense of Szulkin, if every sequence $\{u_n\} \subset X$ for which $I(u_n) \longrightarrow c \in \mathbb{R}$ and

$$\langle \Phi'(u_n); v - u_n\rangle + \psi(v) - \psi(u_n) \geq -\epsilon\|v - u_n\|, \text{ for all } v \in X, \tag{5.19}$$

where $\epsilon_n \longrightarrow 0$ possesses a convergent subsequence. □

**Proposition 5.2.1 ([22], Proposition 1.1)** *If I satisfies (**H**), then each local minimum point of I is necessarily a critical point of I.*

**Theorem 5.2.1 ([20], Theorem 23.2)** *Let $f$ be a convex function, and let $x$ be a point where $f$ is finite. Then $x^*$ is a subgradient of $f$ at $x$ if and only if $f'(x^*, y) \geq \langle x^*; y\rangle$ for all $y \in X$. In fact, the closure of $f'(x^*, y)$ as a convex function of $y$ is the support function of the closed convex set $\partial f(x)$.* □

**Theorem 5.2.2 ([21], Theorem 3.2)** *Assume that I satisfies (**H**), the (PS) condition and the following:*

*(i) $I(0) = 0$, and there exist $\alpha, \rho \geq 0$ such that $I(u) \geq \alpha$ if $\|u\| = \rho$.*
*(ii) $I(e) \leq 0$ for some $e \in X$ with $\|e\| \geq \rho$.*

*Then, I has a critical value $c \geq \alpha$ that can be characterized by*

$$c = \inf_{f\in\Gamma} \sup_{t\in[0,1]} I(f(t)),$$

*where $\Gamma = \{f \in C([0, 1], X) : f(0) = 0, f(1) = e\}$.* □

**Proposition 5.2.2** *Assume that (5.3)–(5.10) hold. Then:*

*(i) $\varphi$ is convex and is in $C^1(X; \mathbb{R})$.*
*(ii) $J$ is proper, convex, and l.s.c.*
*(iii) $\psi$ is proper, convex, and l.s.c.*
*(iv) $\Phi \in C^1(X; \mathbb{R})$.* □

***Proof***

(i) $\varphi$ is well-defined. $A$ is convex with respect to the second variable according to (5.5) and (5.6). According to [16], $\varphi$ is convex on X and $C^1(X, \mathbb{R})$, with derivative given by

$$\langle \varphi'(u), v\rangle = M\left(\sum_{k=1}^{T+1} A(k-1, \Delta u(k-1))\right) \times \sum_{k=1}^{T+1} a(k-1, \Delta u(k-1))\Delta v(k-1), \quad \forall u, v \in X.$$

The continuity of the derivative comes from the continuity of $a(k, .)$. Hence, $\varphi$ is in $C^1(X; \mathbb{R})$.

(ii) Note that as $j$ is proper, convex, and l.s.c., the same properties hold for $J$.

(iii) Since $\varphi$ and $J$ are convex, then $\psi$ is convex.

Suppose that $\psi$ can take the value $-\infty$; then, in this case, $J = \psi - \varphi$ can take the value $-\infty$ that is not possible. Therefore, $\psi$ cannot take the value $-\infty$. Hence, $\psi$ is proper.

Note also that

$$J(u) \leq \liminf_{y \to u} J(y).$$

Then,

$$\begin{aligned}\varphi(u) + J(u) &\leq \liminf_{y \to u} J(y) + \varphi(u)\\ &\leq \liminf_{y \to u} J(y) + \liminf_{y \to u} \varphi(y)\\ &\leq \liminf_{y \to u} \psi(y).\end{aligned}$$

Therefore, $\psi(u) \leq \liminf_{y \to u} \psi(y)$. Hence, $\psi$ is l.s.c.

(iv) $|\Phi(u)| = |\sum_{k=1}^{T} F(k, u(k))| < \infty$. Then, $\Phi$ is well-defined.

By definition, $\Phi$ is derivable, and his derivative is continuous; hence, $\Phi \in C^1(X; \mathbb{R})$. Moreover,

$$\begin{aligned}\langle \Phi'(u); y \rangle &= \lim_{\delta \to 0^+} \frac{\Phi(u + \delta y) - \Phi(u)}{\delta}\\ &= -\lim_{\delta \to 0^+} \sum_{k=1}^{T} \frac{F(k, u(k) + \delta y(k)) - F(k, u(k))}{\delta}\\ &= -\sum_{k=1}^{T} \lim_{\delta \to 0^+} \frac{F(k, u(k) + \delta y(k)) - F(k, u(k))}{\delta}\\ &= -\sum_{k=1}^{T} f(k, u(k))y(k), \quad \text{for all} \quad u, y \in X.\end{aligned}$$

Now, let us claim the following important result.

**Proposition 5.2.3** *If $u \in X$ is a critical point of the functional $I$ in the sense that*

$$\langle \Phi'(u); y - u \rangle + \psi(y) - \psi(u) \geqslant 0, \quad \textit{for all} \quad y \in X, \tag{5.20}$$

*then $u$ is a classical solution of problem (5.1).* □

***Proof*** Since $\langle \Phi'(u); y - u\rangle + \psi(y) - \psi(u) \geqslant 0$, then we can take $y = u + sw$ for all $s > 0$ in (5.20). Dividing (5.20) by $s$ and letting $s \longrightarrow 0^+$, we get

$$\langle \Phi'(u); w\rangle + \langle \varphi'(u); w\rangle + J'(u, w) \geq 0, \ \forall \ w \ \in X, \tag{5.21}$$

where $J'(u; w)$ is the directional derivative of the convex function $J$ at $u$ in the direction of $w$.

Since

$$J(u) = j(u(0), u(T + 1)),$$

then we get from (5.21),

$$\begin{aligned} &\langle \Phi'(u); w\rangle + \langle \varphi'(u); w\rangle + j'((u(0), u(T + 1)),\\ &(w(0), w(T + 1))) \geqslant 0, \quad \text{for all} \quad w \in X. \end{aligned}$$

Since

$$\langle \Phi'(u); w\rangle = -\sum_{k=1}^{T} f(k, u(k))w(k) \quad \text{for all} \quad u, w \in X$$

and

$$\begin{aligned} \langle \varphi'(u), w\rangle = M &\left(\sum_{k=1}^{T+1} A\,(k-1, \Delta u(k-1))\right)\\ &\times \sum_{k=1}^{T+1} a(k-1, \Delta u(k-1))\Delta w(k-1) \quad \text{for all} \quad u, w \in X, \end{aligned}$$

then, one obtains

$$\begin{aligned} -\sum_{k=1}^{T} f(k, u(k))w(k) &+ M\left(\sum_{k=1}^{T+1} A\,(k-1, \Delta u(k-1))\right)\\ \times &\sum_{k=1}^{T+1} a(k-1, \Delta u(k-1))\Delta w(k-1)\\ &+ j'((u(0), u(T + 1)), (w(0), w(T + 1))) \geqslant 0. \end{aligned}$$

Therefore,

$$
\begin{aligned}
&-\sum_{k=1}^{T} f(k, u(k))w(k) + M\left(\sum_{k=1}^{T+1} A\,(k-1, \Delta u(k-1))\right)\\
&\times \sum_{k=1}^{T+1} a(k-1, \Delta u(k-1))[w(k) - w(k-1)]\\
&+j'((u(0), u(T+1)), (w(0), w(T+1)))\\
&\quad \geqslant 0.
\end{aligned}
$$

Then,

$$
\begin{aligned}
&-\sum_{k=1}^{T} f(k, u(k))w(k) + M\left(\sum_{k=1}^{T+1} A\,(k-1, \Delta u(k-1))\right)\\
&\times \sum_{k=1}^{T+1} a(k-1, \Delta u(k-1))w(k)\\
&-M\left(\sum_{k=1}^{T+1} A\,(k-1, \Delta u(k-1))\right)\sum_{k=1}^{T+1} a(k-1, \Delta u(k-1))w(k-1)\\
&+j'((u(0), u(T+1)), (w(0), w(T+1)))\\
&\quad \geqslant 0;
\end{aligned}
$$

thus,

$$
\begin{aligned}
&-\sum_{k=1}^{T} f(k, u(k))w(k) + M\left(\sum_{k=1}^{T+1} A\,(k-1, \Delta u(k-1))\right) a(T, \Delta u(T))w(T+1)\\
&+M\left(\sum_{k=1}^{T+1} A\,(k-1, \Delta u(k-1))\right)\sum_{k=1}^{T} a(k-1, \Delta u(k-1))w(k)\\
&-M\left(\sum_{k=1}^{T+1} A\,(k-1, \Delta u(k-1))\right)\sum_{k=0}^{T} a(k, \Delta u(k))w(k)\\
&+j'((u(0), u(T+1)), (w(0), w(T+1)))\\
&\quad \geqslant 0,
\end{aligned}
$$

which leads to

$$
\begin{aligned}
&-\sum_{k=1}^{T} f(k, u(k))w(k) + M\left(\sum_{k=1}^{T+1} A(k-1, \Delta u(k-1))\right) a(T, \Delta u(T))w(T+1) \\
&-M\left(\sum_{k=1}^{T+1} A(k-1, \Delta u(k-1))\right) a(0, \Delta u(0))w(0) \\
&+M\left(\sum_{k=1}^{T+1} A(k-1, \Delta u(k-1))\right) \sum_{k=1}^{T} a(k-1, \Delta u(k-1))w(k) \\
&-M\left(\sum_{k=1}^{T+1} A(k-1, \Delta u(k-1))\right) \sum_{k=1}^{T} a(k, \Delta u(k))w(k) \\
&+j'((u(0), u(T+1)), (w(0), w(T+1))) \\
&\quad \geqslant 0.
\end{aligned}
$$

Therefore,

$$
\begin{aligned}
&-\sum_{k=1}^{T} f(k, u(k))w(k) + M\left(\sum_{k=1}^{T+1} A(k-1, \Delta u(k-1))\right) a(T, \Delta u(T))w(T+1) \\
&-M\left(\sum_{k=1}^{T+1} A(k-1, \Delta u(k-1))\right) a(0, \Delta u(0))w(0) \\
&-M\left(\sum_{k=1}^{T+1} A(k-1, \Delta u(k-1))\right) \sum_{k=1}^{T} [a(k, \Delta u(k)) - a(k-1, \Delta u(k-1))]w(k) \\
&+j'((u(0), u(T+1)), (w(0), w(T+1))) \\
&\quad \geqslant 0;
\end{aligned}
$$

thus

$$
\begin{aligned}
&-\sum_{k=1}^{T} f(k, u(k))w(k) - M\left(\sum_{k=1}^{T+1} A(k-1, \Delta u(k-1))\right) \\
&\times \sum_{k=1}^{T} \Delta a(k-1, \Delta u(k-1))w(k) \\
&+M\left(\sum_{k=1}^{T+1} A(k-1, \Delta u(k-1))\right) a(T, \Delta u(T))w(T+1)
\end{aligned}
$$

$$
\begin{aligned}
&-M\left(\sum_{k=1}^{T+1} A\,(k-1,\Delta u(k-1))\right) a(0,\Delta u(0))w(0)\\
&+j'((u(0),u(T+1)),(w(0),w(T+1)))\\
&\quad \geqslant 0,
\end{aligned}
$$

for all $w \in X$. Thus we infer

$$
\begin{aligned}
&M\left(\sum_{k=1}^{T+1} A\,(k-1,\Delta u(k-1))\right)\sum_{k=1}^{T}\Big(-\Delta a(k-1,\Delta u(k-1)\Big)w(k)\\
&\quad -\sum_{k=1}^{T} f(k,u(k))w(k)=0.
\end{aligned}
$$

As $w \in X$ is arbitrarily chosen, thus if $w(0)=w(T+1)=0$, we obtain

$$
\begin{aligned}
&M\left(\sum_{k=1}^{T+1} A\,(k-1,\Delta u(k-1))\right)\sum_{k=1}^{T}(-\Delta a(k-1,\Delta u(k-1)))\,w(k)\\
&\quad =\sum_{k=1}^{T} f(k,u(k))w(k).
\end{aligned}
$$

Hence, it follows that

$$
\begin{aligned}
&-M(A(k-1,\Delta u(k-1)))\Delta(a(k-1,\Delta u(k-1)))\\
&\quad = f(k,u(k)) \quad \text{for all} \quad k\in\mathbb{Z}[1,T].
\end{aligned}
\tag{5.22}
$$

It remains to show that $(a(0,\Delta u(0)),-a(T,\Delta u(T)))\in\partial j(u(0),u(T+1))$. One has

$$
\begin{aligned}
&-\sum_{k=1}^{T} f(k,u(k))w(k)-M\left(\sum_{k=1}^{T+1} A\,(k-1,\Delta u(k-1))\right)\\
&\quad\times\sum_{k=1}^{T}\Delta a(k-1,\Delta u(k-1))w(k)\\
&\quad+M\left(\sum_{k=1}^{T+1} A\,(k-1,\Delta u(k-1))\right)[a(T,\Delta u(T))w(T+1)-a(0,\Delta u(0))w(0)]\\
&\quad+j'((u(0),u(T+1)),(w(0),w(T+1)))\geqslant 0
\end{aligned}
$$

and

$$M\left(\sum_{k=1}^{T+1} A\,(k-1, \Delta u(k-1))\right)\sum_{k=1}^{T}\Big(-\Delta a(k-1, \Delta u(k-1)\Big)w(k)$$
$$= \sum_{k=1}^{T} f(k, u(k))w(k).$$

Let

$$C = \begin{cases} B_1 & \text{if} \quad -a(T, \Delta u(T)) + a(0, \Delta u(0)) \geq 0 \\ B_2 & \text{if} \quad -a(T, \Delta u(T)) + a(0, \Delta u(0)) \leq 0. \end{cases}$$

As $M(.)$ is positive, from (5.10), it follows that

$$j'((u(0), u(T+1)), (w(0), w(T+1))) \geqslant C\left(\sum_{k=1}^{T+1} A\,(k-1, \Delta u(k-1))\right)^{\alpha-1}$$
$$\times [-a(T, \Delta u(T))w(T+1) + a(0, \Delta u(0))w(0)].$$

Now, let us take $s = C\left(\sum_{k=1}^{T+1} A\,(k-1, \Delta u(k-1))\right)^{\alpha-1}$.

Thus,

$$j'((u(0), u(T+1)), (w(0), w(T+1))) \geqslant -a(T, \Delta u(T))(sw(T+1))$$
$$+ a(0, \Delta u(0))(sw(0)).$$

Finally, for all $w \in X$, taking $sw(0) = p$ and $sw(T+1) = q$, where $p, q \in \mathbb{R}$ are arbitrarily chosen, it follows that

$$j'((u(0), u(T+1)), (w(0), w(T+1))) \geqslant -a(T, \Delta u(T))q$$
$$+a(0, \Delta u(0))p, \quad \text{for} \quad p, q \in \mathbb{R}.$$

Hence, $(a(0, \Delta u(0)), -a(T, \Delta u(T))) \in \partial j(u(0), u(T+1))$. □

Now, we have the following lemma that will be used later.

**Lemma 5.2.2 (See [16])** *Let $u \in X$ and $p^+ < \infty$; then, the following properties hold:*

$$(i)\ \|u\|_{p(.)} < 1 \Longrightarrow \|u\|_{p(.)}^{p^+} \leq \sum_{k=1}^{T} \frac{|u(k)|^{p(k)}}{p(k)} \leq \|u\|_{p(.)}^{p^-};$$

$$(ii)\ \|u\|_{p(.)} > 1 \Longrightarrow \|u\|_{p(.)}^{p^-} \leq \sum_{k=1}^{T} \frac{|u(k)|^{p(k)}}{p(k)} \leq \|u\|_{p(.)}^{p^+}.$$

In the following section, we turn out to the existence result by using ground state method.

## 5.3 Proof of the Existence of Classical Solutions by Ground State Method

In this section, we prove an existence result of classical solutions. This result shows that the energy functional $I$ has a minimum in $X$, and for the proof, we use the positive constant

$$\lambda_1 := \inf \left\{ \frac{\left( \sum_{k=1}^{T+1} \frac{1}{p(k-1)} |\Delta u(k-1)|^{p(k-1)} \right)^{\alpha-1}}{\sum_{k=1}^{T} \frac{1}{p(k)} |u(k)|^{p(k)}} : \quad u \in X - \{0\} \right.$$
$$\left. \text{and} \quad \big(u(0), u(T+1)\big) \in D(j) \right\},$$

with $\alpha \geq 1$, $\alpha$ is given by (5.10).

It is obvious that $\lambda_1 > 0$.

**Theorem 5.3.3** *Assume that (5.3)–(5.8) hold. Moreover, suppose that*

$$\lim_{|t| \to \infty} \sup \frac{p(k) F(k,t)}{|t|^{p(k)}} < \lambda_1, \quad \textit{for all} \quad k \in \mathbb{Z}[1, T]. \tag{5.23}$$

*Then, problem (5.1) has at least one classical solution that minimizes $I$ on $X$.* □

***Proof***

**Step 1**: We first show that $I$ is sequentially lower semicontinuous on $X$.

Indeed, from Proposition 5.2.2, the functional $\psi$ is lower semicontinuous, and the function $\Phi$ is $C^1$ on $X$. Therefore, the functional $I$ is lower semicontinuous on $X$.

**Step 2**: We prove that $I$ is bounded from below and coercive on $X$.
Using (5.23), one obtains the existence of some constants $\alpha' > 0$ and $\rho > 1$ such that

$$F(k,t) \leq \frac{\lambda_1 - \alpha'}{p(k)}|t|^{p(k)} \quad \text{for all } k \in \mathbb{Z}[1,T] \quad \text{and for all } t \in \mathbb{R} \text{ with } |t| > \rho.$$

On the other hand, by the continuity of $F(k,.)$ over $[-\rho, \rho]$, there is a constant $N_\rho > 0$ such that

$$|F(k,t)| \leq N_\rho \quad \text{for all} \quad k \in \mathbb{Z}[1,T] \quad \text{and} \quad t \in [-\rho, \rho].$$

Hence, we infer that

$$F(k,t) \leq N_\rho + \frac{\lambda_1 - \alpha'}{p(k)}|t|^{p(k)} \quad \text{for all} \quad (k,t) \in \mathbb{Z}[1,T] \times \mathbb{R}.$$

To prove the coercivity of $I$, we use the above inequality to obtain for all $(k,t) \in \mathbb{Z}[1,T] \times \mathbb{R}$,

$$-\sum_{k=1}^{T} F(k, u(t)) \geq -N_\rho T - (\lambda_1 - \alpha') \sum_{k=1}^{T} \frac{|u(k)|^{p(k)}}{p(k)}.$$

It follows that

$$\begin{aligned}
I(u) &\geq \varphi(u) - N_\rho T - (\lambda_1 - \alpha') \sum_{k=1}^{T} \frac{|u(k)|^{p(k)}}{p(k)} + J(u) \\
&\geq \varphi(u) - N_\rho T - \lambda_1 \sum_{k=1}^{T} \frac{|u(k)|^{p(k)}}{p(k)} + \alpha' \sum_{k=1}^{T} \frac{|u(k)|^{p(k)}}{p(k)} + J(u) \\
&\geq \varphi(u) - N_\rho T - \left( \sum_{k=1}^{T+1} \frac{1}{p(k-1)} \mid \Delta u(k-1) \mid^{p(k-1)} \right)^{\alpha - 1} \\
&\quad + \alpha' \sum_{k=1}^{T} \frac{|u(k)|^{p(k)}}{p(k)} + J(u).
\end{aligned}$$

From (5.8) and (5.10), we get

$$A(k-1, \Delta u(k-1)) \geq \frac{1}{p(k-1)} |\Delta u(k-1)|^{p(k-1)},$$

which leads to

$$\begin{aligned}\varphi(u) &= \widehat{M}\left(\sum_{k=1}^{T+1} A\left(k-1, \Delta u(k-1)\right)\right)\\ &\geq \frac{B_1}{\alpha}\left(\sum_{k=1}^{T+1}\frac{1}{p(k-1)}|\Delta u(k-1)|^{p(k-1)}\right)^{\alpha}.\end{aligned} \tag{5.24}$$

Hence, if $\|u\|_{p(.)} > 1$, one uses Lemma 5.2.2 and (5.24) to obtain

$$\begin{aligned}I(u) \geq{}& \frac{B_1}{\alpha}\left(\sum_{k=1}^{T+1}\frac{1}{p(k-1)}|\Delta u(k-1)|^{p(k-1)}\right)^{\alpha} - N_\rho T\\ &-\left(\sum_{k=1}^{T+1}\frac{1}{p(k-1)}|\Delta u(k-1)|^{p(k-1)}\right)^{\alpha-1} + \alpha'\sum_{k=1}^{T}\frac{|u(k)|^{p(k)}}{p(k)} + J(u)\\ \geq{}& \left(\frac{B_1}{\alpha}\sum_{k=1}^{T+1}\frac{1}{p(k-1)}|\Delta u(k-1)|^{p(k-1)} - 1\right)\\ &\times e\left(\sum_{k=1}^{T+1}\frac{1}{p(k-1)}|\Delta u(k-1)|^{p(k-1)}\right)^{\alpha-1}\\ &+\alpha'\sum_{k=1}^{T}\frac{|u(k)|^{p(k)}}{p(k)} - N_\rho T + J(u)\\ \geq{}& \left(\frac{B_1}{\alpha}\|\Delta u\|_{p(.)}^{p^-} - 1\right)\|\Delta u\|_{p(.)}^{(\alpha-1)p^-} - N_\rho T + \alpha'\|u\|_{p(.)}^{p^-} + J(u).\end{aligned}$$

Since $j$ is convex and l.s.c., it is bounded from below by an affine functional. Therefore, using $J(u) = j(u(0), u(T))$, there are constants $m_1, m_2, m_3 \geq 0$ such that

$$\begin{aligned}I(u) \geq{}& \left(\frac{B_1}{\alpha}\|\Delta u\|_{p(.)}^{p^-} - 1\right)\|\Delta u\|_{p(.)}^{(\alpha-1)p^-} - N_\rho T + \alpha'\|u\|_{p(.)}^{p^-}\\ &- m_1|u(0)| - m_2|u(T+1)| - m_3\\ \geq{}& \left(\frac{B_1}{\alpha}\|\Delta u\|_{p(.)}^{p^-} - 1\right)\|\Delta u\|_{p(.)}^{(\alpha-1)p^-} + \alpha'\|u\|_{p(.)}^{p^-} - m_1|u(0)|\\ &- m_2|u(T+1)| - C_1,\end{aligned}$$

where $C_1 = N_\rho T + m_3 - N$; thus

$$I(u) \geq \left(\frac{B_1}{\alpha}\|\Delta u\|_{p(.)}^{p^-} - 1\right)\|\Delta u\|_{p(.)}^{(\alpha-1)p^-} + \alpha'\|u\|_{p(.)}^{p^-} - C_2\|u\|_\infty$$
$$- C_1, \quad where \ C_2 = m_1 + m_2.$$

By the equivalence of the norms on $X$, there is some $C_3 > 0$ such that

$$I(u) \geq \left(\frac{B_1}{\alpha}\|\Delta u\|_{p(.)}^{p^-} - 1\right)\|\Delta u\|_{p(.)}^{(\alpha-1)p^-} + \alpha'\|u\|_{p(.)}^{p^-} - C_3\|u\|_{p(.)} - C_1.$$

Consequently, $I(u) \longrightarrow \infty$ as $\|u\|_{p(.)} \longrightarrow \infty$. Therefore, $I$ is coercive on $X$.

**Step 3:** We now show that the functional $I$ is bounded from below.

For that, let $\|u\|_{p(.)} < 1$. We get by (5.8) and Lemma 5.2.2 the following:

$$\begin{aligned}
I(u) &\geq \varphi(u) - N_\rho T + J(u) \\
&\geq -N_\rho T + \frac{B_1}{\alpha}\|u\|_{p(.)}^{\alpha p^+} + J(u) \\
&\geq -N_\rho T + \frac{B_1}{\alpha}\|u\|_{p(.)}^{\alpha p^+} - m_1|u(0)| - m_2|u(T+1)| - m_3 \\
&\geq \frac{B_1}{\alpha}\|u\|_{p(.)}^{\alpha p^+} - K_1\|u\|_\infty - K', \quad where \ K' = N_\rho T \\
&\quad + m_3 \ and \ K_1 = m_1 + m_2.
\end{aligned}$$

Since any norm on $X$ is equivalent to $\|.\|_{p(.)}$, then there exists $K'' > 0$ such that

$$\begin{aligned}
I(u) &\geq \frac{B_1}{\alpha}\|u\|_{p(.)}^{p^+} - K''\|u\|_{p(.)} - K' \\
&\geq -K''\|u\|_{p(.)} - K' \\
&\geq -K'' - K' \\
&> -\infty.
\end{aligned}$$

Hence, $I$ is bounded from below. Finally, we conclude that $I$ is lower semicontinuous, bounded from below and coercive on the real Banach space $X$. Thus, $I$ attains its infimum at some $u \in X$. Using now Propositions 5.2.1 and 5.2.3, one obtains that the problem (5.1) has at least one solution on $X$. □

Now, we show the existence of solution of problem (5.1) by using mountain pass technique.

## 5.4 Proof of the Existence of Classical Solutions by Mountain Pass Method

In this section, we are concerned with the existence of non-trivial solutions for problem (5.1). The main tool in obtaining such results is Theorem 3.2 in [12].

**Theorem 5.4.4** *Assume that (5.3)–(5.10) and* $(H)$ *hold. Moreover, suppose that there exist constants* $\theta > p^+$, $K$, $M > 0$ *such that:*

($\mathbf{A}_1$) $j(0,0) = 0$.
($\mathbf{A}_2$) $j'(z,z) \leq \theta j(z) + K, \forall z \in D(j)$.
($\mathbf{A}_3$) $\limsup_{|t|\to 0} \frac{p(k)F(k,t)}{|t|^{p(k)}} < \lambda_1$, *for all* $k \in \mathbb{Z}[1,T]$.
($\mathbf{A}_4$) $0 < \theta F(k,t) \leq tf(k,t)$ *for all* $k \in \mathbb{Z}[1,T]$ *with* $|t| > N$.

*Then, there exists a non-trivial solution* $u \in X$ *of problem (5.1).* □

***Proof***

**Step 1**: We show that the functional $I$ verifying (**H**) satisfies the (PS) condition in the sense of Szulkin on $(X, \|.\|_{p(.)})$. So, let $\{u_n\} \subset X$ be a sequence for which $I(u_n) \longrightarrow c \in \mathbb{R}$ and (5.19) holds with $\epsilon_n \longrightarrow 0$.
For this purpose, since $X$ is finite-dimensional, it is sufficient to prove that $\{u_n\}$ is bounded. We may assume that $\{u_n\} \subset D(I) = D(J)$ and $\|u_n\|_{p(.)} > 1$ for all $n \in \mathbb{N}$. By ($\mathbf{A}_2$) and (5.15), it follows that

$$J(v) - \frac{1}{\theta}J'(v; v) \geq -K_1, \quad \text{for all} \quad v \in D(J), \tag{5.25}$$

with $K_1 = \frac{K}{\theta}$. Using the relation ($\mathbf{A}_4$), one deduces that for all $n \in \mathbb{N}$,

$$\sum_{k=1,|u(k)|>N}^{T} [\theta F(k, u_n(k)) - u_n(k)f(k, u_n(k))] \leq 0.$$

Consequently,

$$\begin{aligned}
&-\Phi(u_n) + \frac{1}{\theta} < \Phi'(u_n); u_n > \\
&= \frac{1}{\theta}\sum_{k=1}^{T}[\theta F(k, u_n(k)) - u_n(k)f(k, u_n(k))] \\
&= \frac{1}{\theta}\sum_{k=1,|u_n(k)|>N}^{T} [\theta F(k, u_n(k)) - u_n(k)f(k, u_n(k))]
\end{aligned}$$

$$
\begin{aligned}
&+\frac{1}{\theta}\sum_{k=1,|u_n(k)|\leq N}^{T}[\theta F(k,u_n(k))-u_n(k)f(k,u_n(k))]\\
&\leq \frac{1}{\theta}\sum_{k=1,|u_n(k)|\leq N}^{T}[\theta F(k,u_n(k))-u_n(k)f(k,u_n(k))]\\
&\leq \frac{1}{\theta}\sum_{k=1}^{T}\max_{|x|\leq N}|\theta F(k,x)-xf(k,x)| =: C_3,
\end{aligned}
$$

where $C_3$ is some positive constant. Therefore, one can write

$$
-\Phi(u_n)+\frac{1}{\theta}<\Phi'(u_n);u_n>\leq C_3. \tag{5.26}
$$

Since the real sequence $(I(u_n))_{n\in\mathbb{N}}$ converges toward the real number $c$, it is clear that there is a constant $C_4>0$ such that

$$
|I(u_n)|\leq C_4, \quad \text{for all} \quad n\in\mathbb{N}. \tag{5.27}
$$

Furthermore, setting $v=u_n+su_n$ in (2.6), dividing by $s>0$, and letting $s\longrightarrow 0^+$, one obtains

$$
<\Phi'(u_n);u_n>+<\varphi'(u_n);u_n>+J'(u_n;u_n)\geq -\epsilon_n\|u_n\| \quad \text{for all} \quad n\in\mathbb{N}. \tag{5.28}
$$

Using (5.27) and (5.28), we deduce that

$$
\begin{aligned}
C_4+\frac{\epsilon_n}{\theta}\|u_n\|_{p(.)} &\geq \Phi(u_n)+\varphi(u_n)+J(u_n)+\frac{\epsilon_n}{\theta}\|u_n\|_{p(.)}\\
&\geq \Phi(u_n)-\frac{1}{\theta}<\Phi'(u_n);u_n>+\varphi(u_n)-\frac{1}{\theta}\\
&<\varphi'(u_n);u_n>+J(u_n)-\frac{1}{\theta}J'(u_n;u_n),
\end{aligned}
$$

and by virtue of (5.25), (5.26), and (5.27), it follows that

$$
K_1+C_3+C_4+\frac{\epsilon_n}{\theta}\|u_n\|_{p(.)}\geq \varphi(u_n)-\frac{1}{\theta}<\varphi'(u_n);u_n>.
$$

According to (5.10), it follows that

$$
\varphi(u_n)-\frac{1}{\theta}<\varphi'(u_n);u_n>=-\frac{1}{\theta}M\left(\sum_{k=1}^{T+1}A(k-1,\Delta u_n(k-1))\right)
$$

$$\times \sum_{k=1}^{T+1} a(k-1, \Delta u_n(k-1))\Delta u_n(k-1)$$

$$+ \widehat{M}\left(\sum_{k=1}^{T+1} A(k-1, \Delta u_n(k-1))\right)$$

$$\geqslant \left(\sum_{k=1}^{T+1} A(k-1, \Delta u_n(k-1))\right)^{\alpha-1}$$

$$\times \left[B_1 - \frac{1}{\theta}B_2(\alpha-1)\right]$$

$$\geqslant \lambda_1 p^- \sum_{k=1}^{T} \frac{1}{p(k)}|u(k)|^{p(k)} \left[B_1 - \frac{1}{\theta}B_2(\alpha-1)\right],$$

and from Lemma 5.2.2, we deduce that

$$K_1 + C_3 + C_4 + \frac{\epsilon_n}{\theta}\|u_n\|_{p(.)} \geqslant \lambda_1 p^- \left(B_1 - \frac{1}{\theta}B_2(\alpha-1)\right)\|u\|_{p(.)}^{p^-}.$$

Moreover, $\theta > p^+$. Then, we infer that the sequence $(u_n)_{n\in\mathbb{N}}$ is bounded.

**Step 2**: We show that $I$ has a mountain pass geometry.

From $(\mathbf{A}_1)$, it is clear that

$$I(0) = \Phi(0) + \varphi(0) + J(0) = 0.$$

Using $(\mathbf{A}_3)$, we have

$$\lim_{|u|\to 0} sup \frac{p(k)F(k, u(k))}{|u(k)|^{p(k)}} < \lambda_1.$$

That leads to the existence of $\epsilon, \beta > 0$ such that

$$F(k,t) < \frac{\lambda_1 - \epsilon}{p(k)}|t|^{p(k)} \quad \text{with} \quad |t| < \beta.$$

Consequently,

$$\Phi(u) \geq -(\lambda_1 - \epsilon)\sum_{k=1}^{T} \frac{1}{p(k)}|u(k)|^{p(k)}, \quad \text{for all} \quad u \in X - \{0\},$$

$$u(0) = u(T+1) \ and \ |u| < \beta. \tag{5.29}$$

Using again hypothesis (5.8), we get

$$\Phi(u)+\varphi(u)\geq-(\lambda_1-\epsilon)\sum_{k=1}^{T}\frac{1}{p(k)}|u(k)|^{p(k)}$$
$$+\frac{B_1}{\alpha}\left(\sum_{k=1}^{T+1}\frac{1}{p(k-1)}|\Delta u(k-1)|^{p(k-1)}\right)^{\alpha}$$
$$\geq\epsilon\sum_{k=1}^{T}\frac{1}{p(k)}|u(k)|^{p(k)}-\left(\sum_{k=1}^{T+1}\frac{1}{p(k-1)}|\Delta u(k-1)|^{p(k-1)}\right)^{\alpha-1}$$
$$+\frac{B_1}{\alpha}\left(\sum_{k=1}^{T+1}\frac{1}{p(k-1)}|\Delta u(k-1)|^{p(k-1)}\right)^{\alpha}.$$

According to (5.4) and $(A_1)$, we have $J(u)=j(u(0),u(T+1))=j(0,0)=0$. Therefore, for $\beta<1$,

$$\Phi(u)+\varphi(u)+J(u)\geq\epsilon\|u\|_{p(.)}^{p^+}-\|\Delta u\|_{p(.)}^{(\alpha-1)p^+}+\frac{B_1}{\alpha}\|\Delta u\|_{p(.)}^{(\alpha-1)p^+}+0.$$

Hence, choosing $\|u\|_{p(.)}^{p^+}=\beta$, which is equivalent to $\|u\|_{p(.)}=\beta^{\frac{1}{p^+}}$, and as there exists a positive constant $\gamma$ such that $\|\Delta u\|_{p(.)}^{p^+}=\gamma\|u\|_{p(.)}^{p^+}$, then $I(u)\geq L$ with $L=\beta(\epsilon-\gamma^{\alpha-1}+\frac{B_1}{\alpha}\gamma^{\alpha})$.
Coming back to relation $(A_4)$ and taking $|u|$ big enough, we have

$$\frac{f(k,u(k))}{F(k,u(k))}\geq\frac{\theta}{u}.$$

So, $F(k,u(k))\geq cu^{\theta}$ for $|u|$ big enough. Thus, $F(k,u(k))\geq cu^{\theta}-K,$ for all $u>0$.
One can use (5.4) to say that

$$A(k,\xi)=\int_0^{\xi}a(k,\lambda)d\lambda.$$

Using (5.7), we have the existence of a real $C_1>0$ such that

$$|a(k,\xi)|\leqslant C_1(1+|\xi|^{p(k)-1})\quad\text{for all}\quad k\in\mathbb{Z}[0,T]\quad\text{and for all}\quad\xi\in\mathbb{R}.$$

Therefore,

$$\begin{aligned}\int_0^{\xi} |a(k,\lambda)|d\lambda &\leq C_1 \int_0^{\xi} (1+|\lambda|^{p(k)-1})d\lambda \\ &\leq C_1[\lambda]_0^{\xi} + C_1[\frac{\lambda^{p(k)}}{p(k)}]_0^{\xi} \\ &\leq C_1|\xi| + C_1\frac{|\xi|^{p(k)}}{p(k)}.\end{aligned}$$

One deduces that

$$\sum_{k=1}^{T+1} A(k-1,\Delta u(k-1)) \leq C_1 \sum_{k=1}^{T+1} |\Delta u(k-1)| + C_1 \sum_{k=1}^{T+1} \frac{|\Delta u(k-1)|^{p(k)}}{p(k)}.$$

Thus, according to (5.10), we can write

$$\begin{aligned}\varphi(u) &\leq \frac{B_2 C_1^{\alpha}}{\alpha}\left[\left(\sum_{k=1}^{T+1} |\Delta u(k-1)|\right) + \left(\sum_{k=1}^{T+1} \frac{|\Delta u(k-1)|^{p(k)}}{p(k)}\right)\right]^{\alpha} \\ &\leq 2\frac{B_2 C_1^{\alpha}}{\alpha}\left(\sum_{k=1}^{T+1} |\Delta u(k-1)|\right)^{\alpha}.\end{aligned}$$

Let $u_0 \in X - \{0\}$ be such that $u_0(0) = u_0(T+1) = 0$ and $\|u_0\|_{p(.)} > 1$. From $(\mathbf{A}_1)$, we have that $J(su_0) = 0$ for all $s \in \mathbb{R}$. Then,

$$\begin{aligned}I(su_0) &= \Phi(su_0) + \varphi(su_0) + J(su_0) \\ &= -\sum_{k=1}^{T} F(k, su_0(k)) + \widehat{M}\left(\sum_{k=1}^{T+1} A(k-1, \Delta su_0(k-1))\right) + 0 \\ &\leq \sum_{k=1}^{T}(K - c|su_0(k)|^{\theta}) + 2\frac{B_2 C_1^{\alpha}}{\alpha}\left(\sum_{k=1}^{T+1} |\Delta u(k-1)|\right)^{\alpha} \\ &= TK - c\sum_{k=1}^{T} |su_0(k)|^{\theta} + 2\frac{B_2 C_1^{\alpha}}{\alpha}\left(\sum_{k=1}^{T} |\Delta su_0(k)|\right)^{\alpha} \\ &\leq TK + C_1''\|su_0\|_{\infty} - cs^{\theta}\|u_0\|_{\infty}^{\theta} \longrightarrow -\infty,\end{aligned}$$

as $s \longrightarrow \infty$.

Hence, we can choose $s$ large enough such that $I(su_0) \leq 0$; therefore, $\|su_0\|_{p(.)} > \beta$. We conclude by using Theorem 3.2 in [12] that the problem (5.1) has at least one non-trivial solution.

## References


1. R.P. Agarwal, K. Perera, D. ORegan, Multiple positive solutions of singular discrete $p$-Laplacian problems via variational methods. Adv. Differ. Equ. **2005**(2), 93–99 (2005)
2. C. Bereanu, P. Jebelean, C. Serban, Periodic and Neumann problems for discrete $p(\cdot)$-Laplacian. J. Math. Anal. Appl. **399**, 75–87 (2013)
3. C. Bereanu, P. Jebelean, C. Serban, Ground state and mountain pass solutions for discrete $p(\cdot)$-Laplacian. Bound. Value Probl. **104**, 1–13 (2012)
4. A. Cabada, A. Iannizzotto, S. Tersian, Multiple solutions for discrete boundary value problems. J. Math. Anal. Appl. **356**, 418–428 (2009)
5. P. Candito, G. D'Agui, Three solutions for a discrete nonlinear Neumann problem involving $p$-Laplacian. Adv. Differ. Equ. **2010**, 862016 (2010)
6. M. Galewski, S. Gab, On the discrete boundary value problem for anisotropic equation. J. Math. Anal. Appl. **386**, 956–965 (2012)
7. M. Galewski, S. Gab, Positive solutions for anisotropic discrete BVP. arXiv:1203.4504v1
8. L. Gasinski, N.S. Papageorgiou, Nonlinear second-order multivalued boundary value problems. Proc. Indian Acad. Sci. Math. Sci. **113**, 293–319 (2003)
9. A. Guiro, I. Nyanquini, S. Ouaro, On the solvability of discrete nonlinear Neumann problems involving the $p(x)$-Laplacian. Adv. Differ. Equ. **2011**, 32 (2011)
10. T. He, W. Chen, Periodic solutions of second order discrete convex systems involving the $p$-Laplacian. Appl. Math. Comput. **206**, 124–132 (2008)
11. V.-M. Hokkanen, G. Morosanu, *Functional Methods in Differential Equations* (Chapman and Hall/CRC, 2002)
12. P. Jebelean, C. Şerban, Ground state periodic solutions for difference equations with discrete $p$-Laplacian. Appl. Math. Comput. **217**, 9820–9827 (2011)
13. G. Kirchhoff, *Vorlesungen über mathematische Physik: Mechanik* (Teubner, Leipzig, 1876)
14. B. Koné, S. Ouaro, Weak solutions for anisotropic discrete boundary value problems. J. Differ. Equ. Appl. **16**(2), 1–11 (2010)
15. B. Koné, I. Nyanquini, S. Ouaro, Weak solutions to discrete nonlinear two-point boundary-value problem of Kirchhoff type. Electron. J. Differ. Equ. **2015**(105), 110 (2015)
16. B.A. Kyelem, S. Ouaro, M. Zoungrana, Classical solutions for discrete potential boundary value problems with generalized Leray-Lions type operator and variable exponent. Electron. J. Differ. Equ. **2017**(109), 1–16 (2017)
17. R. Mashiyev, Z. Yucedag, S. Ogras, Existence and multiplicity of solutions for a Dirichlet problem involving the discrete $p(x)$-Laplacian operator. Electron. J. Qual. Theory Differ. Equ. **67**, 1–10 (2011)
18. M. Mihailescu, V. Radulescu, S. Tersian, Eigenvalue problems for anisotropic discrete boundary value problems. J. Differ. Equ. Appl. **15**(6), 557–567 (2009)
19. M. Mihailescu, V. Radulescu, S. Tersian, Homoclinic solutions of difference equations with variable exponents. Topol. Methods Nonlinear Anal. **38**, 277–289 (2011)
20. R.T. Rockafellar, *Convex Analysis* (Princeton University Press, Princeton, 1972)
21. A. Szulkin, Minimax principles for lower semicontinuous functions and applications to nonlinear boundary value problems. Ann. Inst. Henri Poincaré, Anal. Non Linéaire **3**(2), 77–109 (1986)
22. Y. Tian, W. Ge, The existence of solutions for a second-order discrete Neumann problem with a p-Laplacian. J. Appl. Math. Comput. **26**, 333–340 (2008)
23. Z. Yucedag, Existence of solutions for anisotropic discrete boundary value problems of Kirchhoff type. Int. J. Differ. Equ. Appl. **13**(1), 1–15 (2014)

# Chapter 6
# From Calculus of Variation to Exterior Differential Calculus: A Presentation and Some New Results

**Ngalla Djitte and Mariama Ndiaye**

**Abstract** In this chapter, we provide an introduction to exterior differential calculus. In detail, the Cartan–Kähler theorem is revisited. Using this, we give necessary and sufficient conditions for a second-order differential equation to be equivalent to some Euler–Lagrange equation.

**Keywords** Exterior differential calculus · Integral element · Manifold · calculus of variation

## 6.1 Introduction

Let $F_i : \mathbb{R}^n \times \mathbb{R}^n \to \mathbb{R}$, $n \geq 1$, $1 \leq i \leq n$, be real-valued functions defined on $\mathbb{R}^n \times \mathbb{R}^n$. Consider the following system of second-order differential equations:

$$\frac{d^2 x_i}{dt^2} = F_i\left(x_1, x_2, \cdots, x_n, \frac{dx_1}{dt}, \frac{dx_2}{dt}, \cdots, \frac{dx_n}{dt}\right) \forall i = 1, \ldots, n. \tag{6.1}$$

A natural question, initially raised by Douglas [3], is the following: When is it possible to find a Lagrangian $L : \mathbb{R}^n \times \mathbb{R}^n \to \mathbb{R}$ such that the solutions of (6.1) correspond to that of the Euler–Lagrange equations

$$\frac{d}{dt}\frac{\partial L}{\partial y_i}\left(x_1, x_2, \cdots, x_n, \frac{dx_1}{dt}, \frac{dx_2}{dt}, \cdots, \frac{dx_n}{dt}\right) - \frac{\partial L}{\partial x_i}\left(x_1, x_2, \cdots, x_n, \frac{dx_1}{dt}, \frac{dx_2}{dt}, \cdots, \frac{dx_n}{dt}\right) = 0, \quad i = 1, \ldots, n, \tag{6.2}$$

N. Djitte (✉) · M. Ndiaye
Gaston Berger University, Saint Louis, Senegal
e-mail: ngalla.djitte@ugb.edu.sn; mariama-ndiaye.diakhaby@ugb.edu.sn



T. Diagana et al. (eds.), *Partial Differential Equations and Applications*,
Springer Proceedings in Mathematics & Statistics 420,
https://doi.org/10.1007/978-3-031-27661-3_6

corresponding to $L$? In other words, given $F_i$, $1 \leq i \leq n$, can we find $L : \mathbb{R}^n \times \mathbb{R}^n \to \mathbb{R}$ such that $L$ solves the following partial differential equation:

$$\sum_{k=1}^{n} \frac{\partial^2 L}{\partial x_k \partial y_i} y_k + \sum_{h=1}^{n} \frac{\partial^2 L}{\partial y_h \partial y_i} F_h = \frac{\partial L}{\partial x_i} \quad i = 1, \ldots, n? \tag{6.3}$$

It is our purpose in this chapter to propose an introduction to *exterior differential calculus*. In particular, we describe in some details two very powerful theorems, respectively, due to Darboux and to Cartan and Kähler. These theorems have recently been used in a very sophisticated way in economic theory of demand by Ekeland and Chiappori (see, e.g., [5–8]) and further by Ekeland and Djitté [9, 10]. We strongly believe that these theorems will be extremely helpful in many contexts, and they should profitably be included in economist's toolbox. Here, we show how this approach can be useful for the solvability of some inverse problems of calculus of variation. In fact, we present an application of this approach to the Douglas problem.

## 6.2 Exterior Differential Calculus

In this section, we introduce the basic notions of exterior differential calculus. For a much more exhaustive presentation, the interested reader is referred to Cartan's book [2] or to Bryant et al. [1].

### *6.2.1 Linear and Differential Forms*

A linear form (or 1-form) on $E = \mathbb{R}^n$ is a linear mapping from $E$ to $\mathbb{R}$:

$$\omega : \xi \in \mathbb{R}^n \mapsto \langle \omega, \xi \rangle = \sum_{i=1}^{n} \omega^i \xi_i .$$

The set of linear forms on $E$ is the dual $E^*$ of $E$. Basic examples of linear forms are the projection $\pi_i : \xi \mapsto \xi_i$, which, to any vector, associated its $i$-th coordinate. These form a basis for $E^*$ in the sense that any form $\omega$ can be decomposed as

$$\omega = \sum_{i=1}^{n} \omega^i \pi_i .$$

In what follows, we are specifically interested in *differential forms*. Consider a smooth manifold $M$ of dimension $n \geq 1$ and for $p \in M$; let $T_pM$ denote its tangent space at the point $p$. A differential 1-form on $M$ , $\omega$, is a map defined on $M$ such that for every $p \in M$, $\omega(p)$ is a 1-form on the tangent space $T_pM$ to $M$ at $p$ with, say, $\langle \omega(p), \xi \rangle = \sum_{i=1}^{n} \omega^i(p)\xi_i)$, where the coefficient $\omega^i(p)$ depends smoothly on $p$. A local coordinate system at $p$ provides $T_pM$ with a coordinate system as well. If $M$ is a $n$-dimensional manifold, then $T_pM$ is a copy of $\mathbb{R}^n$, and the projection maps $\pi_i : T_pM \to \mathbb{R}$, which associate with a tangent vector $\xi$ its $i$-th coordinate $\xi_i$, will be denoted by $dp_i$.

As a simple example of a differential 1-form, we may, for any smooth mapping $V$ from $E$ to $\mathbb{R}$, consider the differential form $dV$ defined at any point $p$ by

$$dV(p) = \sum_{i=1}^{n} \frac{\partial V}{\partial p_i}(p)dp_i,$$

so that

$$dV(p) : \xi \mapsto \langle dV(p), \xi \rangle = \sum_{i=1}^{n} \frac{\partial V}{\partial p_i}(p)\xi_i.$$

Of course, this form is extremely specific, for the following reason. Consider the hypersurface (that is the $(n-1)$-dimensional submanifold) $N \subset M$ defined by

$$N = \{p \in M \mid V(p) = a\},$$

where $a$ is a constant. Then for any $p \in N$, the form $dV(p)$ or any form $\omega(p) = \lambda(p)dV(p)$ (proportional to $dV(p)$) vanishes on the tangent space $T_pN$:

$$\forall\ p \in N,\ \forall \xi \in T_pN,\ \langle \omega(p), \xi \rangle = 0.$$

This is exactly the *integration problem*: starting from some given differential form $\omega(p)$, when is it possible to find a hypersurface $N$ such that, for any $p \in N$, the restriction of $\omega(p)$ to $T_pN$ is zero? Such a submanifold will be called an *integrating submanifold* or an *integral element* for $\omega$.

One point must, however, be emphasized. When $\omega$ is proportional to some total differential $dV$, the submanifold $N$ can be found of (maximum) dimension $n-1$. But of course, life is not always that easy. Starting from an arbitrary form, it is in general impossible to find such an integrating submanifold of dimension $n-1$.

### 6.2.2 Exterior k-form

Before addressing the integration problem in detail, we must generalize our basic concept.

**Definition 6.2.1** An exterior $k$-form is a mapping $\omega : E^k \to \mathbb{R}$ that is:

- Multilinear, that is, linear with respect to each component
- Antisymmetric, that is, the sign is changed when two vectors are permuted □

Note that if $k = 1$, we are back to the definition of linear forms.

Consider, for instance, the case $k = 2$. A 2-form is defined by a matrix:

$$\omega(\xi, \eta) = \sum_{i,j=1}^{n} \omega^{i,j} \xi_i \eta_j = \xi' \Omega \eta.$$

Additional restrictions are usually imposed upon the matrix $\Omega$. A standard one is symmetric, i.e., $\Omega = \Omega'$. In exterior differential calculus, on the contrary, since one considers exterior forms, *antisymmetry* is imposed. This gives $\Omega = -\Omega'$, i.e., $\omega^{i,j} = -\omega^{j,i}$ for all $i$, $j$. Hence,

$$\omega(\xi, \eta) = \sum_{i<j}^{n} \omega^{i,j} (\xi_i \eta_j - \xi_j \eta_i).$$

Another case of interest is $k = n$, where $n$ is the dimension of the space $E$. Then the space of exterior $n$-forms is of dimension one and includes the *determinant*. That is, any $n$-form $\omega$ is collinear to the determinant:

$$\omega(\xi_1, \cdots, \xi_n) = \lambda \det(\xi_1, \cdots, \xi_n).$$

Some well-known properties of the determinant are in fact due exclusively to multilinearity together with antisymmetry and can thus be generalized to forms of any order. For instance, take any $k$-form $\omega$, and take $k$ vectors $(\xi_1, \cdots, \xi_k)$ that are not linearly independent; then $\omega(\xi_1, \cdots, \xi_k) = 0$. An important consequence is that, for any $k > n$, any exterior $k$-form must be zero.

### 6.2.3 Exterior Product

The set of exterior forms on $M$ is an algebra, on which the multiplication, called the *exterior product*, is formally defined as follows:

**Definition 6.2.2** Let $\alpha$ be a $k$-form and $\beta$ be an $l$-form; then $\alpha \wedge \beta$ is the $(k+l)$-form such that:

$$\alpha\wedge\beta(\xi_1, \cdots, \xi_{k+l})=\sum_{\sigma}\frac{1}{k!l!}(-1)^{\mathrm{sign}(\sigma)}\alpha(\xi_{\sigma(1)}, \cdots, \xi_{\sigma(k)})\beta(\xi_{\sigma(k+1)}, \cdots, \xi_{\sigma(k+l)}),$$

where the sum is over all permutations $\sigma$ of $\{1, \cdots, k+l\}$. □

This formula may seem complex. Note, however, that it satisfies two basic requirements: $\alpha \wedge \beta$ is multilinear and antisymmetry. To grasp the intuition, consider the case of two linear forms ($k = l = 1$). Then

$$\alpha \wedge \beta(\xi, \eta) = \alpha(\xi)\beta(\eta) - \alpha(\eta)\beta(\xi).$$

Obviously, this is the simplest exterior 2-form related to $\alpha$ and $\beta$ and satisfies the two requirements above. A few consequences of this definition must be kept in mind:

- If $w$ is linear (or of odd order), then $w \wedge w = 0$. More generally, let $\omega_1 \cdots, \omega_s$ be 1-forms. If the forms are linearly dependent, then $\omega \wedge \cdots \wedge \omega_s = 0$.
- If $\omega$ is a two-form (or a form of even order), $\omega \wedge \omega$ need not to be zero.
- For any $k$-form $\omega$, $(\omega)^k = \omega \wedge \omega \cdots \wedge \omega$ is a $ks$-form. In particular, $(\omega)^s = 0$ as soon as $ks > n$.
- Any $k$-form can be decomposed into exterior product of 1-forms. If $\omega$ is a $k$-form, then

$$\omega = \sum_{\sigma} = \omega_{\sigma(1),\cdots,\sigma(k)} dp_{\sigma(1)} \cdots \sigma(k), \tag{6.4}$$

where the sum is over all ordered maps $\sigma : \{1, \cdots, k\} \to \{1, \cdots, n\}$.

### 6.2.4 Differential Forms and Exterior Differentiation

A differential $k$-form is, for every $p \in M$, an exterior $k$-form $\omega(p)$ on $(T_pM)^k$, depending smoothly on $p$. Exterior differentiation sends differential $k$-forms into differentials $(k+1)$-forms. We first define it on 1-forms. For this, set

$$\omega(p) = \sum_{j=1}^{n} \omega^j(p) dp_j.$$

To define the exterior differential of $\omega$, we may first remark that the $\omega^j(p)$ are standard functions from $M$ to $\mathbb{R}$. As such, they admit differentials:

$$d\omega^j(p) = \sum_{i=1}^{n} \frac{\partial w^j}{\partial p_i} dp_i.$$

Then the exterior differential $d\omega(p)$ of $\omega$ is the differential 2-form defined by

$$d\omega(p) = \sum_{j=1}^{n} d\omega^j(p) \wedge dp_j \tag{6.5}$$
$$= \sum_{i,j=1}^{n} \frac{\partial w^j}{\partial p_i} dp_i \wedge dp_j$$
$$= \sum_{i<j}^{n} \left( \frac{\partial w^j}{\partial p_i} - \frac{\partial w^i}{\partial p_j} \right) dp_i \wedge dp_j.$$

Generally, if

$$\omega = \sum_{i_1<i_2<\ldots<i_k} \omega_{i_1,\ldots,i_k} dp_{i_1} \wedge \ldots \wedge dp_{i_k}$$

is a $k$-form, then $d\omega$ is the $(k+1)$-form defined by

$$d\omega = \sum_{i_1<i_2<\ldots<i_k} d\omega_{i_1,\ldots,i_k} \wedge dp_{i_1} \wedge \ldots \wedge dp_{i_k}. \tag{6.6}$$

Note that this formula guarantees that $d\omega(p)$ is bilinear and antisymmetric.

**Proposition 6.2.1** *Exterior differentiation is a linear operation, and it satisfies the following product formula: if $\alpha$ is a differential $p$-form and $\beta$ is a differential $q$-form, we have*

$$d[\alpha + \beta] = d\alpha + d\beta; \tag{6.7}$$
$$d[\alpha \wedge \beta] = d\alpha \wedge \beta + (-1)^p \alpha \wedge d\beta. \tag{6.8}$$

□

### 6.2.5 Pullback

Let $\varphi : \mathbb{R}^m \to \mathbb{R}^n$ be a smooth mapping. To any smooth function $f : \mathbb{R}^n \to \mathbb{R}$, we can associate the function $\varphi^*(f) := fo\varphi$, which is a smooth function on $\mathbb{R}^m$.

Similarly to $df$, which is a differential 1-form on $\mathbb{R}^n$, we associate $\varphi^*(df) := d(fo\varphi)$, which is a differential 1-form on $\mathbb{R}^m$, called the pullback of $df$. More generally, we have the following:

**Definition 6.2.3** Let $\omega$ be a $p$-form on $\mathbb{R}^n$. The pullback of $\omega$ w.r.t. $\varphi$ is the $p$-form $\varphi^*(\omega)$ on $\mathbb{R}^m$ defined by

$$(\varphi^*\omega)(p).(v_1, \ldots, v_p) = \omega(\varphi(p)).(d\varphi(p)v_1, \ldots, d\varphi(p)v_p), \tag{6.9}$$

for all $p \in \mathbb{R}^m$ and $(v_1, \ldots, v_p) \in (T_p\mathbb{R}^m)^p$. By convention, if $p = 0$, then the formula (6.9) is reduced to

$$\varphi^* f(x) = f(\varphi(x)). \tag{6.10}$$

Finally, we note that the *pullback* is natural with respect to *exterior product* and *exterior differentiation* in the following sense.

**Proposition 6.2.2** *With the preceding definition, we have*

$$\varphi^*(\alpha \wedge \beta) = \varphi^*(\alpha) \wedge \varphi^*(\beta); \tag{6.11}$$

$$\varphi^*(d\omega) = d\varphi^*(\omega). \tag{6.12}$$

□

### 6.2.6 Poincaré Theorem

The construction detailed above has strong implications for the resolution of the type of equations we are interested in. Let us start with a simple problem: what are the conditions for a given exterior form $\omega$ to be the tangent form of some twice continuously differentiable function $V$? An immediate, necessary condition is given by the following result:

**Theorem 6.2.1** *Let $U$ be an open subset of $\mathbb{R}^n$ and $\omega$ be an exterior form on $U$. Assume there exists a twice continuously differentiable function $V$ such that $\omega(p) = dV(p)$. Then*

$$d\omega = 0.$$

***Proof*** Just note that

$$d\omega = \sum_{i<j} \left( \frac{\partial^2 V}{\partial x_i \partial x_j} - \frac{\partial^2 V}{\partial x_j \partial x_i} \right) dp_i \wedge dp_j = 0.$$

□

This proposition admits a converse, due to Poincaré, that requires some topological condition upon $U$. We have the following result.

**Theorem 6.2.2 (Poincaré)** *Let $\omega$ be a differential $k$-form on $U$ such that $d\omega = 0$. Assume that $U$ is convex. Then, there exists a differential $(k-1)$-form on $U$, say $\alpha$, such that*

$$\omega = d\alpha.$$

***Proof*** See Bryant et al. [1]. □

**Corollary 6.2.1** *Let $U$ be a nonempty convex subset of $\mathbb{R}^n$, and let $\omega^1, \cdots, \omega^n$ be given differentiable functions on $U$. There exists a differentiable function $V$ on $U$ such that*

$$\omega^i = \frac{\partial V}{\partial p_i}, \quad i = 1, \cdots, n,$$

*if and only if*

$$\frac{\partial \omega^i}{\partial p_j} = \frac{\partial \omega^j}{\partial p_i}.$$

***Proof*** Define $\omega(p) := \sum_{i=1}^{n} \omega^i dp_i$ and apply Theorem 6.2.2. □

### 6.2.7 Darboux Theorem

Poincaré theorem provides necessary and sufficient conditions for a differential 1-form to be a total differential. In this case, the integration problem is straightforward, as illustrated above. But, at the same time, these conditions are very strong. We now generalize this result, by giving necessary and sufficient conditions for a differential 1-form to be a linear combination of $k$ tangent forms. In this case, the integration problem can be solved, but only with an integration manifold of dimension at least $n - k$.

**Proposition 6.2.3** *Let $U$ be an open subset of $\mathbb{R}^n$ and $\omega$ be an exterior form on $U$. Assume there exist twice continuously differentiable functions $V^i$ and functions $\lambda_i$, $1 \leq i \leq k$, such that*

$$\omega(p) = \sum_{i=1}^{k} \lambda_i(p) dV^i(p), \ \forall\, p \in U.$$

*Then,*

$$\omega \wedge (d\omega)^k = 0.$$

This simple necessary condition admits an important converse.

**Theorem 6.2.3 (Darboux)** *Let $\omega$ be a linear form defined on some neighbourhood $U_0$ of $p$. Let $k \geq 1$ be such that:*

$$\omega \wedge (d\omega)^{k-1} \neq 0, \ \forall\, p \in U_0;$$
$$\omega \wedge (d\omega)^{k} = 0, \ \forall\, p \in U_0.$$

□

*Then there exists a neighbourhood $U_1$ of $p$ and $2k$ functions $V^i$ and $\lambda_i$, $1 \leq i \leq k$, such that:*

1. *The $V^i$ are linearly independent.*
2. *None of the $\lambda_i$ vanishes on $U_1$.*
3. 

$$\omega(p) = \sum_{i=1}^{k} \lambda_i(p) dV^i(p), \ \forall \, p \in U_1.$$

***Proof*** See Bryant et al. [1]. □

Before ending this section, let us give some applications of Darboux theorem. In demand theory, many problems take one of the following forms: Given a smooth vector function $x : \mathbb{R}^n \to \mathbb{R}^n$, a natural number $k \geq 1$:

**Q1.** When is it possible to find scalar functions $V^1, \cdots, V^k$ and $\lambda_1, \cdots, \lambda_k$ such that

$$x(p) = \sum_{i=1}^{k} \lambda_i(p) \nabla V^i(p)? \tag{6.13}$$

**Q2.** Can we choose in the decomposition (6.13) the $V^i$ convex and the $\lambda_i$ positive?

**Q3.** Can we require that $V^i$ and $\lambda_i$ satisfy some additional equations of the following type:

$$\Phi_j(p, \lambda_i(p), \nabla V^i(p)) = 0, \ \ 1 \leq j \leq m?$$

where the $\Phi_j$ are prescribed given functions.

For **Q1.**, if $k = 1$ and $\lambda_1$ is a constant function, then the answer follows from Poincaré theorem. If $\lambda_1$ is not constant, then this is Frobenius theorem, and it requires $\omega$ to satisfy the so-called *Frobenius condition*:

$$\omega \wedge d\omega \neq 0,$$

where $\omega$ is the differential form given by

$$\omega(p) := \sum_{i=1}^{k} x_i(p) dp_i.$$

Still for **Q1.**, in the general case, this is an application of Darboux theorem, and the conditions are

$$\omega \wedge (d\omega)^{k-1} \neq 0;$$
$$\omega \wedge (d\omega)^{k} = 0.$$

Ekeland and Chiappori [6] studied **Q2.** in the framework of *demand theory*. In fact, they give a positive answer in the case where $k = n$.

Motivated by the work of Ekeland and Chiappori [6], still in the framework of *demand theory*, Ekeland and Djitte [10] investigated the case $k \leq n$, and they got a positive answer provided that the differential form $\omega$ satisfies the Darboux condition

$$\omega \wedge (d\omega)^{k-1} \neq 0;$$
$$\omega \wedge (d\omega)^{k} = 0$$

and the *Slutsky symmetry condition*, that is, the Jacobian matrix

$$D_p x(p) := \left( \frac{\partial x_i}{\partial p_j}(p) \right) \tag{6.14}$$

is a sum of a symmetric definite positive matrix $S$ and a matrix $R_k$ of rank $k$.

## 6.3 Exterior Differential System

We now present the key result upon which our approach relies. This theorem, due to Cartan and Kähler, solves the following general problem.

*Given a certain family of differential forms (not necessarily* 1*-form, nor even of the same degree), a point* $\bar{p}$*, and an integer* $m \geq 1$*, can one find some* $m$*-dimensional submanifold* $M$ *containing* $\bar{p}$ *and on which all the given forms vanish on the tangent space* $T_{\bar{p}}M$*?*

### 6.3.1 Introductive Examples

**Cauchy–Lipschitz Theorem** Let us start from a simple version of our problem, namely the Cauchy–Lipschitz Theorem for ordinary differential equations. It states that, given a point $\bar{p} \in \mathbb{R}^n$ and a $C^1$ function $f$, defined from some neighbourhood $U$ of $\bar{p}$ into $\mathbb{R}^{n-1}$, there exist some $\epsilon > 0$ and a $C^1$ function $\varphi : (-\epsilon, \epsilon) \to U$ such that:

$$\begin{cases} \dfrac{d\varphi}{dt}(t) = f(\varphi(t)), \quad \forall\, t \in (-\epsilon, \epsilon) \\ \\ \varphi(0) = \bar{p}. \end{cases} \tag{6.15}$$

It follows that $\frac{d\varphi}{dt}(0) = f(\bar{p})$. If $f(\bar{p}) = 0$, the solution is trivially, $\varphi(t) = \bar{p}$ for all $t$. So, we assume that $f(\bar{p})$ does not vanish.

This theorem can be rephrased in a geometric way. Consider the graph $M$ of $\varphi$:

$$M := \{(t, \varphi(t)) \ : \ t \in (-\epsilon, \epsilon)\},$$

which is a 1-dimensional submanifold of $(-\epsilon, \epsilon) \times U$. Let us introduce the 1-forms $\omega^i$ defined by

$$\omega^i := f^i(p)dt - dp^i, \quad 1 \leq i \leq n. \tag{6.16}$$

Clearly, $\varphi$ solves the differential equation (6.15) if and only if the $\omega^i$ all vanish on $M$. More precisely, substituting $p^i = \varphi^i(t)$ into formula (6.16) yields the pullbacks:

$$\varphi^* \omega_i = \left[ f^i(\varphi(t)) - \frac{d\varphi^i}{dt}(t) \right] dt,$$

which vanish if and only if $\varphi$ solves the differential equation (6.15). So the Cauchy–Lipschitz theorem tells us how to find a 1-dimensional submanifold of $\mathbb{R} \times \mathbb{R}^n$ on which certain 1-forms vanish.

**First-Order Partial Differential Equations** Consider the following partial differential equations of order one, with unknown function $u$ of variables $x_1, \cdots, x_n$:

$$F\left(x_1, \ldots, x_n, \frac{\partial u}{\partial x_1}, \ldots, \frac{\partial u}{\partial x_n}\right) = 0, \tag{6.17}$$

where $F$ is a $C^\infty$ function defined on an open subset $U$ of $\mathbb{R}^{2n+1}$. We denote by $M_0$ the submanifold of $U$ defined by

$$M_0 := \{(x_1, \cdots, x_n, u, p_1, \cdots, p_n) \in U \ : \ F(x_1, \cdots, x_n, u, p_1, \cdots, p_n) = 0\}. \tag{6.18}$$

Let us associate to Eq. (6.17) the exterior differential system generated by the 0-form $F$, the 1-forms

$$dF = \sum_{i=1}^{n} \frac{\partial F}{\partial x_i} dx_i + \frac{\partial F}{\partial u} du + \sum_{i=1}^{n} \frac{\partial F}{\partial p_i} dp_i \tag{6.19}$$

$$\omega = du - \sum_{i=1}^{n} p_i dx_i, \tag{6.20}$$

and the 2-form

$$d\omega = -\sum_{i=1}^{n} dp_i \wedge dx_i. \tag{6.21}$$

Now, the geometric point of view is to find a submanifold $M$ of $\mathbb{R}^{2n+1}$ of dimension $n$ for which all the above forms vanish.

### 6.3.2 *The General Problem*

The Cauchy–Lipschitz theorem deals with 1-forms, while the first-order partial differential equations deal with 0-form, 1-forms, and 2-form. By extension, the general problem can formally be stated as follows.

**Definition 6.3.4** Let $\omega^k$, $1 \le k \le K$, be differential forms on an open subset of $\mathbb{R}^n$, and $M \subset \mathbb{R}^n$ a submanifold. $M$ is called an *integral submanifold* of the exterior differential system:

$$\omega^1 = 0, \dots, \omega^K = 0 \tag{6.22}$$

if the pullbacks of the $\omega^k$ to $M$ all vanish, that is, if

$$\omega^k(p)(\xi^1, \cdots, \xi^{d_k}) = 0, \ 1 \le k \le K, \tag{6.23}$$

whenever $p \in M$, $\omega^k$ has degree $d_k$, and $\xi^i \in T_pM$ for $1 \le i \le d_k$.

Given $\bar{p} \in \mathbb{R}^n$, the Cartan–Kähler theorem will give necessary and sufficient conditions for the existence of an integral manifold containing $\bar{p}$. Necessary conditions are easy to find. Assume that an integration manifold $M$ containing $\bar{p}$ exists, and let $m$ be its dimension. Then the tangent space at $\bar{p}$, denoted by $T_pM$, is an $m$-dimensional space, and all the $\omega^k(p)$ must vanish on $T_pM$, because of (6.23). Any subspace $E \subset T_pM$ with this property will be called an *integral element* of system (6.22) at $\bar{p}$. The set of all $m$-dimensional integral elements at $\bar{p}$ will be denoted by

$$G^m_{\bar{p}} = \left\{ E \mid \begin{array}{l} E \subset T_{\bar{p}}M \text{ and } \mathbf{dim} E = m, \\ \omega^1(\bar{p}), \dots, \omega^K(\bar{p}) \text{ all vanish on E.} \end{array} \right\}$$

So the first necessary condition follows:

$$G^m_{\bar{p}} \neq \emptyset. \tag{6.24}$$

To get a second one, let us ask a strange question: have we written all the equations? In other words, does the system:

$$\omega^1 = 0, \dots, \omega^K = 0 \tag{6.25}$$

exhibit all the relevant information? The answer may be no. To see this, recall that $M$ is a submanifold of $\mathbb{R}^n$. Denote by $\varphi_M : M \to \mathbb{R}^n$ the standard embedding $\varphi_M(x) = x$ for all $x \in M$. Then $M$ is an integral manifold of the system (6.22) if

$$\varphi_M^* \omega^1 = 0, \dots, \varphi_M^* \omega^K = 0. \tag{6.26}$$

But we know that exterior differentiation is natural with respect to pullbacks, that is, that $d$ commutes with $\varphi_M^*$. So (6.26) implies that

$$\varphi_M^*(d\omega^1) = 0, \dots, \varphi_M^*(d\omega^K) = 0. \tag{6.27}$$

In other words, $M$ is also an integral manifold of the larger system:

$$\begin{cases} \omega^1 = 0, \dots, \omega^K = 0; \\ d\omega^1 = 0, \dots, d\omega^K = 0, \end{cases} \tag{6.28}$$

which is different from (6.22). If integral elements of (6.28) are different from those of (6.22), it is not clear which ones we should be working with. To resolve this quandary, we shall assume that the systems (6.22) and (6.28) have the same integral elements. In other words, the second equations in (6.28) must be algebraic consequences of the first ones. The precise statement for this is as follows:

**Definition 6.3.5** The family $\{\omega^k,\ 1 \le k \le K\}$ is said to generate a differential ideal if there are forms $\{\alpha_j^k,\ 1 \le k, j \le K\}$ such that:

$$\forall\, k, \quad d\omega^k = \sum_{j=1}^{K} \alpha_j^k \wedge \omega^j. \tag{6.29}$$

Our second necessary condition is that the $\omega^k$, $1 \le k \le K$, must generate a differential ideal. If this is the case, we say that the exterior differential system is closed.

Note that if the given family $\{\omega^k,\ 1 \le k \le K\}$ does not satisfy this condition, the enlarged family $\{\omega^k, d\omega^k,\ 1 \le k \le K\}$ certainly will (because $d(d\omega)) = 0$). So the condition that the system is closed can be understood as saying that the enlargement procedure has already taken place.

Unfortunately, conditions (6.24) and (6.29) are not sufficient. We give two counterexamples to show that an additional condition is needed:

- **A first counterexample.** Consider two functions $f$ and $g$ from $\mathbb{R}^{n-1}$ into itself, with $f(0) = g(0) = v \neq 0$ and $f(p) \neq g(p)$ for $p \neq 0$. Define $\alpha^k$ and $\beta^k$, $1 \le k \le n-1$, by

$$\alpha^k(p,t) = f^k(p)dt - dp^k;$$
$$\beta^k(p,t) = g^k(p)dt - dp^k$$

and consider the exterior differential system in $\mathbb{R}^n$:

$$\alpha^k = 0, \quad 1 \leq k \leq n-1;$$
$$\beta^k = 0, \quad 1 \leq k \leq n-1.$$

The $\alpha^k$ and $\beta^k$ generate a differential ideal, and there is an integral element at 0, namely the line carried by $(1, v)$, so $G^1(0) \neq \emptyset$. However, finding an integral manifold of the initial system containing 0 amounts to finding a common solution of the two Cauchy problems:

$$\frac{dp}{dt} = f(p), \quad p(0) = 0, \tag{6.30}$$

$$\frac{dp}{dt} = g(p), \quad p(0) = 0, \tag{6.31}$$

which does not exist in general. The problem, clearly, is that the equality $f(p) = g(p)$ holds only at $p = 0$. So we need a regularity condition that will exclude such situations, which guarantees that the required equality holds true at ordinary, a concept we now formally define.

- **A second counterexample.** Let us work in $\mathbb{R}^2$, and let us find all functions $f = f(x, y)$ that can be written as

$$f(x, y) = u(x) + v(y). \tag{6.32}$$

It is well known that a necessary and sufficient condition for such a decomposition to be possible, at least for smooth function $f$, is that the cross derivative vanishes:

$$\frac{\partial^2 f}{\partial x \partial y} \equiv 0. \tag{6.33}$$

Consider the exterior differential system in $\mathbb{R}^4 = (x, y, u, v)$

$$\begin{cases} du + dv - \dfrac{\partial f}{\partial x} dx - \dfrac{\partial f}{\partial y} dy = 0, \\ du \wedge dx \qquad\qquad\qquad\quad = 0, \\ dv \wedge dy \qquad\qquad\qquad\quad = 0. \end{cases} \tag{6.34}$$

Any 2-dimensional integral submanifold $M$ of this system will be the graph of a pair of functions $(u, v)$ that solve the problem, provided only that it is not vertical, that is, that neither $dx$ nor $dy$ vanishes on $M$. Let us try to find such an integral submanifold. The system is obviously closed. We then look for non-vertical integral elements, at $(x, y, u, v) \in \mathbb{R}^4$ say. They are defined by a set of linear equations:

$$du = A_1 dx + B_1 dy,$$
$$dv = A_2 dx + B_2 dy.$$

Plugging into the system, we get

$$B_1 = 0,\ A_2 = 0,\ A_1 = \frac{\partial f}{\partial x}(x, y),\ B_2 = \frac{\partial f}{\partial y}(x, y).$$

So there is an integral element. However, there is no 2-dimensional integral submanifold, unless (6.33) is satisfied.

### 6.3.3 The Regularity Condition

- If all the $\omega^k$ are 1-forms, the regularity condition is clear enough: *the dimension of the space spanned by the $\omega^k(p)$ should be constant on a neighbourhood of $\bar{p}$* (which is not the case in the first counterexample).
- If some of the $\omega^k$ have higher degree, the regularity condition is more complicated. It is expressed as follows. Let $\bar{p} \in \mathbb{R}^n$; from now, we work on the tangent space $V := T_p\mathbb{R}^n$. Let $E \subset V$ be an $m$-dimensional integral element at $\bar{p}$. Let $\bar{\alpha}_1, \cdots, \bar{\alpha}_n$ be a basis of the dual $V^*$ such that

$$E = \{\zeta \in V \mid\ <\zeta, \bar{\alpha}_i>= 0\ \forall\, i \geq m+1\}.$$

For $n' \leq n$, denote by $I(n', d)$ the set of all ordered subsets of $\{1, \cdots, n'\}$ with $d$ elements. Denote by $d_k$, the degree of $\omega^k$. For every $k$, writing $\omega^k(p)$ in the $\bar{\alpha}_i$ basis, we get

$$\omega_k(\bar{p}) = \sum_{I \in I(n,d_k)} C_I^k \bar{\alpha}_{i_1} \wedge \ldots \wedge \bar{\alpha}_{i_{d_k}}. \tag{6.35}$$

In this summation, it is understood that $I = \{i_1, \cdots, i_{d_k}\}$. Since $\omega^k(\bar{p})$ vanishes on $E$, each monomial must contain some $\bar{\alpha}_i$ with $i \geq m+1$. Let us single out the monomials containing one such term only. Regrouping and writing, we get

$$\omega_k(\bar{p}) = \sum_{J \in I(m,d_k-1)} \bar{\beta}_J^k \wedge \bar{\alpha}_{j_1} \wedge \ldots \wedge \bar{\alpha}_{j_{d_k-1}} + R, \tag{6.36}$$

where $\beta_J^k$ is a linear combination of the $\bar{\alpha}_i$ for $i \geq m+1$, and all the monomials in the remainder $R$ contain $\bar{\alpha}_i \wedge \bar{\alpha}_j$ for some $i > j \geq m+1$. Define an increasing sequence of linear subspaces $H_0^* \subset H_1^* \subset \ldots \subset H_m^* \subset V^*$ of $V^*$ by

$$H_m^* = \text{span}\{ \bar{\beta}_J^k \,|1 \le k \le K, \; J \in \mathcal{I}(m, d_k - 1)\};$$
$$H_{m-1}^* = \text{span}\{ \bar{\beta}_J^k \,|1 \le k \le K, \; J \in \mathcal{I}(m-1, d_k - 1)\};$$
$$H_0^* = \text{span}\{ \bar{\beta}_J^k \,|1 \le k \le K, \; J \in \mathcal{I}(0, d_k - 1)\}.$$

The later subspace is just the linear space generated by those of the $\omega^k(\bar{p})$ that happen to be 1-forms. We define an increasing sequence of integers $0 \le c_0(\bar{p}, E) \le \ldots \le c_m(\bar{p}, E) \le n$ by

$$c_i(\bar{p}, E) = \mathbf{dim} H_i^*, \quad \forall i = 1, \ldots, m.$$

We are finally able to express *Cartan's regularity condition*. Denote by $\P^m(\mathbb{R}^n)$ the set of all $m$-dimensional subspaces of $\mathbb{R}^n$ with the standard (Grassmannian) topology. It is known to be a manifold of dimension $m(n-m)$. Denote by $G^m$ the set of all $(p, E)$ such that $E$ is an $m$-dimensional integral element at $p$. Note that $G^m$ is a subset of $\mathbb{R}^n \times \P^m(\mathbb{R}^n)$.

**Definition 6.3.6** Let $(\bar{p}, \bar{E}) \in G^m$. We say that $(\bar{p}, \bar{E})$ is *ordinary* if there is some neighbourhood $U$ of $(\bar{p}, \bar{E})$ in $\mathbb{R}^n \times \P^m(\mathbb{R}^n)$ such that $G^m \cap U$ is a submanifold of co-dimension

$$c_0(\bar{p}, \bar{E}) + \ldots + c_{m-1}(\bar{p}, \bar{E}).$$

If all the $\omega^k$ are 1-form, denote by $d(p)$ the dimension of the space spanned by the $\omega^k(p)$. Then $c_i(p, E) = d(p)$ for every $i$, and $(\bar{p}, \bar{E})$ is ordinary if $G^m \cap U$ is a submanifold of co-dimension $md(\bar{p})$ in $\mathbb{R}^n \times \P^m(\mathbb{R}^n)$. This implies that, for every $p$ in a neighbourhood of $\bar{p}$, the set of $E \in G^m(p)$ ($m$-dimensional integral element at $p$) has co-dimension $md(\bar{p})$ in $\mathbb{R}^n \times \P^m(\mathbb{R}^n)$. It can be seen directly to have co-dimension $md(p)$. So $d(p) = d(\bar{p})$ in a neighbourhood of $\bar{p}$. This is exactly the regularity condition we would like to have for 1-forms.

In the general case, if $(\bar{p}, \bar{E})$ is ordinary, the numbers $c_i$ will also be locally constant on a neighbourhood of $U$ of $(\bar{p}, \bar{E})$, that is,

$$c_i(p, E) = c_i(\bar{p}, \bar{E}) = c_i \;\; \forall \; p \in U.$$

The nonnegative numbers

$$\begin{aligned} s_0 &= c_0 \\ s_i &= c_i - c_{i-1}, \quad 1 \le i < m \\ s_m &= n - m - c_{m-1} \end{aligned}$$

are called the *Cartan characters*. We shall use them later.

### 6.3.4 The Main Theorem

We are now in a position to state the Cartan–Kähler theorem. Recall that a real-valued function on $\mathbb{R}^n$ is called *analytic* if its Taylor series at every point is absolutely convergent.

**Theorem 6.3.4 (Cartan–Kähler)** *Consider the exterior differential system:*

$$\omega^k = 0, \quad 1 \leq k \leq K. \tag{6.37}$$

*Assume that the $\omega^k$ are real analytic and that they generate a differential ideal. Let $\bar{p}$ be a point and $\bar{E}$ be an integral element at $\bar{p}$ such that $(\bar{p}, \bar{E})$ is ordinary. Then there is a real analytic integral manifold $M$, containing $\bar{p}$ such that*

$$T_{\bar{p}}M = \bar{E}. \tag{6.38}$$

*Remark 6.3.1* Nothing should come as a surprise in this statement, except the real analyticity. It comes from the generality of the Cartan–Kähler theorem. Indeed, every system of partial differential equations, linear or not, can be written as an exterior differential system, and there is a famous example, due to Hans Lewy, of a system of two first-order nonhomogeneous linear partial differential equations (with nonconstant coefficients) for two unknown functions , which has no solution if the right-hand side is $C^\infty$ but not analytic.

Let us mention the question of uniqueness. There is no uniqueness in the Caratn–Kähler theorem: there may be infinitely many analytic integral manifolds going through the point $\bar{p}$ and having $E$ as a tangent space at $\bar{p}$. However, the theorem describes in a precise way the set

$$T_U = \left\{ M \mid \begin{array}{l} \text{M is an integral manifold and there exists} \\ (p, E) \in U \text{ such that } p \in M \text{ and } T_pM = E \end{array} \right\},$$

where $U$ is a suitable chosen neighbourhood of $(\bar{p}, \bar{E})$. Each $M$ in $T_U$ is completely determined by the (arbitrary) choice of $s_m$ analytic functions of $m$ variables, the $s_m$ being the Cartan character. □

Let us illustrate the Cartan–Kähler theorem with an example.

### 6.3.5 An Example

Let us go back to the second counterexample. There is only one integral element at every point $a \in \mathbb{R}^4$, so $G_a^2$ is a point in $\P^2(\mathbb{R}^4)$ that has dimension 4. So its co-dimension is 4. Let us compute the Cartan characters. For this, let $\bar{E}$ be the integral element at a point $\bar{a} = (\bar{x}, \bar{y}, \bar{u}, \bar{v}) \in \mathbb{R}^4$, and define the 1-forms $\bar{\alpha}_1, \cdots, \bar{\alpha}_4$ as follows:

$$\bar{\alpha}_1 = dx,$$
$$\bar{\alpha}_2 = dy,$$
$$\bar{\alpha}_3 = du - A_1 dx,$$
$$\bar{\alpha}_4 = dv - A_2 dy.$$

These 1-forms define a basis of the tangent space of $(T_{\bar{a}}\mathbb{R}^4)^*$. Furthermore, we have that

$$\bar{E} = \{\xi \in \mathbb{R}^4 \mid \langle \xi, \bar{\alpha}_i \rangle = 0,\ i = 3, 4\}.$$

Plugging the $\bar{\alpha}_i$ in the system (6.34), it follows that

$$\omega_1 = \bar{\alpha}_3 + \bar{\alpha}_4,$$
$$\omega_2 = \bar{\alpha}_3 \wedge \bar{\alpha}_1,$$
$$\omega_3 = \bar{\alpha}_4 \wedge \bar{\alpha}_2.$$

So

$$H_0^* = \text{span}\{\bar{\alpha}_3 + \bar{\alpha}_4\},\ \ c_0(\bar{a}, \bar{E}) = 1,$$
$$H_1^* = \{0\}\text{span}\{\bar{\alpha}_3, \bar{\alpha}_4\},\ \ c_1(\bar{a}, \bar{E}) = 2.$$

Hence, $c_0 + c_1 = 1 + 2 = 3 \neq 4$, which is the co-dimension of $G^2$. Therefore, $(\bar{a}, \bar{E})$ is not ordinary.

## 6.4 Main Result: Douglas Problem

We now come back to Douglas' problem described in the introduction. To apply our approach to this problem, let us take $n = 1$. So, given an analytic function $F : \mathbb{R} \times \mathbb{R} \to \mathbb{R}$, we are looking for a Lagrangian $L : \mathbb{R} \times \mathbb{R} \to \mathbb{R}$ of class $C^2$ such that the solutions of the second-order differential equation:

$$\frac{d^2x}{dt^2} = F\left(x, \frac{dx}{dt}\right),\ \ t \in (-\epsilon, \epsilon) \tag{6.39}$$

correspond to those of the Euler–Lagrange equation corresponding to $L$:

$$\frac{d}{dt}\frac{\partial L}{\partial y}\left(x, \frac{dx}{dt}\right) - \frac{\partial L}{\partial x}\left(x, \frac{dx}{dt}\right) = 0\ t \in (-\epsilon, \epsilon), \tag{6.40}$$

for some positive real number $\epsilon$. In other words, given an analytic function $F : \mathbb{R} \times \mathbb{R} \to \mathbb{R}$, can we find $L : \mathbb{R}^2 \times \mathbb{R} \to \mathbb{R}$ such that $L$ solves the following partial differential equation:

$$\frac{\partial^2 L}{\partial x \partial y}(x, y)y + \frac{\partial^2 L}{\partial y^2}(x, y)F(x, y) - \frac{\partial L}{\partial x}(x, y) = 0? \tag{6.41}$$

We now describe the basic strategy used throughout the proof. Consider the space

$$E = \{(x, y, u, v, q, r)\} = \mathbb{R}^6,$$

where $u, v, q,$ and $r$ will later be interpreted as $\frac{\partial L}{\partial x}, \frac{\partial L}{\partial y}, \frac{\partial^2 L}{\partial y^2},$ and $\frac{\partial^2 L}{\partial x \partial y}$, respectively.

*Remark 6.4.2* Clearly, if a solution exists, then the system

$$\begin{cases} \frac{\partial L}{\partial x}(x, y) = u; \\ \frac{\partial L}{\partial y}(x, y) = v; \\ \frac{\partial^2 L}{\partial y^2}(x, y) = q; \\ \frac{\partial^2 L}{\partial x \partial y}(x, y) = r \end{cases} \tag{6.42}$$

defines a 2-dimensional manifold $S$ in $E$ included in the 5-dimensional manifold $M$ defined by

$$ry + F(x, y)q - u = 0. \tag{6.43}$$

Conversely, assume that we have found the functions $u = u(x, y)$, $v = v(x, y)$, $q = q(x, y)$, and $r = r(x, y)$ such that:

- For every $(x, y)$ holds $(x, y, u(x, y), v(x, y), q(x, y), r(x, y)) \in M$.
- $d(udx + vdy) = du \wedge dx + dv \wedge dy = 0$.
- $dv - rdx - qdy = 0$. □

Then by Theorem 6.2.2, there exists the function $L = L(x, y)$ of class $C^2$ such that

$$u = \frac{\partial L}{\partial x}(x, y), \quad v = \frac{\partial L}{\partial y}(x, y), \quad q = \frac{\partial^2 L}{\partial y^2}(x, y), \quad r = \frac{\partial^2 L}{\partial x \partial y}(x, y)$$

and

$$\frac{\partial^2 L}{\partial x \partial y}(x, y)y + \frac{\partial^2 L}{\partial y^2}(x, y)F(x, y) - \frac{\partial L}{\partial x}(x, y) = 0.$$

In the language of Sect. 6.3, we are looking for a 2-dimensional integral submanifold in $M$ of the exteriors differential system:

$$\begin{cases} \omega_1 := du \wedge dx + dv \wedge dy = 0, \\ \omega_2 := dv - rdx - qdy = 0, \\ \omega_3 := dr \wedge dx + dq \wedge dy = 0. \end{cases} \tag{6.44}$$

Finally, the solution must be parametrized by $(x, y)$. The formal translation of that is

$$dx \wedge dy \neq 0. \tag{6.45}$$

From Remark 6.4.2, we have the following result:

**Lemma 6.4.1** *Any integral manifold of this system is the graph of a map:*

$$(x, y) \mapsto (u, v, q, r),$$

*where the functions $u, v, q$, and $r$ satisfy Eq.* (6.43).

We now prove the following theorems.

**Theorem 6.4.5** *Assume that the function $F$ is analytic. Let $(\bar{x}, \bar{y}, \bar{u}, \bar{v}, \bar{q}, \bar{r})$ be a point of $M$ such that $F(\bar{x}, \bar{y}) \neq 0$. Then there exists a real analytic 2-dimensional integral manifold $N$ containing the point $(\bar{x}, \bar{y})$.*

***Proof*** It is obvious from the system (6.44) that the differential forms $\omega_i$, $i = 1, 2, 3$ generate a differential ideal. The proof is in two steps:

**Step 1**: *Finding integral elements.* We linearize $u, v, q$, and $r$ (as functions of $(x, y)$) around $(\bar{x}, \bar{y})$ by setting:

$$\begin{aligned} dv &= V_1 dx + V_2 dy, \\ dq &= Q_1 dx + Q_2 dy, \\ dr &= R_1 dx + R_2 dy. \end{aligned}$$

Solving the linearized system is equivalent to finding all the $V_i, Q_i, R_i$, $i = 1, 2$, that satisfy the system (6.44) and (6.45), plus Eq. (6.43) expressing that $(x, y, u, v, q, r)$ remains on the manifold $M$.

Substituting $du, dv, dq$, and $dr$ in the system (6.44) and differentiating (6.43), we get

$$\begin{cases} \bar{y}R_2 + F(\bar{x},\bar{y})Q_2 = -\bar{q}\dfrac{\partial F}{\partial y}(\bar{x},\bar{y}), \\ V_1 \qquad\qquad\qquad = \bar{r}, \\ V_2 \qquad\qquad\qquad = \bar{q}, \\ Q_1 - R_2 \qquad\quad\; = 0. \end{cases} \tag{6.46}$$

Note that all these equations are linearly independent. So, the set of integral elements has co-dimension 4 in the Grassmannian.

**Step 2**: *Cartan's test.* Set

$$\begin{aligned} \bar{\alpha}_1 &= dx, \\ \bar{\alpha}_2 &= dy, \\ \bar{\alpha}_3 &= dv - V_1 dx - V_2 dy, \\ \bar{\alpha}_4 &= dq - Q_1 dx - Q_2 dy, \\ \bar{\alpha}_5 &= dr - R_1 dx - R_2 dy, \end{aligned}$$

where the $V_i$, $Q_i$, and $R_i$ satisfy (6.46). Note that because of (6.43), (6.44) and relations of (6.46), we have

$$\begin{aligned} \omega_1 &= [\bar{y}\bar{\alpha}_5 + F(\bar{x},\bar{y})\bar{\alpha}_4] \wedge \bar{\alpha}_1 + \bar{\alpha}_3 \wedge \bar{\alpha}_2, \\ \omega_2 &= \bar{\alpha}_3, \\ \omega_3 &= \bar{\alpha}_5 \wedge \bar{\alpha}_1 + \bar{\alpha}_4 \wedge \bar{\alpha}_2. \end{aligned}$$

We then apply the Cartan procedure, as described in Sect. 6.3. We have

$$\begin{aligned} H_0^* &= \text{Span}\,\{\bar{\alpha}_3\}, \\ H_1^* &= \text{Span}\,\{\bar{\alpha}_3, \bar{y}\bar{\alpha}_5 + F(\bar{x},\bar{y})\bar{\alpha}_4, \bar{\alpha}_5\}. \end{aligned}$$

Hence, $c_0 = 1$, $c_1 = 3$. So $C := c_0 + c_1 = 1 + 3 = 4$. Which is exactly the co-dimension of the set of integral elements in the Grassmannian $G^2(M)$. There the exterior differential system (6.44) passes the Cartan test. So the conclusion follows from Cartan–Kähler theorem. □

**Theorem 6.4.6** *Let $F : \mathbb{R} \times \mathbb{R} \to \mathbb{R}$ be a real analytic function. Let $(\bar{x},\bar{y}) \in \mathbb{R}^2$ such that $F(\bar{x},\bar{y}) \neq 0$. Then, there exist an open subset of $\mathbb{R}^2$ containing $(\bar{x},\bar{y})$ and a real analytic function $L : U \to \mathbb{R}$ such that*

$$\frac{\partial^2 L}{\partial x \partial y}(x,y)y + \frac{\partial^2 L}{\partial y^2}(x,y)F(x,y) - \frac{\partial L}{\partial x}(x,y) = 0 \;\forall (x,y) \in U. \tag{6.47}$$

***Proof*** The proof follows from Theorem 6.4.5 and Lemma 6.4.1. □

**Corollary 6.4.2** *Let $F : \mathbb{R} \times \mathbb{R} \to \mathbb{R}$ be a real analytic function and $(\bar{x}, \bar{y}) \in \mathbb{R}^2$ such that $F(\bar{x}, \bar{y}) \neq 0$. Then, there exists a real analytic function $L : \mathbb{R} \times \mathbb{R} \to \mathbb{R}$, a positive real number $\epsilon$ such that any solution of the second-order differential equation:*

$$\frac{d^2x}{dt^2} = F\left(x, \frac{dx}{dt}\right), \ t \in (-\epsilon, \epsilon), \tag{6.48}$$

*corresponds to that of the Euler–Lagrange equation corresponding to $L$:*

$$\frac{d}{dt}\frac{\partial L}{\partial y}\left(x, \frac{dx}{dt}\right) - \frac{\partial L}{\partial x}\left(x, \frac{dx}{dt}\right) = 0 \ t \in (-\epsilon, \epsilon). \tag{6.49}$$

*Furthermore, if $\frac{\partial^2 F}{\partial y^2}(x, y) \neq 0$ for all $(x, y) \in U$, then Eqs. (6.48) and (6.49) have the same solutions.* □

***Proof*** The proof follows from Theorem 6.4.6 and the chain rule for derivation. □

**Acknowledgments** The authors thank the referees for their work and their valuable suggestions that helped to improve the presentation of this paper.

## References

1. R.L. Bryant, S.S. Chern, R.B. Gardner, H.L. Goldschmidt, P.A. Griffiths, *Exterior Differential Systems* (Springer, New York, 1991)
2. E. Cartan, *Les systèmes différentiels extérieurs et leurs applications géométriques* (Hermann, 1945)
3. J. Douglas, Solution of the inverse problem of the calculus of variations. Trans. Am. Math. Soc. **50**, 71–128 (1941)
4. P.A. Chiappori, I. Ekeland, *Problèmes d'agrègation en théorie du consommateur et calcul différentiel extérieur*, vol. 323 (CRAS Paris, 1996), pp. 565–570
5. P.A. Chiappori, I. Ekeland, Restrictions on the equilibrium manifold: the differential approach (to appear)
6. P.A. Chiappori, I. Ekeland, Aggregation and market demand: an exterior differential calculus viewpoint. Econometrica **67**(6), 1435–1458 (1999)
7. P.A. Chiappori, I. Ekeland, *Disaggregation of Collective Excess Demand Functions in Incomplete Markets* (mimeo 1997)
8. P.A. Chiappori, I. Ekeland, *A Convex Darboux Theorem*. Annali della Scuola Normale di Pisa, Cl. Sci (4), XXV (1997), pp. 287–297
9. I. Ekeland, N. Djitte, An inverse problem in the economic theory of demand. Ann. I.H. Poincaré - AN **23**, 269–281 (2006)
10. I. Ekeland, N. Djitte, Inverse problem in calculus of variations (submitted)
11. I. Ekeland, L. Nirenberg, A convex Darboux theorem. Methods Appl. Anal. **9**(3), 329–344 (2002)

# Chapter 7
# Existence of Local and Maximal Mild Solutions for Some Non-autonomous Functional Differential Equation with Finite Delay

**Khalil Ezzinbi, Bila Adolphe Kyelem, and Stanislas Ouaro**

**Abstract** This chapter is devoted to the study of the existence results of local and maximal solutions on the one hand and the existence and uniqueness results of mild solutions on the second hand, for the non-autonomous evolution equation with finite delay $\frac{d}{dt}u(t) = A(t)u(t) + f(t, u_t), \quad t \in [0, T]$, subjected to the initial datum $u_0 = \phi$, where $T > 0$ is some positive constant. The unbounded operators associated to the non-autonomous system are assumed to be stable family that generates $C_0$-semigroups, while the nonlinear part is supposed to be continuous. Using some boundedness assumptions on the delayed nonlinear continuous part, we prove the local existence of solution that blows up at the finite time. Under some Lipschitz condition on the nonlinear term, we establish the existence and uniqueness of mild solution. Finally, an example of reaction–diffusion non-autonomous partial functional differential equations is used to illustrate our theoretical obtained results.

**Keywords** Non-autonomous equation · Evolution system · Delayed differential equations · Local solution · Mild solution · Maximal solution

K. Ezzinbi (✉)
Faculté des Sciences Semlalia, Département de Mathématiques, Université Cadi Ayyad, Marrakech, Marocco
e-mail: ezzinbi@ucam.ac.ma

B. A. Kyelem
Département de Mathématiques et Informatique, Université de Ouahigouya, Unité de Formation et de Recherche en Sciences et Technologies, Ouahigouya, Burkina Faso

S. Ouaro
Département de Mathématiques, Université Joseph KI-ZERBO, Unité de Formation et de Recherche en Sciences Exactes et Appliquées, Ouagadougou, Burkina Faso



T. Diagana et al. (eds.), *Partial Differential Equations and Applications*,
Springer Proceedings in Mathematics & Statistics 420,
https://doi.org/10.1007/978-3-031-27661-3_7

## 7.1 Introduction

The main purpose of this chapter is to outline the existence results of local and maximal solutions for the following class of non-autonomous partial functional differential equations with finite delay

$$\begin{cases} \dfrac{d}{dt}u(t) = A(t)u(t) + f(t, u_t), & t \in [0, T] \\ \\ u(\theta) = \varphi(\theta), \quad \theta \in [-r, 0], \end{cases} \tag{7.1}$$

where $f : \mathbb{R}^+ \times C \to X$ is a continuous function with value in the Banach space denoted by $X$. $A(t) : D(A(t)) \subseteq X \to X$ is a closed linear operator and generates a $C_0$-semigroup. $C$ is the space of continuous functions from $[-r, 0]$ to $X$, which will be defined later. Also, we denote by $u_t$ for $t \in [0, T]$, the historic function defined on $[-r, 0]$ by

$$u_t(\theta) = u(t + \theta) \text{ for } \theta \in [-r, 0],$$

where $u$ is a function from $[-r, T]$ into $X$.

Following the work done in [17], we make some sufficient conditions on the following family $\{-A(t) : 0 \le t \le T\}$ of closed linear operators to obtain the existence of the associated evolution system $\{U(t, s) : 0 \le s \le t \le T\}$, which is used to express the mild solution of (7.1). Observe that the evolution system introduced for the first time in 1974 by Howland in [13] remains an important tool in the study of the quantitative and qualitative results for some non-autonomous evolution equations.

For more information related to some non-autonomous evolution equations, one can see the book of Friedman in [8] in which he imposed that the family $\{A(t) : 0 \le t \le T\}$ of linear operators verifies the following conditions:

**($\mathbf{B}_1$)** The domain $D(A(t))$ of the closed linear operator $A(t)$ is dense in $X$ and is also independent of $t \in [0, T]$.

**($\mathbf{B}_2$)** For each $t \in [0, T]$, the resolvent $R(\lambda, A(t))$ exists for all $\lambda$ with $Re\lambda \le 0$, and there exists $K > 0$ such that $\|R(\lambda, A(t))\| \le \dfrac{K}{(|\lambda| + 1)}$.

**($\mathbf{B}_3$)** There exists $0 < \delta \le 1$ and $K > 0$ such that $\|(A(t) - A(s))A^{-1}(r)\| \le K|t - s|^\delta$ for all $t, s, r \in [0, T]$.

Under those assumptions, he showed that the family $\{A(t) : 0 \le t \le T\}$ generates a unique linear evolution system $\{U(t, s) : 0 \le s \le t \le T\}$. Moreover, there exists a family of bounded linear operators $\{R(t, \mu) : 0 \le \mu \le t \le T\}$ with $\|R(t, \mu)\| \le K|t - \mu|^{\delta - 1}$ such that $U(t, s)$ has the following representation:

$$U(t,s) = e^{-(t-s)A(t)} + \int_s^t e^{-(t-\xi)A(\xi)} R(\xi,s)d\xi,$$

where $e^{-\tau A(t)}$ denotes the analytic semigroup having the infinitesimal generator $-A(t)$.

Observe that the study of the problem (7.1) has many interactions in applied sciences. The delayed non-autonomous models are naturally appeared in many branches of biological modelling for the first time. For details, they have been used to describe the quantitative and qualitative behaviours of dynamic infection diseases such primary infection, drug therapy, and immune response. For more information, the reader can see [4, 16] and the related references therein. The delayed terms can also be seen in the study of chemosynthesis models, circadian rhythms, epidemiology, the respiratory system, tumour growth, and statistical analysis of ecological data of many biological species. For more lightening related to those cases, we direct the reader to the works done in [3, 5, 18, 19, 22].

Also, one can mention that the study of non-autonomous abstract evolution equations was the subject of many works, and among others, we cite explicitly [2, 7, 10, 12, 17]. Besides, in the autonomous case where $A(t) = A$, the problem (7.1) has been the subject of various quantitative and qualitative studies (see [11, 21]).

In the similar setting, Acquistapace and Terreni in [1] proved some regularity results for the following non-autonomous evolution equation without delay in a Banach space $E$, under the so-called classical Kato–Tanabe assumptions:

$$\begin{cases} u'(t) - A(t)u(t) = f(t), & t \in [0,T] \\ u(0) = x \\ x \in E,\ f \in C([0,T];E) & \text{prescribed}, \end{cases} \tag{7.2}$$

where for each $t \in [0,T]$, the operator $A(t)$ is assumed to generate an analytic semigroup on $E$, and the domain $D(A(t))$ of $A(t)$ varies with $t \in [0,T]$ and is not necessarily dense in $E$. It is important to note that the case of variable domains was first studied by Kato in [15].

As we are concerned, it is to assume that the domain $D(A(t))$ of the operator $A(t)$ is time-independent, and for each $t \in [0,T]$, the operator $A(t)$ generates only a $C_0$-semigroup not necessarily an analytic semigroup. Consequently, it is worth in our case to work following the arguments developed by Pazy in [17] to study some non-autonomous evolution equations in the hyperbolic case.

This chapter is organized as follows: in Sect. 7.2, we recall some preliminaries that will play an important role in the study of the problems such as (7.1). In Sect. 7.3, the local and maximal existence of mild solution of Eq. (7.1) is proved. The last section is devoted to applying our theoretical results to the study of some example of non-autonomous partial functional differential equations of the form (7.1).

## 7.2 Preliminary

Let us denote by $(X, \|.\|)$ the Banach space $X$ endowed with the norm $\|.\|$. For the convenience, we assume that there exists a Banach space $Y$ densely and continuously embedded in $X$. The space $C = C([-r, 0]; X)$ endowed with the uniform norm topology

$$\|\phi\|_C = \sup_{-r\leq\theta\leq 0} \|\phi(\theta)\|$$

is a Banach space.

We also make the following definitions given in [17] that will be used in this chapter.

**Definition 7.2.1** Let $\{S(t)\}_{t\geq 0}$ be a $C_0$-semigroup, and let $A$ be its infinitesimal generator. A subspace $Y$ of $X$ is said to be $A$-admissible if it is an invariant subspace of $\{S(t)\}_{t\geq 0}$, and the restriction of $\{S(t)\}_{t\geq 0}$ to $Y$ is a $C_0$-semigroup in $Y$ (i.e., it is strongly continuous in the norm $\|.\|_Y$). □

**Definition 7.2.2** Let $X$ be a Banach space. A family $\{A(t)\}_{t\in[0,T]}$ of infinitesimal generators of $C_0$-semigroups on $X$ is said to be stable if there are constants $M \geq 1$ and $\omega$ (called the stability constants) such that

$$(\omega, +\infty[ \subset \rho(A(t)) \quad \text{for} \quad t \in [0, T] \tag{7.3}$$

and

$$\left\| \prod_{j=1}^{k} R(\lambda; A(t_j)) \right\| \leq M(\lambda - \omega)^{-k} \quad \text{for} \quad \lambda > \omega \tag{7.4}$$

and any sequence $0 \leq t_1 \leq t_2 \leq \cdots \leq t_k \leq T, k = 1, 2, \cdots$

Here, $\rho(A(t))$ is the resolvent set of the operator $A(t)$, and $R(\lambda; A(t))$ defines the resolvent operator associated to $A(t)$ at the point $\lambda$. □

Observe that the stability of a family $\{A(t)\}_{t\in[0,T]}$ of infinitesimal generators of $C_0$-semigroups on $X$ is preserved when we replace the norm in $X$ by an equivalent norm.

In [17], the existence and uniqueness of evolution system associated to the family of the unbounded operators $\{A(t)\}_{t\in[0,T]}$ are obtained under the following assumptions:

**(H$_1$)** $\{A(t)\}_{t\in[0,T]}$ is a stable family with the stability constants $M, \omega$.

**(H$_2$)** $Y \subset X$ is $A(t)$-admissible for $t \in [0, T]$, and the family $\{\tilde{A}(t)\}_{t\in[0,T]}$ of parts $\tilde{A}(t)$ of $A(t)$ in $Y$ is a stable family in $Y$ with the stability constants $\tilde{M}, \tilde{\omega}$.

**(H$_3$)** For $t \in [0, T]$, $Y \subset D(A(t))$, $A(t)$ is a bounded operator from $Y$ into $X$ and $t \mapsto A(t)$ is continuous in the space of bounded linear operators from $Y$ into $X$ denoted by $\mathcal{L}(Y, X)$ equipped with the uniform norm topology $\|.\|_{\mathcal{L}(Y,X)}$.

**Proposition 7.2.1 ([17])** *Let $\{A(t)\}_{t\in[0,T]}$ be the infinitesimal generator of a $C_0$-semigroup $\{S_t(s)\}_{s\geq 0}$ on $X$. If the family $\{A(t)\}_{t\in[0,T]}$ satisfies the conditions* **($H_1$)**, **($H_2$)**, *and* **($H_3$)**, *then there exists a unique evolution system $\{U(t,s): \ 0\leq s\leq t\leq T\}$ in $X$ verifying:*

**($E_1$)** $\|U(t,s)\|\leq Me^{\omega(t-s)}$ *for* $0\leq s\leq t\leq T$;

**($E_2$)** $\frac{\partial^+}{\partial t}U(t,s)v\mid_{t=s}= A(s)v$ *for* $v\in Y,\ 0\leq s\leq t\leq T$;

**($E_3$)** $\frac{\partial}{\partial s}U(t,s)v=-U(t,s)A(s)v$ *for* $v\in Y,\ 0\leq s\leq t\leq T$.

**Proposition 7.2.2 ([17])** *The evolution system of linear operator $\{U(t,s): \ 0\leq s \leq t\leq T\}$ generated by the family $\{A(t)\}_{t\in[0,T]}$ satisfies the following properties:*

(*a*) *$U(t,s)\in\mathcal{L}(X)$, the space of bounded linear transformations on $X$, whenever $0\leq s\leq t\leq T$, and for all $x\in X$, the mapping $(t,s)\mapsto U(t,s)x$ is continuous.*
(*b*) *$U(t,s)U(s,\mu)=U(t,\mu)$ for $0\leq\mu\leq s\leq t\leq T$.*
(*c*) *$U(t,t)=I$.*
(*d*) *$\dfrac{\partial}{\partial t}U(t,s)=A(t)U(t,s)$, for $s<t$.*
(*e*) *$\dfrac{\partial}{\partial s}U(t,s)=-U(t,s)A(s)$, for $s<t$.* □

Now, we can give the notion of solutions that will be studied in this chapter.

**Definition 7.2.3** Let $\phi\in C$. A continuous function $u:[-r,T]\to X$ is called a mild solution of Eq. (7.1) associated to $\phi$ if:

$$\begin{cases} u(t)=U(t,s)u(0)+\displaystyle\int_0^t U(t,s)f(s,u_s)ds \quad \text{for } t\in[0,T] \\ \\ u_0=\phi \qquad \text{on} \quad [-r,0]. \end{cases} \tag{7.5}$$

For the study of Eq. (7.1), we will make the following assumptions that give us some sufficient conditions to obtain the local and maximal solutions:

**($C_1$)** The domain $D(A(t))=D$ is independent of $t\in[0,T]$.
In this case, we define on $D$ a norm $\|.\|_Y$ by

$$\|y\|_Y=\|y\|+\|A(0)y\| \quad \text{for all} \quad y\in D=Y. \tag{7.6}$$

Using the closedness of $A(0)$, then $Y=(D,\|.\|_Y)$ is a Banach space.

**($C_2$)** The application $t\mapsto A(t)x$ for all $x\in D$ is continuously differentiable on $\mathbb{R}^+$.

To prove the local and maximal existence of mild solution of (7.1), we need the following compactness hypothesis.

**($C_3$)** The evolution system verifies the following property:

$$U(t,s) \quad \text{is compact for} \quad t>s.$$

*Remark 7.2.1* It is well-known that the operator $A(0) \in \mathcal{L}(Y, X)$. Using the fact that the family $\{A(t)\}_{t\in[0,T]}$ has the common closed domain, the closed graph theorem gives that $A(t) \in \mathcal{L}(Y, X)$. If the application $t \mapsto A(t)x$ is continuously differentiable on $[0, T]$, then this condition leads to $\sup_{t\in[0,T]} \|A(t)\|_{\mathcal{L}(Y,X)} < +\infty$ via the principle of uniform boundedness. □

Now, we are able to make our first result that is the local and maximal existence of mild solutions to the problem (7.1).

## 7.3 Existence of Local and Maximal Mild Solutions

Often in this chapter, $u(., \phi)$ denotes the mild solution associated to the initial data $\phi$, and we simply denote it by $u$ if there is no confusion. Let us give the first existence result.

**Theorem 7.3.1** *Let $\{A(t)\}_{t\in[0,T]}$ be a stable family of infinitesimal generators of $C_0$-semigroups on $X$, and assume that the conditions* ($\mathbf{C}_1$), ($\mathbf{C}_2$), *and* ($\mathbf{C}_3$) *hold. Moreover, suppose that the function $f : [0, T] \times O \to X$ is continuous where $O$ is an open subset of $C$. Then, for all $\phi \in C$, there exists at least a local mild solution $u(., \phi)$ associated to* (7.1). □

***Proof*** The proof will essentially be based on the Schauder's fixed-point theorem. Let us consider $\phi \in O$. Using the fact that $O$ is an open set on $C$ and the continuity of the function $f : [0, a] \times O \to X$ with $a \in (0, T)$, then there exist some positive constants $\gamma_1$ and $\gamma_2$ such that $\overline{B}_{\gamma_1}(\phi) = \left\{\psi \in C : \left\|\phi - \psi\right\|_C \le \gamma_1\right\} \subset O$ and $\|f(t, \psi)\| \le \gamma_2$ for all $(t, \psi) \in [0, \gamma_1] \times \overline{B}_{\gamma_1}(\phi)$. Consider the function $z \in C([-r, a]; X)$ be defined by

$$z(t) = \begin{cases} U(t, 0)\phi(0) & \text{for } t \in [0, a] \\ \phi(t) & \text{for } t \in [-r, 0]. \end{cases}$$

From definition of $z$, it follows that $z_t \in C$.

For some fixed positive constant $\gamma$ with $\gamma < \gamma_1$, one can choose $b_\phi \in (0, \gamma)$ such that

$$M\gamma_2 \int_0^{b_\phi} e^{\omega s} ds \le \gamma$$

and

$$\left\|z_t - \phi\right\|_C \le \gamma_1 - \gamma \quad \text{for all} \quad t \in [0, b_\phi]. \tag{7.7}$$

Let us set

$$\mathcal{K}_\phi = \Big\{ y \in C([-r, b_\phi]; X) : \quad y_0 = \phi \quad \|y_t - \phi\|_C \leq \gamma_1 \quad \text{for all} \quad t \in [0, b_\phi] \Big\}$$

provided with the uniform norm topology. It is clear that the restriction of $z$ on $[-r, b_\phi]$ belongs to $\mathcal{K}_\phi$ and $\mathcal{K}_\phi \neq \emptyset$. Moreover, $\mathcal{K}_\phi$ is closed, bounded, and convex subset of $C([-r, b_\phi]; X)$.

Now, consider the mapping $\mathcal{T} : \mathcal{K}_\phi \to C([-r, b_\phi]; X)$ defined by

$$(\mathcal{T}y)(t) = \begin{cases} U(t, 0)\phi(0) + \displaystyle\int_0^t U(t, s) f(s, y_s) ds \ \text{ for } \ t \in [0, b_\phi] \\ \\ \phi(t) \ \text{for} \ t \in [-r, 0]. \end{cases}$$

First, we have to prove that $\mathcal{T}(\mathcal{K}_\phi) \subset \mathcal{K}_\phi$. Since for all $y \in \mathcal{K}_\phi$, one has that $s \mapsto U(t, s) f(s, y_s)$ is continuous on $[0, t]$ with $t \in [0, b_\phi]$, then $\mathcal{T}y \in C([-r, b_\phi]; X)$. Setting $u = \mathcal{T}y$ and $h = u - z$, we obtain for $t \in [0, b_\phi]$ via (7.7),

$$\begin{aligned} \|u_t - \phi\|_C &= \|h_t + z_t - \phi\|_C \\ &\leq \|h_t\|_C + \|z_t - \phi\|_C \\ &\leq \|h_t\|_C + \gamma_1 - \gamma. \end{aligned}$$

Also, we can write

$$h(t) = \begin{cases} \displaystyle\int_0^t U(t, s) f(s, y_s) ds \ \text{ for } \ t \in [0, b_\phi] \\ \\ 0 \ \text{for} \ t \in [-r, 0]. \end{cases}$$

Hence,

$$\begin{aligned} \|h(t)\| &= \Big\| \int_0^t U(t, s) f(s, y_s) ds \Big\| \\ &\leq \int_0^t \|U(t, s) f(s, y_s)\| ds \\ &\leq \int_0^t \|U(t, s)\| \|f(s, y_s)\| ds \\ &\leq M\gamma_2 \int_0^{b_\phi} e^{\omega s} ds \leq \gamma. \end{aligned}$$

Since for all $s \in [-r, 0]$, $h(s) = 0$, one can see that for all $t \in [0, b_\phi]$

$$\|h_t\|_C = \sup_{\theta \in [-r, 0]} \|h(t + \theta)\| \leq \sup_{s \in [0, b_\phi]} \|h(s)\| \leq \gamma.$$

Consequently,

$$\begin{aligned}\|(\mathcal{T}y)_t - \phi\|_C = \|u_t - \phi\|_C &\le \|h_t\|_C + \gamma_1 - \gamma \\ &\le \gamma + \gamma_1 - \gamma = \gamma_1.\end{aligned}$$

Also, the definition of $\mathcal{T}$ gives for all $y \in \mathcal{K}_\phi$, $(\mathcal{T}y)_0 = \phi$. So

$$\mathcal{T}(\mathcal{K}_\phi) \subset \mathcal{K}_\phi.$$

Let us show that the family $\{(\mathcal{T})(y) : y \in \mathcal{K}_\phi\}$ is equicontinuous. Let $y \in \mathcal{K}_\phi$, $t_1, t_2 \in [0, b_\phi]$ with $t_1 < t_2$. Then,

$$\begin{aligned}(\mathcal{T}y)(t_2) - (\mathcal{T}y)(t_1) &= U(t_2, 0)\phi(0) + \int_0^{t_2} U(t_2, s) f(s, y_s) ds \\ &\quad -\Big(U(t_1, 0)\phi(0) + \int_0^{t_1} U(t_1, s) f(s, y_s) ds\Big) \\ &= \Big(U(t_2, 0)\phi(0) - U(t_1, 0)\phi(0)\Big) \\ &\quad + \int_0^{t_1} \Big(U(t_2, s) - U(t_1, s)\Big) f(s, y_s) ds \\ &\quad + \int_{t_1}^{t_2} U(t_2, s) f(s, y_s) ds.\end{aligned}$$

On the one hand, for any $0 < \epsilon < t_1$, it follows

$$\begin{aligned}&\Big\| \int_0^{t_1} \Big(U(t_2, s) - U(t_1, s)\Big) f(s, y_s) ds \Big\| \\ &= \Big\| \int_0^{t_1-\epsilon} \Big(U(t_2, s) - U(t_1, s)\Big) f(s, y_s) ds \Big\| \\ &\quad + \Big\| \int_{t_1-\epsilon}^{t_1} \Big(U(t_2, s) - U(t_1, s)\Big) f(s, y_s) ds \Big\| \\ &\le \Big\| \Big(U(t_2, t_1 - \epsilon) - U(t_1, t_1 - \epsilon)\Big) \int_0^{t_1-\epsilon} U(t_1 - \epsilon, s) f(s, y_s) ds \Big\| \\ &\quad + \Big\| \int_{t_1-\epsilon}^{t_1} \Big(U(t_2, s) - U(t_1, s)\Big) f(s, y_s) ds \Big\| \\ &\le \Big\| \Big(U(t_2, t_1 - \epsilon) - U(t_1, t_1 - \epsilon)\Big) \int_0^{t_1-\epsilon} U(t_1 - \epsilon, s) f(s, y_s) ds \Big\| \\ &\quad + M\gamma_2 \Big[ \int_0^{\epsilon} e^{\omega s} ds + \int_{t_2-t_1}^{t_2-t_1+\epsilon} e^{\omega s} ds \Big].\end{aligned}$$

We can also note that $\left\{ \int_0^{t_1-\epsilon} U(t_1-\epsilon, s) f(s, y_s) ds \ : \quad y \in \mathcal{K}_\phi \right\}$ is uniformly bounded. On the other hand,

$$\begin{aligned}
\left\| \int_{t_1}^{t_2} U(t_2, s) f(s, z_s) ds \right\| &\leq \int_{t_1}^{t_2} \left\| U(t_2, s) f(s, y_s) \right\| ds \\
&\leq M\gamma_2 \int_{t_1}^{t_2} e^{\omega(t_2-s)} ds \\
&= M\gamma_2 \int_0^{t_2-t_1} e^{\omega s} ds.
\end{aligned}$$

Consequently,

$$\begin{aligned}
\left\| \Big( (\mathcal{T}y)(t_2) - (\mathcal{T}y)(t_1) \Big) \right\| \leq & \left\| \Big( U(t_2, 0)\phi(0) - U(t_1, 0)\phi(0) \Big) \right\| + M\gamma_2 \int_0^{t_2-t_1} e^{\omega s} ds \\
& + \left\| \Big( U(t_2, t_1-\epsilon) - U(t_1, t_1-\epsilon) \right) \\
& \times \int_0^{t_1-\epsilon} U(t_1-\epsilon, s) f(s, y_s) ds \Big) \Big\| \\
& + M\gamma_2 \Big[ \int_0^{\epsilon} e^{\omega s} ds + \int_{t_2-t_1}^{t_2-t_1+\epsilon} e^{\omega s} ds \Big].
\end{aligned}$$

Using the fact that the compactness of $U(t, s)$ for $t > s$ implies the continuity of $U(t, s)$ in the uniform norm topology, then it follows that the family $\{(\mathcal{T})(y) : y \in \mathcal{K}_\phi\}$ is equicontinuous.

To end the proof, we have to show that the set $\overline{\mathcal{T}(\mathcal{K}_\phi)}$ is compact. The collection $\mathcal{T}(\mathcal{K}_\phi)$ is equicontinuous; therefore, it remains to prove via Arzela–Ascoli theorem that the set $\{(\mathcal{T}y)(t) : y \in \mathcal{K}_\phi\}$ is precompact in $X$-norm for some fixed $t \in [0, b_\phi]$. The precompactness of the set $\{(\mathcal{T}y)(t) : y \in \mathcal{K}_\phi\}$ is a consequence of the fact that $\{(\mathcal{T}y)(t) - U(t, 0)\phi(0) : y \in \mathcal{K}_\phi\}$ is a precompact set. Let $0 < \epsilon < t$, and then

$$\begin{aligned}
\int_0^t U(t, s) f(s, y_s) ds &= \int_0^{t-\epsilon} U(t, s) f(s, y_s) ds + \int_{t-\epsilon}^t U(t, s) f(s, y_s) ds \\
&= U(t, t-\epsilon) \int_0^{t-\epsilon} U(t-\epsilon, s) f(s, y_s) ds \\
&\quad + \int_{t-\epsilon}^t U(t, s) f(s, y_s) ds.
\end{aligned}$$

One can write for all $y \in \mathcal{K}_\phi$,

$$\begin{aligned}\left\| \int_{t-\epsilon}^{t} U(t,s)f(s,y_s)ds \right\| &\leq \int_{t-\epsilon}^{t} \left\| U(t,s)f(s,y_s) \right\| ds \\ &\leq M\gamma_2 \int_{t-\epsilon}^{t} e^{\omega(t-s)} ds \\ &= M\gamma_2 \int_0^{\epsilon} e^{\omega s} ds \\ &\leq \alpha\epsilon,\end{aligned}$$

where $\alpha > 0$ is some constant real number. Also, using the fact that $\left\{ \int_0^{t-\epsilon} U(t-\epsilon,s)f(s,y_s)ds \;:\; y \in \mathcal{K}_\phi \right\}$ is uniformly bounded and the operator $U(t, t-\epsilon)$ is compact, then one can find a compact subset $W_\epsilon$ of $X$ such that

$$U(t,t-\epsilon)\left\{ \int_0^{t-\epsilon} U(t,s)f(s,y_s)ds : \quad y \in \mathcal{K}_\phi \right\} \subset W_\epsilon.$$

Hence, for each $t \in [0, b_\phi]$, $\left\{ (\mathcal{T}y)(t) - U(t,0)\phi(0) : y \in \mathcal{K}_\phi \right\}$ is totally bounded and the set $\{(\mathcal{T}y)(t) : y \in \mathcal{K}_\phi\}$ is precompact in $X$-norm.

To finish, let us prove that the application $\mathcal{T}$ is continuous on $\mathcal{K}_\phi$. For this aim, let $\epsilon > 0$ be given. Since the application $f : [0, b_\phi] \times \overline{B}_{\gamma_1}(\phi) \to X$ is continuous, then there exists $\delta > 0$ such that for all $y^1, y^2 \in \mathcal{K}_\phi$, $\left\| y^1 - y^2 \right\|_C \leq \delta$ implies $\left\| f(s, y_s^1) - f(s, y_s^2) \right\| < \epsilon$. So, for all $t \in [0, b_\phi]$,

$$\begin{aligned}\left\| (\mathcal{T}y^1)(t) - (\mathcal{T}y^2)(t) \right\| &\leq \int_0^t \left\| U(t,s) \right\| \left\| f(s,y_s^1) - f(s,y_s^2) \right\| ds \\ &\leq \epsilon M \int_0^t e^{\omega s} ds,\end{aligned}$$

which yields the continuity of $\mathcal{T}$ on $\mathcal{K}_\phi$.

The conditions of Schauder's fixed-point theorem are satisfied for the application $\mathcal{T}$ on $\mathcal{K}_\phi$. Consequently, the function $\mathcal{T}$ has a fixed point $y = u$ on $\mathcal{K}_\phi$, which solves the problem (7.1) on $[-r, b_\phi]$. □

Now, we have to prove that a mild solution to (7.1) can be defined on its maximal interval $[-r, T]$ of existence.

**Theorem 7.3.2** *Let $\{A(t)\}_{t\in[0,T]}$ be a stable family of infinitesimal generators of $C_0$-semigroups on $X$, and assume that the conditions* ($\mathbf{C}_1$), ($\mathbf{C}_2$) *and* ($\mathbf{C}_3$) *hold.*

*Furthermore, suppose that* $f : \mathbb{R}^+ \times C \to X$ *is a continuous function and takes bounded sets of* $[0, +\infty) \times C$ *into bounded sets of* $X$. *Then, Eq.* (7.1) *has at least one mild solution* $y(., \phi)$ *on the maximal interval* $[0, b_\phi)$. *Moreover, either* $b_\phi = T$ *or* $b_\phi < T$ *and* $\overline{\lim_{t \to b_\phi^-}} \|y(t, \phi)\| = +\infty$. □

***Proof*** Using Theorem 7.3.1, then we have the existence of mild solution $y(., \phi)$ that is defined on $[-r, b_1]$. Moreover, one can extend the solution $y(., \phi)$ to the interval $[-r, b_2]$ with $b_1 < b_2$. To do this, we consider the following equation:

$$\begin{cases} \dfrac{d}{dt}u(t) = A(t)u(t) + f(t, u_t) \quad \text{for} \quad t \in [b_1, b_2] \\ \\ u_{b_1} = y_{b_1}(., \phi). \end{cases} \tag{7.8}$$

To prove that Eq. (7.8) has a mild solution, we consider the operator $\mathcal{T}$ defined on

$$\mathcal{K}_{b_2}(\phi) = \Big\{ u \in C([-r, b_2]; X) : \ u_{b_1} = y_{b_1} \ \|u_t - y_{b_1}\|_C \le \gamma_1 \ \text{for all} \ t \in [b_1, b_2] \Big\},$$

as follows:

$$\mathcal{T}(u)(t) = U(t, b_1)y(b_1, \phi) + \int_{b_1}^{t} U(t, s) f(s, u_s) ds \ \text{for} \ t \in [b_1, b_2].$$

Using the similar argument, one obtains that $\mathcal{T}$ verifies the Schauder's fixed-point theorem that solves Eq. (7.8). This solution gives a mild solution $u$ of (7.8) on $[-r, b_2]$ that is an extension of $y(., \phi)$. Proceeding inductively, the solution $y(., \phi)$ is continuously extended to a maximal interval $[-r, b_\phi)$.

Assume that $b_\phi < T$ and the conclusion of Theorem 7.3.2 is false. Then, there exists some constant $R > 0$ such that $\overline{\lim_{t \to b_\phi}} \|y(t, \phi)\| < R$.

Also, since $f : \mathbb{R}^+ \times C \to X$ is a continuous function and takes bounded sets of $[0, +\infty) \times C$ into bounded sets of $X$, then there exists a positive constant $\delta$ such that

$$\Big\| f(t, y_t(., \phi)) \Big\| \le \delta, \quad \text{for all} \quad t \in [0, b_\phi).$$

Let $t_0 \in (0, b_\phi)$ be fixed and $y : [t_0, b_\phi) \to X$ be the restriction of $y(., \phi)$ to $[t_0, b_\phi)$. Consider $t_0 \le t_1 \le t_2 < b_\phi$ and $0 < \epsilon < t_0$. One can use the fact that $U(t, s)$ is strongly continuous to obtain a positive constant $\eta_1$ such that

$$\Big\| (U(t_2, 0) - U(t_1, 0))y(0) \Big\| < \epsilon, \quad \text{for} \ |t_2 - t_1| \le \eta_1.$$

Moreover,

$$\begin{aligned}\left\|y(t_2)-y(t_1)\right\| &\le \left\|(U(t_2,0)-U(t_1,0))y(0)\right\| + \left\|\int_0^{t_1}[U(t_2,s)\right.\\ &\quad \left.-U(t_1,s)]f(s,y_s)ds\right\|\\ &\quad +\left\|\int_{t_1}^{t_2} U(t_2,s)f(s,y_s)ds\right\|.\end{aligned}$$

Otherwise,

$$\int_0^{t_1}[U(t_2,s)-U(t_1,s)]f(s,y_s)ds=\Big[U(t_2,t_1)-I\Big]\int_0^{t_1} U(t_1,s)f(s,y_s)ds.$$

Let us set

$$S_{b_\phi}=\left\{\int_0^t U(t,s)f(s,y_s)ds:\quad t\in[0,b_\phi)\right\}.$$

Obviously, we can observe that the application $F:[0,b_\phi)\to X$ defined by

$$F(t)=\int_0^t U(t,s)f(s,y_s)ds$$

is continuous on the interval $[0,b_\phi)$. Moreover, for all $t\in[0,b_\phi)$,

$$\begin{aligned}\left\|F(t)\right\| &\le M\delta\int_0^t e^{\omega(t-s)}ds = M\delta\int_0^t e^{\omega s}ds\\ &\le M\delta\int_0^{b_\phi} e^{\omega s}ds.\end{aligned}$$

Therefore, $F(t)$ is bounded in the $X$-norm in the neighbourhood of $b_\phi$. Hence, there exists $0<\tau<t_0$ such that $F(t)$ is bounded for all $t\in(b_\phi-\tau,b_\phi)$. Also, $F$ is continuous on the compact set $[0,b_\phi-\tau]$. Then, $F([0,b_\phi-\tau])$ is the compact set of $X$. It follows that the set $S_{b_\phi}$ is included in a compact subset $\Sigma$ of the Banach space $X$. Using Banach–Steinhaus theorem, one claims the existence of a positive constant $\eta_2$ verifying

$$\sup_{x\in\Sigma}\left\|[U(t_2,t_1)-I]x\right\|<\epsilon \quad \text{for } |t_1-t_2|\le\eta_2.$$

Moreover,

$$\left\| \int_{t_1}^{t_2} U(t_2, s) f(s, y_s) \right\| ds \leq M\delta \int_{t_1}^{t_2} e^{\omega(t_2-s)} ds$$
$$\leq M\delta \int_{0}^{t_2-t_1} e^{\omega s} ds$$
$$\leq (t_2 - t_1)\delta M \max\{1, e^{\omega(t_2-t_1)}\}.$$

Hence, taking $\eta = \inf\{\eta_1, \eta_2, \epsilon\}$ such that for all $t_1, t_2 \in [0, b_\phi)$ with $|t_2 - t_1| < \eta$, it follows

$$\|y(t_2) - y(t_2)\| < (2\epsilon + \epsilon\delta M \max\{1, e^{\omega(t_2-t_1)}\}).$$

Therefore, using similar argument, we can conclude that

$$\lim_{|t_1-t_2|\to 0} \left\| y(t_2) - y(t_2) \right\| = 0.$$

Hence, $u$ is uniformly continuous on $[t_0, b_\phi)$. Consequently, $\lim\limits_{t\to b_\phi} y(t, \phi)$ exists. Let define $y(b_\phi, \phi) := \lim\limits_{t\to b_\phi} y(t, b_\phi)$. Then, the function $\Psi : [-r, b_\phi] \to X$ defined by

$$\Psi(t) = \begin{cases} y(t, \phi) & \text{if} \quad t < b_\phi \\ y(b_\phi, \phi) & \text{if} \quad t = b_\phi \end{cases}$$

extends $y$. This contradicts the existence of the maximal interval $[-r, b_\phi)$, and the proof is complete. □

In order to obtain the existence and uniqueness of mild solution, we have to take the nonlinear term of the problem (7.1) to be Lipschitz function with respect to its second argument. Moreover, in the rest of this chapter, to prove the existence of global solution, we will assume that:

**(H)** $A(t)$ is defined for each $t \geq 0$, and $\{A(t)\}_{t\geq 0}$ is a stable family of infinitesimal generators of $C_0$-semigroups on $X$ with stability constants $M, \omega$.

**Theorem 7.3.3** *Assume that the conditions* **(H)**, **(C$_1$)**, *and* **(C$_2$)** *hold. Furthermore, suppose that* $f : \mathbb{R}^+ \times C \to X$ *is a continuous function and verifies the following condition: for all* $t \geq 0$,

$$\left\| f(t, \phi) - f(t, \psi) \right\| \leq L \left\| \phi - \psi \right\|_C, \quad \forall \phi, \psi \in C \tag{7.9}$$

*when* $L > 0$ *is some positive constant. Then, Eq.* (7.1) *has a unique mild solution* $y(., \phi)$ *on* $[-r, +\infty)$. □

***Proof*** Let $a > 0$ and $\mathcal{M}_a = C([0, a]; X)$ be the space of continuous functions from $[0, a]$ to $X$ endowed with the uniform norm topology. Let us set for $\phi \in C$

$$K(\phi) = \{z \in \mathcal{M}_a : \; z(0) = \phi(0)\}.$$

For $z \in K(\phi)$, we introduce the extension $\tilde{z}$ of $z$ on $[-r, a]$ by

$$\tilde{z}(t) = \begin{cases} z(t) & \text{for } t \in [0, a] \\ \phi(t) & \text{for } t \in [-r, 0]. \end{cases}$$

Let $\mathcal{T}$ be a mapping defined on $K(\phi)$ by

$$\mathcal{T}(z)(t) = U(t, 0)\phi(0) + \int_0^t U(t, s) f(s, \tilde{z}_s) ds \;\; \text{for } \; t \in [0, a].$$

Consider $z \in K(\phi)$, $t_1, t \in [0, a]$ with $t_1 < t$. Then

$$\begin{aligned} \mathcal{T}(z)(t) - \mathcal{T}(z)(t_1) &= U(t, 0)\phi(0) + \int_0^t U(t, s) f(s, \tilde{z}_s) ds \\ &\quad -\Big(U(t_1, 0)\phi(0) + \int_0^{t_1} U(t_1, s) f(s, \tilde{z}_s) ds\Big) \\ &= \Big(U(t, 0)\phi(0) - U(t_1, 0)\phi(0)\Big) \\ &\quad + \int_0^{t_1} \Big(U(t, s) - U(t_1, s)\Big) f(s, \tilde{z}_s) ds \\ &\quad + \int_{t_1}^t U(t, s) f(s, \tilde{z}_s) ds. \end{aligned}$$

Immediately,

$$\Big(U(t, 0)\phi(0) - U(t_1, 0)\phi(0)\Big) \to 0 \text{ as } t \to t_1 \quad \text{in } X\text{-norm}$$

since $(t, 0) \mapsto U(t, 0)\phi(0)$ is continuous.

Also, since $f$ is continuous and verifies (7.9), then

$$\sup_{s\in[0,t]} \|f(s, \tilde{z}_s)\| \leq L \sup_{s\in[0,t]} \|\tilde{z}_s - \phi\| + \sup_{s\in[0,t]} \|f(s, \phi)\| < R,$$

where $R > 0$ is some positive constant. Therefore, for each $\epsilon > 0$ and $\epsilon \leq t_1 \leq t$,

$$\begin{aligned}
&\Big\| \int_0^{t_1} \Big(U(t,s) - U(t_1,s)\Big) f(s,\tilde{z}_s)ds \Big\| \\
&\leq \Big\| \int_0^{t_1-\epsilon} \Big(U(t,s) - U(t_1,s)\Big) f(s,\tilde{z}_s)ds \Big\| \\
&\quad + \Big\| \int_{t_1-\epsilon}^{t_1} \Big(U(t,s) - U(t_1,s)\Big) f(s,\tilde{z}_s)ds \Big\| \\
&\leq \Big\| (U(t,t_1-\epsilon) - U(t_1,t_1-\epsilon)) \int_0^{t_1-\epsilon} U(t_1-\epsilon,s) f(s,\tilde{z}_s)ds \Big\| \\
&\quad + \Big\| \int_{t_1-\epsilon}^{t_1} \Big(U(t,s) - U(t_1,s)\Big) f(s,\tilde{z}_s)ds \Big\| \\
&\leq \Big\| (U(t,t_1-\epsilon) - U(t_1,t_1-\epsilon)) \int_0^{t_1-\epsilon} U(t_1-\epsilon,s) f(s,\tilde{z}_s)ds \Big\| \\
&\quad + MR\Big[ \int_0^{\epsilon} e^{\omega s} ds + \int_{t-t_1}^{t-t_1+\epsilon} e^{\omega s} ds \Big].
\end{aligned}$$

Since $\epsilon$ is arbitrary chosen, then

$$\int_0^{t_1} \Big(U(t,s) - U(t_1,s)\Big) f(s,\tilde{z}_s)ds \to 0, \quad \text{as } t \to t_1.$$

Also, we have

$$\begin{aligned}
\Big\| \int_{t_1}^{t} U(t,s) f(s,\tilde{z}_s)ds \Big\| &\leq \int_{t_1}^{t} \Big\| U(t,s) f(s,\tilde{z}_s) \Big\| ds \\
&\leq RM \int_{t_1}^{t} e^{\omega(t-s)} ds \\
&= RM \int_0^{t-t_1} e^{\omega s} ds.
\end{aligned}$$

Then,

$$\int_{t_1}^{t} U(t,s) f(s,\tilde{z}_s)ds \to 0, \quad \text{as} \quad t \to t_1.$$

Consequently,

$$\mathcal{T}(z)(t) - \mathcal{T}(z)(t_1) \to 0 \text{ as } t \to t_1 \text{ and } a \geq t > t_1.$$

Using a similar argument, one obtains that for $t_1, t \in [0, a]$ with $t_1 > t$

$$\mathcal{T}(z)(t_1) - \mathcal{T}(z)(t) \to 0 \text{ as } t \to t_1.$$

Hence, $\mathcal{T}(z) \in K(\phi)$ for all $z \in K(\phi)$.

Now let show that $\mathcal{T}(z)$ is a strict contraction on $K(\phi)$. For that, let $z, u \in K(\phi)$, and $t \in [0, a]$.

$$\Big(\mathcal{T}(z)(t) - \mathcal{T}(u)(t)\Big) = \int_0^t U(t,s)\Big[f(s,\tilde{z}_s) - f(s,\tilde{u}_s)\Big]ds.$$

Then, we can write

$$\Big\|(\mathcal{T}(z)(t) - \mathcal{T}(u)(t))\Big\| \leq \int_0^t \Big\|U(t,s)[f(s,\tilde{z}_s) - f(s,\tilde{u}_s)]\Big\|ds.$$

$$\Big\|(\mathcal{T}(z)(t) - \mathcal{T}(u)(t))\Big\| \leq ML\int_0^t e^{\omega(t-s)}\Big\|\tilde{z}_s - \tilde{u}_s\Big\|_C ds.$$

Since $\tilde{z}(\theta) - \tilde{u}(\theta) = 0$ for all $\theta \in [-r, 0]$, then

$$\Big\|\tilde{z}_s - \tilde{u}_s\Big\|_C \leq \sup_{0\leq\tau\leq s}\Big\|z(\tau) - u(\tau)\Big\|.$$

So,

$$\Big\|\mathcal{T}(z)(t) - \mathcal{T}(u)(t)\Big\| \leq \Big(ML\int_0^t e^{\omega s}ds\Big)\Big\|z - u\Big\|_C,$$

where $\Big\|z - u\Big\|_C$ denotes the supremum norm in $C([0, a]; X)$. One can choose $a$ small enough such that

$$\Big(ML\int_0^a e^{\omega s}ds\Big) < 1.$$

Then, $\mathcal{T}$ is a strict contraction on $K(\phi)$. Therefore, $\mathcal{T}$ has a unique fixed point $u$ that is the unique mild solution of Eq. (7.1) on $[0, a]$. Moreover, one can extend the solution $u$ to $[a, 2a]$. Therefore, we consider the following equation:

$$\begin{cases} \dfrac{d}{dt}z(t) = A(t)z(t) + f(t, z_t) \quad \text{for} \quad t \in [a, 2a] \\ z_a = u_a. \end{cases} \tag{7.10}$$

To show that Eq. (7.10) has a unique mild solution, we consider the operator $\mathcal{T}_a$ defined on $K_a(\phi) = \{z \in C([a, 2a]; X) : z(a) = u(a)\}$ by

$$\mathcal{T}_a(z)(t) = U(t, a)u(a) + \int_a^t U(t, s)f(s, \tilde{z}_s)ds \text{ for } t \in [a, 2a],$$

where the function $\tilde{z}$ is defined by

$$\tilde{z}(t) = \begin{cases} z(t) \text{ for } t \in [a, 2a] \\ u(t) \text{ for } t \leq a. \end{cases}$$

Using the similar argument, one obtains that $\mathcal{T}_a$ is a strict contraction on $[a, 2a]$ that gives a unique mild solution of (7.10) on $[a, 2a]$ that is an extension of $u$. Proceeding inductively, the solution $u$ is uniquely and continuously extended to $[na, (n + 1)a]$ for all $n \geq 1$. Finally, we obtain that Eq. (7.1) has a unique mild solution on $[-r, +\infty)$. □

## 7.4 Application

$$\begin{cases} \frac{\partial}{\partial t}u(t, x) = \alpha(t)\frac{\partial^2}{\partial x^2}u(t, x) + \beta(t)\int_{-r}^{0} u(t + \theta, x) \sin\Big(u(t + \theta, x)\Big)d\theta, \\ t \geq 0, \ x \in [0, \pi], \\ u(t, 0) = u(t, \pi) = 0, \quad t \geq 0, \\ u(\theta, x) = \phi_0(\theta, x), \quad \theta \in [-r, 0], \end{cases} \tag{7.11}$$

where $\beta : \mathbb{R}^+ \to \mathbb{R}$ is a bounded continuous function and $\alpha : \mathbb{R}^+ \to \mathbb{R}^*_+$ is a bounded continuously differentiable function. The given function $\phi_0 : [-r, 0] \times [0, \pi] \to \mathbb{R}$ will be specified later.

In order to write the system (7.11) in an abstract form, we introduce the space $X = L^2([0, \pi]; \mathbb{R})$. Let $A$ be the operator defined on $X$ by

$$\begin{cases} D(A) = H^2((0, \pi); \mathbb{R}) \cap H_0^1((0, \pi); \mathbb{R}), \\ Ay = y'', \quad y \in D(A). \end{cases}$$

Let us note $A(t)$ the operator defined on $X$ as follows:

$$A(t)y = \alpha(t)Ay = \alpha(t)y''.$$

Its domain $D(A(t))$ is independent of $t \in [0, T]$ and is given by

$$D(A(t)) = \{y \in X : \; y, \; y' \text{ are absolutely continuous } y'' \in X, \;\; y(0) = y(\pi) = 0\}.$$

Consequently, the assumption $(\mathbf{C}_1)$ is verified.

We equipped a subspace $D = D(A(t))$ with the graph norm

$$\left\|x\right\|_Y = \left\|x\right\| + \left\|A(0)x\right\| \quad \text{for every} \quad x \in Y = D,$$

which is a Banach space. Also, since $t \mapsto \alpha(t)$ is continuously differentiable on $\mathbb{R}^+$, then $t \mapsto A(t)y = \alpha(t)A$ is continuously differentiable on $\mathbb{R}^+$. Hence, the condition $(\mathbf{C}_2)$ is satisfied.

Note also that $Ay = y''$ for $y \in D(A)$. In [20], it is well-known that $A$ generates an analytic semigroup $(T(t))_{t\in\mathbb{R}^+}$ on $X$. Moreover, $T(t)$ is compact on $X$ for all $t > 0$. Furthermore, the operator $A$ has a discrete spectrum, and the eigenvalues are $\{-n^2 \; : \; n \in \mathbb{N}^*\}$ with the corresponding normalized eigenvectors $z_n(\xi) = \sqrt{\frac{2}{\pi}} sin(n\xi)$. Thus for $y \in D(A) = D$, there holds

$$Ay = \sum_{n=1}^{+\infty} -n^2(y, z_n)z_n,$$

$$A(t)y = \sum_{n=1}^{+\infty} (-\alpha(t)n^2)(y, z_n)z_n,$$

where $(.,.)$ is the usual inner product on $X$. It is clear that the common domain of $A(t)$, $t \geq 0$, coincides with that of the operator $A$. In the one hand, we have

$$\overline{D} = \overline{Y} = \overline{D(A(t))} = X, \quad A(t) \quad \text{is closed} \quad \text{and} \quad \mathbb{R}^+ \subset \rho(A(t)) \quad \forall t \in [0, T].$$

In the other hand, it is well-known that for all $\lambda > 0$ such that $R(\lambda, A)$ exists, $\left\|R(\lambda, A)\right\| \leq \frac{1}{\lambda}$. Consequently,

$$\begin{aligned} \left\|R(\lambda, A(t))\right\| &= \left\|\frac{1}{\alpha(t)} R(\frac{\lambda}{\alpha(t)}, A)\right\| \\ &\leq \frac{1}{\alpha(t)} \frac{\alpha(t)}{\lambda} \leq \frac{1}{\lambda}. \end{aligned}$$

Hence, using Hille–Yosida theorem, one obtains that $\{A(t)\}_{t\geq 0}$ is a family of generators of $C_0$-semigroups $\{S_t(s)\}_{s\geq 0}$ on $X$.

Therefore, for all $t \geq 0$, there exists $\omega = 0$ such that

$$(0, +\infty) \subset \rho(A(t)), \quad \forall \lambda > 0. \tag{7.12}$$

Better, it is obvious that

$$\left\| \prod_{j=1}^{k} R(\lambda; A(t_j)) \right\|_{\mathcal{L}(X)} \leq \frac{1}{\lambda^k} \quad \text{for } \lambda > 0 \tag{7.13}$$

and any finite sequence $0 \leq t_1 \leq t_2 \leq \cdots \leq t_k < +\infty$, $j = 1, 2, \cdots$

Consequently, one claims via Definition 2.1, p.130 of [17] that $\{A(t)\}_{t\geq 0}$ is a stable family.

It remains to show that the assumption ($\mathbf{C}_3$) is insured. Following the work done in [9], it suffices to prove that

$$\text{for each} \quad t \in [0, T], \quad \text{and some } \lambda \in \rho(A(t)), \quad \text{the resolvent}$$
$$R(\lambda, A(t)) \text{ is a compact operator.}$$

Since for all $\lambda > 0$,

$$R(\lambda, A(t)) = \Big(\lambda - \alpha(t)A\Big)^{-1} = \frac{1}{\alpha(t)}\Big(\frac{\lambda}{\alpha(t)} - A\Big)^{-1}$$

and

$$\Big(\frac{\lambda}{\alpha(t)} - A\Big)^{-1} = \int_0^{+\infty} e^{-\frac{\lambda}{\alpha(t)}s}\, T(s)ds$$

with $\{T(s)\}_{s\geq 0}$ compact, then $\Big(\frac{\lambda}{\alpha(t)} - A\Big)^{-1}$ is compact. Consequently, the operator $U(t, s), t > s$, is a compact operator.

To complete the abstract form of Eq. (7.11), let us define the initial data function $\phi \in C = C([-r, 0]; X)$ by

$$\phi(\theta)(x) = \phi_0(\theta, x) \quad \text{for all} \quad (\theta, x) \in [-r, 0] \times [0, \pi].$$

Moreover, let us define the following function $f : \mathbb{R}^+ \times C \to X$

$$f(t, \psi)(x) = \beta(t) \int_{-r}^{0} \psi(\theta)(x) \sin\Big(\psi(\theta)(x)\Big) d\theta \quad \text{for all } x \in [0, \pi] \text{ and } \quad \psi \in C.$$

Using the above notations and setting $v(t)(x) = u(t,x)$, Eq. (7.11) can be written as the following abstract form:

$$\begin{cases} \dfrac{d}{dt}v(t) = A(t)v(t) + f(t, v_t), & t \geq 0, \\ v(\theta) = \phi(\theta), & \theta \in [-r, 0]. \end{cases} \tag{7.14}$$

**Proposition 7.4.3** *The function $f$ is continuous on $\mathbb{R}^+ \times C \to X$ and takes bounded sets of $\mathbb{R}^+ \times C$ into bounded sets of $X$.* □

***Proof*** For all $(t, \psi) \in \mathbb{R}^+ \times C$, one has for all $x \in [0, \pi]$

$$\begin{aligned} \left| f(t,\psi)(x) \right| &= \left| \beta(t) \int_{-r}^{0} \psi(\theta)(x) \sin\left(\psi(\theta)(x)\right) d\theta \right| \\ &\leq |\beta(t)| \int_{-r}^{0} \left| \psi(\theta)(x) \right| d\theta. \end{aligned}$$

Using Hölder inequality, we can write

$$\begin{aligned} \left| f(t,\psi)(x) \right| &\leq |\beta(t)| \int_{-r}^{0} \left| \psi(\theta)(x) \right| d\theta \\ &\leq |\beta(t)| \left( \int_{-r}^{0} \left| \psi(\theta)(x) \right|^2 d\theta \right)^{\frac{1}{2}} \left( \int_{-r}^{0} \left| 1 \right|^2 d\theta \right)^{\frac{1}{2}} \\ &= r^{\frac{1}{2}} |\beta(t)| \left( \int_{-r}^{0} \left| \psi(\theta)(x) \right|^2 d\theta \right)^{\frac{1}{2}}. \end{aligned}$$

Consequently,

$$\begin{aligned} \int_0^{\pi} \left| f(t,\psi)(x) \right|^2 dx &\leq r|\beta(t)|^2 \int_0^{\pi} \int_{-r}^{0} \left| \psi(\theta)(x) \right|^2 d\theta dx \\ &\leq r|\beta(t)|^2 \int_{-r}^{0} \left( \int_0^{\pi} \left| \psi(\theta)(x) \right|^2 dx \right) d\theta \\ &\leq r|\beta(t)|^2 \int_{-r}^{0} \left\| \psi(\theta) \right\|^2 d\theta \\ &\leq r^2 |\beta(t)|^2 \left\| \psi \right\|_C^2. \end{aligned}$$

So,

$$\left\| f(t,\psi) \right\| \leq r|\beta(t)| \left\| \psi \right\|_C < +\infty,$$

for all $(t, \psi)$ in the bounded set of $\mathbb{R}^+ \times C$ since $\beta$ is continuous on $\mathbb{R}^+$. Therefore, $f$ takes the bounded sets of $\mathbb{R}^+ \times C$ into bounded sets of $X$.

Now, let us show that $f$ is continuous on $\mathbb{R}^+ \times C$. Let $(t_n)_n$ be a sequence of $\mathbb{R}^+$ such that $\lim_{n\to+\infty} t_n = t$. Then, for each $\phi \in C$,

$$\begin{aligned}
\Big|f(t_n,\psi)(x) - f(t,\psi)(x)\Big| &= \Big|(\beta(t_n)-\beta(t))\int_{-r}^{0}\psi(\theta)(x)\sin\Big(\psi(\theta)(x)\Big)d\theta\Big| \\
&\leq \Big|\beta(t_n)-\beta(t)\Big|\Big|\int_{-r}^{0}\psi(\theta)(x)\sin\Big(\psi(\theta)(x)\Big)d\theta\Big| \\
&\leq \Big|\beta(t_n)-\beta(t)\Big|\int_{-r}^{0}\Big|\psi(\theta)(x)\Big|d\theta \\
&\leq \Big|\beta(t_n)-\beta(t)\Big|\Big(\int_{-r}^{0}\Big|\psi(\theta)(x)\Big|^2 d\theta\Big)^{\frac{1}{2}}\Big(\int_{-r}^{0}\Big|1\Big|^2 d\theta\Big)^{\frac{1}{2}} \\
&\leq \Big|\beta(t_n)-\beta(t)\Big|r^{\frac{1}{2}}\Big(\int_{-r}^{0}\Big|\psi(\theta)(x)\Big|^2 d\theta\Big)^{\frac{1}{2}}.
\end{aligned}$$

Thus,

$$\Big\|f(t_n,\psi) - f(t,\psi)\Big\| \leq \Big|\beta(t_n)-\beta(t)\Big|r\Big\|\psi\Big\|_C.$$

Since $\beta$ is continuous on $\mathbb{R}^+$, then

$$\lim_{n\to+\infty} f(t_n,\psi) = f(t,\psi) \quad \text{in} \quad X\text{-norm}.$$

Also, let $(\psi_n)_{n\in\mathbb{N}}$ be a sequence of $C$ such that $\lim_{n\to+\infty}\psi_n = \psi$. Then, for $t \in \mathbb{R}^+$,

$$\begin{aligned}
\Big|f(t,\psi_n)(x) - f(t,\psi)(x)\Big| &\leq \Big|\beta(t)\int_{-r}^{0}\Big(\psi_n(\theta)(x)-\psi(\theta)(x)\Big)\sin\Big(\psi_n(\theta)(x)\Big)d\theta\Big| \\
&\quad + \Big|\beta(t)\int_{-r}^{0}\psi(\theta)(x)\Big(\sin\Big(\psi_n(\theta)(x)\Big)-\sin\Big(\psi(\theta)(x)\Big)\Big)d\theta\Big| \\
&\leq \Big|\beta(t)\Big|\Big(\int_{-r}^{0}\Big|\psi_n(\theta)(x)-\psi(\theta)(x)\Big|^2 d\theta\Big)^{\frac{1}{2}}\Big(\int_{-r}^{0}\Big|1\Big|^2 d\theta\Big)^{\frac{1}{2}} \\
&\quad + \Big|\beta(t)\Big|\Big(\int_{-r}^{0}\Big|1\Big|^2 d\theta\Big)^{\frac{1}{2}}\Big(\int_{-r}^{0}\Big|\psi(\theta)(x)\Big[\sin\Big(\psi_n(\theta)(x)\Big) \\
&\quad - \sin\Big(\psi(\theta)(x)\Big)\Big]\Big|^2 d\theta\Big)^{\frac{1}{2}}.
\end{aligned}$$

Therefore,

$$\left\| f(t,\psi_n) - f(t,\psi) \right\| \le \left|\beta(t)\right| r \left\| \psi_n - \psi \right\| + r\left|\beta(t)\right| \left\| \psi \right\| \left\| \sin\left(\psi_n\right) - \sin\left(\psi\right) \right\|.$$

Since the function sin is continuous on $\mathbb{R}$, then

$$\lim_{n\to+\infty} f(t,\psi_n) = f(t,\psi) \quad \text{in X-norm.}$$

□

Consequently, the existence of the local and maximal mild solution for the problem (7.11) is proved.

## References


1. P. Acquistapace, B. Terreni, Some existence and regularity results for abstract non-autonomous parabolic equations. J. Math. Anal. Appl. **99**, 9–64 (1984)
2. T. Akrid, L. Maniar, A. Ouhinou, Periodic solutions of non-densely non-autonomous differential equations with delay. Afr. Diaspora J. Math. **15**(1), 25–42 (2013)
3. S.A. Campbell, R. Edwards, P. van den Driessche, Delayed coupling between two neural network loops. SIAM J. Appl. Math. **65**(1), 316–335 (2004)
4. K. Cooke, Y. Kuang, B. Li, Analyses of an antiviral immune response model with time delays. Can. Appl. Math. Q. **6**(4), 321–354 (1998)
5. K.L. Cooke, P. van den Driessche, X. Zou, Interaction of maturation delay and nonlinear birth in population and epidemic models. J. Math. Biol. **39**, 332–352 (1999)
6. G. DaPrato, E. Sinestrari, Non autonomous evolution operators of hyperbolic type. Semigroup Forum **45**, 302–312 (1992)
7. K. Engel, R. Nagel, *One-Parameter Semigroups for Linear Evolution Equations*. Graduate Texts in Mathematics, vol. 194 (Springer, Berlin, 2001)
8. A. Friedman, *Partial Differential Equation* (Holt, Rinehart and Winston, New York, 1979)
9. W.E. Fitzgibbon, Semilinear functional differential equations in Banach space. J. Diff. Equ. **29**, 1–14 (1978)
10. X. Fu, Existence of solutions for non-autonomous functional evolution equations with nonlocal conditions. Electron. J. Differ. Equ. **2012**(110), 1–15
11. H.R. Henriquez, Periodic solutions of quasi-linear partial functional differential equations with unbounded delay. Funkcial. Ekvac. **37**, 329–343 (1994)
12. D. Henry, *Geometric Theory of Semilinear Parabolic Equations*. Lecture notes in Mathematics, vol. 840 (Springer, Berlin Heidelberg, New York, 1981)
13. J.S. Howland, Stationary scattering theory for time-dependent Hamiltonians. Math. Ann. **207**, 315–335 (1974)
14. T. Kato, Linear evolution equations of hyperbolic type. J. Fac. Sci. Univ. Tokyo **17**, 241–258 (1970)
15. T. Kato, On linear differential equations in Banach spaces. Commun. Pure. Appl. Math. **9**, 479–486 (1956)
16. P.W. Nelson, J.D. Murray, A.S. Perelson, A model of HIV-1 pathogenesis that includes an intracellular delay. Math. Biosci. **163**, 201–215 (2000)
17. A. Pazy, *Semigroups of Linear Operators and Applications to Partial Differential Equations* (Springer, New York, 1993)
18. P. Smolen, D. Baxter, J. Byrne, A reduced model clarifies the role of feedback loops and time delays in the Drosophila circadian oscillator. Biophys. J. **83**, 2349–2359 (2002)

19. B. Vielle, G. Chauvet, Delay equation analysis of human respiratory stability. Math. Biosci. **152**(2), 105–122 (1998)
20. C.C. Travis, G.F. Webb, Existence, stability and compactness in $\alpha$-norm for partial functional differential equations. Trans. Am. Math. Soc. **240**, 129–143 (1978)
21. J. Wu, *Theory and Applications of Partial Functional Differential Equations*. Applied Mathematical Sciences (Springer, Berlin, 1996)
22. T. Zhao, Global periodic solutions for a differential delay system modeling a microbial population in the chemostat. J. Math. Anal. Appl. **193**, 329–352 (1995)

# Chapter 8
# Existence, Regularity, and Stability in the $\alpha$-Norm for Some Neutral Partial Functional Differential Equations in Fading Memory Spaces

**Khalil Ezzinbi, Bila Adolphe Kyelem, and Stanislas Ouaro**

**Abstract** The aim of this chapter is to study the regularity and the stability in the $\alpha$-norm for neutral partial functional differential equations in fading memory spaces. We assume that a linear part is densely defined and generates an analytic semigroup. The delayed part is assumed to be Lipschitzian. For illustration, we provide an example for some reaction–diffusion equation involving infinite delay.

**Keywords** Analytic semigroup · Neutral partial functional differential equations · $\alpha$-norm · Stability · Fading memory space

## 8.1 Introduction

Let $(X, |.|)$ be a Banach space, $(\mathscr{L}(X), |.|_{\mathscr{L}})$ be the space of bounded linear operators on $X$, and $\alpha$ be a constant such that $0 < \alpha < 1$. The aim of this chapter is to study the stability results of the following class of neutral partial functional differential equations in the $\alpha$-norm in fading memory spaces

---

K. Ezzinbi (✉)
Université Cadi Ayyad, Faculté des Sciences Semlalia, Département de Mathématiques, Marrakech, Marocco
e-mail: ezzinbi@ucam.ac.ma

B. A. Kyelem
Université de Ouahigouya, Unité de Formation et de Recherche en Sciences et Technologies, Département de Mathématiques et Informatique, Ouahigouya, Burkina Faso

S. Ouaro
Université Joseph KI-ZERBO, Unité de Formation et de Recherche en Sciences Exactes et Appliquées, Département de Mathématiques, Ouagadougou, Burkina Faso

 
T. Diagana et al. (eds.), *Partial Differential Equations and Applications*, Springer Proceedings in Mathematics & Statistics 420,
https://doi.org/10.1007/978-3-031-27661-3_8

$$
\begin{cases}
\dfrac{d}{dt}\mathcal{D}(u_t) = -A\mathcal{D}(u_t) + f(u_t) \quad \text{for} \quad t \geq 0, \\
\\
u_0 = \phi \in \mathcal{B}_\alpha,
\end{cases}
\tag{8.1}
$$

where $f : \mathcal{B}_\alpha \to X$ is a continuous function and $A : D(A) \subseteq X \to X$ is a linear operator such that $(-A)$ generates an analytic semigroup $(T(t))_{t\geq 0}$ on the Banach space $X$. $D(A)$ is the domain of the operator $A$. We also denote $R(A)$ the range of the operator $A$. For $0 < \alpha < 1$, $A^\alpha$ denotes the fractional power of $A$, and the space $X_\alpha$ will be defined later. The initial function $\phi$ belongs to a Banach space $\mathcal{B}_\alpha$ of functions mapping $(-\infty, 0]$ into $X_\alpha$ and satisfying some axioms to be introduced later. $\mathcal{D}$ is a bounded linear operator defined on $\mathcal{B}_\alpha$ with values in $X$ as follows:

$$
\mathcal{D}(\phi) = \phi(0) - \mathcal{D}_0(\phi) \ \text{for}\ \phi \in \mathcal{B}_\alpha,
\tag{8.2}
$$

where $\mathcal{D}_0$ is also a bounded linear operator defined on $\mathcal{B}_\alpha$ with values in $X$.

We denote by $u_t$ for $t \in \mathbb{R}^+$ the historic function defined on $(-\infty, 0]$ by

$$
u_t(\theta) = u(t + \theta) \quad \text{for all} \quad \theta \leq 0,
$$

where $u$ is a function from $\mathbb{R}$ into $X_\alpha$.

The existence results of neutral partial functional differential equations with delay are an important subject studied by many authors (see [1, 3, 5, 6, 8, 11, 20] and the references therein). One of the qualitative behaviours of solutions of neutral partial functional differential equations with delay developed in many works is the stability (see [2, 4, 7, 9, 10, 15, 21, 22] and the references therein).

One of the most important qualitative results of the functional partial differential equations is the stability, extensively studied by many authors. A mechanical or an electrical device can be constructed to a level of perfect accuracy that is restricted by technical, economic, or environmental constraints. What happens to the expected result if the construction is a little off specifications? Does output remain near design values? How sensitive is the design to variations in fabrication parameters? Stability theory gives some answers to these and similar questions.

Adimy and Ezzinbi in [4] established the stability results in the $\alpha$-norm for the problem of neutral type of the form

$$
\begin{cases}
\dfrac{d}{dt}\mathcal{D}(u_t) = -A\mathcal{D}(u_t) + f(u_t) \quad \text{for} \quad t \geq 0, \\
\\
u_0 = \phi \in C_\alpha,
\end{cases}
$$

where $f : \mathbb{R} \times C_\alpha \to X$ is a continuous function and $A : D(A) \subseteq X \to X$ is a linear operator;

$u_t$ for $t \in \mathbb{R}$ is the historic function defined on $[-r, 0]$ with $r > 0$ by $u_t(\theta) = u(t+\theta)$ for $\theta \in [-r, 0]$, where $u$ is a continuous function from $\mathbb{R}$ into $X_\alpha$; $C_\alpha = C([-r, 0]; D(A^\alpha))$ is the space of continuous functions from $[-r, 0]$ into $D(A^\alpha)$ provided with the uniform norm topology, $\mathcal{D}$ is a bounded linear operator from $C = C([-r, 0]; X)$ into $X$ defined by

$$\mathcal{D}(\phi) = \phi(0) - \mathcal{D}_0(\phi) \quad \text{for } \phi \in C,$$

where the operator $\mathcal{D}_0$ is given by

$$\mathcal{D}_0(\phi) = \int_{-r}^{0} d\eta(\theta)\phi(\theta) \quad \text{for } \phi \in C,$$

and $\eta : [-r, 0] \to \mathcal{L}(X)$ is of bounded variation and non-atomic at zero, that is, there exists a continuous nondecreasing function $\delta : [0, r] \to [0, +\infty)$ such that $\delta(0) = 0$ and

$$\left| \int_{-s}^{0} d\eta(\theta)\phi(\theta) \right| \leq \delta(s)\, |\phi|_C \quad \text{for } \phi \in C \quad \text{and } s \in [0, r].$$

In our work, we study the stability results of Eq. (8.1) following the results obtained in [2, 4, 7, 9, 10, 21].

To get some stability results in the uniform fading memory spaces, we make use of the spectral theory of linear operators, the fractional power operators, and the linear semigroup theory (see [13, 19]).

The organization of this chapter is as follows: In Sect. 8.2, we introduce some preliminary results on analytic semigroups, fractional powers of operator, and axiomatic phase space adapted to the fractional norm space for infinite delay. In Sect. 8.3, the existence and uniqueness of strict solutions is established. In Sect. 8.4, we are concerned with the smoothness results of the solutions. In Sect. 8.5, we investigate the stability near an equilibrium by using the linearized principle. In the last section, an example is provided to illustrate the applications of the main results of this chapter.

## 8.2 Analytic Semigroup, Fractional Power of Its Generator, and Partial Functional Differential Equations

Throughout this chapter, we assume the following:

**(H$_1$)** $(-A)$ is the infinitesimal generator of an analytic semigroup of linear operators $\{T(t)\}_{t\geq 0}$ on a Banach space $X$. Without loss of generality, we suppose that $0 \in \rho(A)$; otherwise, instead of $A$, we take $A - \delta I$, where $\delta$ is chosen such that $0 \in \rho(A - \delta I)$ and where $\rho(A)$ is the resolvent set of $A$.

It is well-known that $|T(t)x| \leq Me^{\omega t}|x|$ for all $t \geq 0$, $x \in X$, where $M \geq 1$ and $\omega \in \mathbb{R}$.

For all $0 < \alpha < 1$, we define (see [19]) the operator $A^{-\alpha}$ by

$$A^{-\alpha}x = \frac{1}{\Gamma(\alpha)} \int_0^{+\infty} t^{\alpha-1} T(t)x dt \text{ for all } x \in X,$$

where $\Gamma(\alpha)$ denotes the well-known gamma function at the point $\alpha$. The operator $A^{-\alpha}$ is bijective, and the operator $A^{\alpha}$ is defined by

$$A^{\alpha} = (A^{-\alpha})^{-1}.$$

We denote by $D(A^{\alpha})$ the domain of the operator $A^{\alpha}$. Then, $D(A^{\alpha})$ endowed with the norm $|x|_{\alpha} = |A^{\alpha}x|$ for all $x \in D(A^{\alpha})$ is a Banach space [19]. We denote it by $X_{\alpha}$. Moreover, we recall the following known results.

**Theorem 8.2.1 ([19], p.69–75)** *Let $0 < \alpha < 1$, and assume that* $(\mathbf{H}_1)$ *holds. Then:*

*(a)* $T(t) : X \to D(A^{\alpha})$ *for each* $t > 0$ *and* $\alpha \geq 0$.
*(b)* *For all* $x \in D(A^{\alpha})$, $T(t)A^{\alpha}x = A^{\alpha}T(t)x$.
*(c)* *For each* $t > 0$, *the linear operator* $A^{\alpha}T(t)$ *is bounded and* $|A^{\alpha}T(t)x| \leq M_{\alpha}t^{-\alpha}e^{\omega t}|x|$, *where* $M_{\alpha}$ *is a positive real constant.*
*(d)* *For* $0 < \alpha \leq 1$ *and* $x \in D(A^{\alpha})$, $|T(t)x - x| \leq N_{\alpha}t^{\alpha}|A^{\alpha}x|$, *for* $t > 0$, *where* $N_{\alpha}$ *is a positive real constant.*
*(e)* *For* $0 < \alpha < \beta < 1$, $X_{\beta} \hookrightarrow X_{\alpha}$.

From now on, we use an axiomatic definition of the phase space $\mathcal{B}$ that was first introduced by Hale and Kato in [16]. We assume that $\mathcal{B}$ is the normed space of functions mapping $(-\infty, 0]$ into $X$ and satisfying the following axioms:

**(A)** There exist a positive constant $N$, a locally bounded continuous function $M(.)$ on $[0, +\infty)$, and a continuous function $K(.)$ on $[0, +\infty)$, such that if $u : (-\infty, a] \to X$ is continuous on $[\xi, a]$ with $u_{\xi} \in \mathcal{B}$ for some $\xi < a$ where $0 < a$, then for all $t \in [\xi, a]$:

*(i)* $u_t \in \mathcal{B}$.
*(ii)* $t \to u_t$ is continuous on $[\xi, a]$.
*(iii)* $N|u(t)| \leq |u_t|_{\mathcal{B}} \leq K(t-\xi) \sup_{\xi \leq s \leq t} |u(s)| + M(t-\xi)|u_{\xi}|_{\mathcal{B}}$.

**(B)** $\mathcal{B}$ is a Banach space.

**Lemma 8.2.1 ([7])** *Let* $C_{00}$ *be the space of continuous functions mapping* $(-\infty, 0]$ *into* $X$ *with compact supports and* $C_{00}^{a}$ *be the subspace of functions in* $C_{00}$ *with supports included in* $[-a, 0]$ *endowed with the uniform norm topology. Then* $C_{00}^{a} \hookrightarrow \mathcal{B}$. □

Let

$$\mathcal{B}_\alpha = \left\{\phi \in \mathcal{B} : \ \phi(\theta) \in D(A^\alpha) \text{ for } \theta \le 0 \text{ and } A^\alpha \phi \in \mathcal{B}\right\}.$$

and provide $\mathcal{B}_\alpha$ with the following norm:

$$|\phi|_{\mathcal{B}_\alpha} = |A^\alpha \phi|_{\mathcal{B}} \text{ for } \phi \in \mathcal{B}_\alpha.$$

We also assume that

**(H$_2$)** $A^{-\alpha}\phi \in \mathcal{B}$ for all $\phi \in \mathcal{B}$, where the function $A^{-\alpha}\phi$ is defined by

$$(A^{-\alpha}\phi)(\theta) = A^{-\alpha}(\phi(\theta)) \text{ for } \theta \le 0$$

and

**(H$_3$)** $K(0)|\mathcal{D}_0| < 1$.

**Lemma 8.2.2 ([7])** *Assume that* (**H**$_1$) *and* (**H**$_2$) *hold. Then,* $\mathcal{B}_\alpha$ *is a Banach space and satisfies the axiom* (**A**). □

For regularity results in the Banach space $X$, consider the following problem:

$$\begin{cases} \dfrac{d}{dt}\mathcal{D}(u_t) = -A\mathcal{D}(u_t) + f(t) \quad \text{for} \quad t \ge 0, \\ u_0 = \phi. \end{cases} \tag{8.3}$$

**Definition 8.2.1** Let $\phi \in \mathcal{B}$. A function $u : (-\infty, a] \to X$ is called a mild solution of Eq. (8.3) associated to $\phi$ if

$$\begin{cases} \mathcal{D}(u_t) = T(t)\mathcal{D}(u_0) + \displaystyle\int_0^t T(t-s)f(s)ds \text{ for } t \in [0, a] \\ u_0 = \phi. \end{cases}$$

**Definition 8.2.2** Let $\phi \in \mathcal{B}$. A function $u : (-\infty, a] \to X$ is called a strict solution of Eq. (8.3) associated to $\phi$ if

$$\begin{cases} t \longmapsto \mathcal{D}(u_t) \quad \text{is continuously differentiable on} \quad [0, a] \\ \mathcal{D}(u_t) \in D(A) \quad \text{for} \quad t \ge 0 \\ u(t) \quad \text{satisfies the system} \quad (8.3) \quad \text{for} \quad t \ge 0. \end{cases}$$

We have the following important result.

**Theorem 8.2.2** *Let $u_0 = \phi$, $\mathcal{D}(\phi) \in D(A)$, and $f \in C^1([0, a]; X)$. The existence of a mild solution $u$ of* (8.3) *on $[0, a]$ implies the existence of a strict solution of* (8.3) *on $[0, a]$.* □

***Proof*** Let $u$ be a mild solution of (8.3). Then,

$$\mathcal{D}(u_t) = T(t)\mathcal{D}(u_0) + \int_0^t T(t-s)f(s)ds \text{ for } t \in [0, a]. \tag{8.4}$$

Show that $t \mapsto \mathcal{D}(u_t)$ is continuously differentiable. We need to only examine the second term of the right-hand side of (8.4), which will be denoted by $v(t)$. It is well-known that $T(t-s) = -\frac{\partial}{\partial s}(T(t-s))(-A)^{-1}$ since $(-A)$ generates the analytic semigroup $(T(t))_{t\geq 0}$. Hence,

$$\begin{aligned} v(t) &= -\int_0^t \frac{\partial}{\partial s}(T(t-s))(-A)^{-1}f(s)ds \\ &= \left[-(T(t-s))(-A)^{-1}f(s)\right]_0^t + \int_0^t T(t-s)(-A)^{-1}f'(s)ds \\ &= -(-A)^{-1}f(t) + T(t)(-A)^{-1}f(0) + \int_0^t T(t-s)(-A)^{-1}f'(s)ds. \end{aligned}$$

Since $\lim_{h\to 0}\Big[\int_0^t \frac{T(t+h-s)-T(t-s)}{h}(-A)^{-1}f'(s)ds + \frac{1}{h}\int_t^{t+h} T(t-s)(-A)^{-1}$ $f'(s)ds\Big] = (-A)^{-1}f'(t) + \int_0^t T(t-s)f'(s)ds$, it is easy to see that

$$\frac{d}{dt}v(t) = T(t)f(0) + \int_0^t T(t-s)f'(s)ds. \tag{8.5}$$

Using Eq. (8.5) and the fact that $f \in C^1([0, a]; X)$ and the semigroup $(T(t))_{t\geq 0}$ is analytic, then $t \mapsto \frac{d}{dt}v(t)$ is continuous. Consequently, $t \mapsto \mathcal{D}(u_t)$ is continuously differentiable on $t \in [0, a]$.

Now, let us show that $\mathcal{D}(u_t) \in D(A)$. Since $T(t)\mathcal{D}(\phi) \in D(A)$, it remains to prove that $v(t) \in D(A)$. We use the relation (8.5) in order to obtain

$$\begin{aligned} \frac{d}{dt}v(t) &= T(t)f(0) + \int_0^t T(t-s)f'(s)ds \\ &= -Av(t) + f(t). \end{aligned}$$

Thus, $Av(t) = -\frac{d}{dt}v(t) + f(t)$ exists and $v(t) \in D(A)$.

To finish, let us prove that $u$ verifies (8.3). Using (8.4), one can write

$$\begin{aligned}
\frac{d}{dt}(\mathcal{D}(u_t)) &= T'(t)\mathcal{D}(\phi) + \int_0^t \frac{\partial}{\partial s}(T(t-s))f(s)ds + f(t) \\
&= -AT(t)\mathcal{D}(\phi) - A\int_0^t T(t-s)f(s)ds + f(t) \\
&= -A\left[T(t)\mathcal{D}(\phi) + \int_0^t T(t-s)f(s)ds\right] + f(t) \\
&= -A\mathcal{D}(u_t) + f(t).
\end{aligned}$$

□

## 8.3 Existence and Uniqueness of Strict Solutions

Now, we give the notions of solutions that will be studied in our work.

**Definition 8.3.1** Let $\phi \in \mathcal{B}_\alpha$. A function $u : (-\infty, +\infty) \to X_\alpha$ is called a mild solution of Eq. (8.1) associated to $\phi$ if:

*(i)* $\mathcal{D}(u_t) = T(t)\mathcal{D}(\phi) + \int_0^t T(t-s)f(u_s)ds$ for $t \geq 0$.
*(ii)* $u_0 = \phi$.

□

**Definition 8.3.2** Let $\phi \in \mathcal{B}_\alpha$. A function $u : (-\infty, +\infty) \to X_\alpha$ is called a strict solution of Eq. (8.1) associated to $\phi$ if:

*(i)* $t \longmapsto \mathcal{D}(u_t)$ is continuously differentiable on $[0, +\infty)$.
*(ii)* $\mathcal{D}(u_t) \in D(A)$ for $t \geq 0$.
*(iii)* $u(t)$ satisfies the system (8.1) for $t \geq 0$.

□

Often in this chapter, $u_t(., \phi)$ and $u_t(\phi)$ denote the mild solution associated to the initial data $\phi$, and we simply denote it by $u_t$ if there is no confusion.

We assume that there exists $k > 0$ such that

**(H$_4$)** $|f(\phi_1) - f(\phi_2)| \leq k|\phi_1 - \phi_2|_{\mathcal{B}_\alpha}$ for all $\phi_1, \phi_2 \in \mathcal{B}_\alpha$.

**Theorem 8.3.1 ([14])** *Assume that* (**H**$_1$), (**H**$_2$), (**H**$_3$), *and* (**H**$_4$) *hold. Then, for each* $\phi \in \mathcal{B}_\alpha$, *there exists a unique mild solution of Eq.* (8.1) *that is defined for* $t \geq 0$.

**Lemma 8.3.1** *Assume that* (**H**$_1$), (**H**$_2$), *and* (**H**$_3$) *hold. Let* $\phi \in \mathcal{B}_\alpha$ *and* $h \in C(\mathbb{R}^+; X_\alpha)$ *such that* $\mathcal{D}(\phi) = h(0)$. *Then, there exists a unique continuous function* $x$ *on* $\mathbb{R}^+$ *that solves the following problem:*

$$\begin{cases} \mathcal{D}(x_t) = h(t) \quad for \quad t \geq 0, \\ x(t) = \phi(t) \ for \ t \in (-\infty, 0]. \end{cases} \tag{8.6}$$

*Moreover, there exist two functions a and b in $L^\infty_{loc}(\mathbb{R}^+; \mathbb{R}^+)$ such that*

$$|x_t|_{\mathcal{B}_\alpha} \leq a(t)|\phi|_{\mathcal{B}_\alpha} + b(t) \sup_{0 \leq s \leq t} |h(s)|_\alpha \quad for \quad t \geq 0. \tag{8.7}$$

***Proof*** We define for $p > 0$ the space

$$W = \{x \in C([0, p]; X_\alpha) : x(0) = \phi(0)\}$$

endowed with the uniform norm topology. For $x \in W$, we define its extension $\tilde{x}$ on $\mathbb{R}^-$ by

$$\tilde{x}(t) = \begin{cases} x(t) & \text{for } t \in [0, p] \\ \phi(t) & \text{for } t \in (-\infty, 0]. \end{cases}$$

Using axiom (**A**), one can see that $t \mapsto \tilde{x}_t$ is continuous from $[0, p]$ to $\mathcal{B}_\alpha$. Let us define the function $\mathcal{K}$ on $W$ by

$$(\mathcal{K}(x))(t) = \mathcal{D}_0(\tilde{x}_t) + h(t) \quad \text{for } t \geq 0.$$

One must show that $\mathcal{K}$ has a unique fixed point on $W$. Since $h \in C(\mathbb{R}^+; X_\alpha)$, then $h \in C([0, p]; X_\alpha)$. Moreover, $h(0) = \mathcal{D}(\phi) = \phi(0) - \mathcal{D}_0(\phi)$. It follows that

$$\mathcal{K}(W) \subset W.$$

We can also write for $x, y \in W$ with their respective extensions $\tilde{x}$ and $\tilde{y}$ associated to $\phi$

$$\begin{aligned} |(\mathcal{K}(x) - \mathcal{K}(y))(t)|_\alpha &\leq |\mathcal{D}_0||\tilde{x}_t - \tilde{y}_t|_{\mathcal{B}_\alpha} \\ &\leq |\mathcal{D}_0|K(t) \sup_{0 \leq s \leq t} |x(s) - y(s)|_\alpha \\ &\leq |\mathcal{D}_0|K(t)|x - y|_W. \end{aligned}$$

Choosing $p > 0$ small enough, one obtains that $\mathcal{K}$ is a strict contraction. Consequently, (8.6) has a unique solution $x$ on $(-\infty, p]$.

It follows for $s \in [0, p]$ that

$$\begin{aligned}
|x_s|_{\mathcal{B}_\alpha} &\le K(s) \sup_{0\le\tau\le s} |x(\tau)|_\alpha + M(s)|\phi|_{\mathcal{B}_\alpha} \\
&\le K(s)\Big(|\mathcal{D}_0| \sup_{0\le\tau\le s} |x_\tau|_{\mathcal{B}_\alpha} + \sup_{0\le\tau\le s} |h(\tau)|_\alpha\Big) + M(s)|\phi|_{\mathcal{B}_\alpha} \\
&\le K_p|\mathcal{D}_0| \sup_{0\le\tau\le s} |x_\tau|_{\mathcal{B}_\alpha} + K_p \sup_{0\le\tau\le s} |h(\tau)|_\alpha + M_p|\phi|_{\mathcal{B}_\alpha},
\end{aligned}$$

where $K_p = \sup_{s\in[0,p]} K(s)$ and $M_p = \sup_{s\in[0,p]} M(s)$.

Therefore,

$$\begin{aligned}
\sup_{0\le s\le t} |x_s|_{\mathcal{B}_\alpha} &\le \sup_{0\le s\le t} \left\{K_p|\mathcal{D}_0| \sup_{0\le\tau\le s} |x_\tau|_{\mathcal{B}_\alpha} + K_p \sup_{0\le\tau\le s} |h(\tau)|_\alpha + M_p|\phi|_{\mathcal{B}_\alpha}\right\} \\
&\le K_p|\mathcal{D}_0| \sup_{0\le s\le t} |x_s|_{\mathcal{B}_\alpha} + K_p \sup_{0\le s\le t} |h(s)|_\alpha + M_p|\phi|_{\mathcal{B}_\alpha}.
\end{aligned}$$

Thus, for $p > 0$ small enough and using (**H**$_3$), one can write for $t \in [0, p]$,

$$\sup_{0\le s\le t} |x_s|_{\mathcal{B}_\alpha} \le \frac{K_p}{1 - K_p|\mathcal{D}_0|} \sup_{0\le s\le t} |h(s)|_\alpha + \frac{M_p}{1 - K_p|\mathcal{D}_0|}|\phi|_{\mathcal{B}_\alpha}.$$

As a consequence, we have the existence of $a, b \in L^\infty_{loc}([0, p]; \mathbb{R}^+)$ such that

$$|x_t|_{\mathcal{B}_\alpha} \le a(t)|\phi|_{\mathcal{B}_\alpha} + b(t) \sup_{0\le s\le t} |h(s)|_\alpha, \quad \text{for} \quad t \in [0, p].$$

Now, to extend the solution $x$ on $[p, 2p]$, we consider the space

$$W_1 = \{u \in C([p, 2p]; X_\alpha) : u(p) = x(p)\}$$

endowed with the uniform norm topology and the following problem:

$$\tilde{u}(t) = \begin{cases} u(t) & \text{for} \quad t \in [p, 2p], \\ x(t) & \text{for} \ \ t \in (-\infty, p]. \end{cases}$$

We define the function $\mathcal{K}_1$ on $W_1$ by

$$(\mathcal{K}_1(u))(t) = \mathcal{D}_0(\tilde{u}_t) + h(t), \qquad \text{for } \ t \in [p, 2p].$$

Using the same arguments as above, we show that $\mathcal{K}_1$ is a strict contraction on $W_1$. That leads to the existence of a unique solution $u$ of (8.6) on $(-\infty, 2p]$, and $u$ is the extension of $x$ on $(-\infty, 2p]$.

Also, we have to extend $a, b$ on $[p, 2p]$. Therefore, let $s \in [p, 2p]$. Then, one can write

$$\begin{aligned} |x_s|_{\mathcal{B}_\alpha} &\le K(s-p) \sup_{p\le\tau\le s} |x(\tau)|_\alpha + M(s-p)\left|x_p\right|_{\mathcal{B}_\alpha} \\ &\le K_p \sup_{p\le\tau\le s} |x(\tau)|_\alpha + M_p\left|x_p\right|_{\mathcal{B}_\alpha} \\ &\le K_p \sup_{p\le\tau\le s} \left\{|\mathcal{D}_0|\,|x_\tau|_{\mathcal{B}_\alpha} + |h(\tau)|_\alpha\right\} + M_p\left|x_p\right|_{\mathcal{B}_\alpha}. \end{aligned}$$

Therefore, for each $t \in [p, 2p]$ such that $s \le t$, we have

$$\begin{aligned} \sup_{p\le s\le t} |x_s|_{\mathcal{B}_\alpha} &\le \sup_{p\le s\le t}\left\{K_p \sup_{p\le\tau\le s}\left\{|\mathcal{D}_0|\,|x_\tau|_{\mathcal{B}_\alpha} + |h(\tau)|_\alpha\right\} + M_p\left|x_p\right|_{\mathcal{B}_\alpha}\right\} \\ &\le K_p|\mathcal{D}_0| \sup_{p\le s\le t}|x_\tau|_{\mathcal{B}_\alpha} + K_p \sup_{p\le s\le t}|h(\tau)|_\alpha + M_p\left|x_p\right|_{\mathcal{B}_\alpha}. \end{aligned}$$

Thus, for $t \in [p, 2p]$,

$$|x_t|_{\mathcal{B}_\alpha} \le \frac{K_p}{1-K_p|\mathcal{D}_0|} \sup_{p\le s\le t}|h(s)|_\alpha + \frac{M_p}{1-K_p|\mathcal{D}_0|}\left|x_p\right|_{\mathcal{B}_\alpha}.$$

Since $p \in [0, p]$, one can write

$$|x_p|_{\mathcal{B}_\alpha} \le a(p)|\phi|_{\mathcal{B}_\alpha} + b(p) \sup_{0\le s\le p}|h(s)|_\alpha.$$

Consequently,

$$\begin{aligned} |x_t|_{\mathcal{B}_\alpha} &\le \frac{K_p}{1-K_p|\mathcal{D}_0|} \sup_{p\le s\le t}|h(s)|_\alpha + \frac{M_p}{1-K_p|\mathcal{D}_0|}\left|x_p\right|_{\mathcal{B}_\alpha} \\ &\le \frac{K_p}{1-K_p|\mathcal{D}_0|} \sup_{p\le s\le t}|h(s)|_\alpha + \frac{M_p a(p)}{1-K_p|\mathcal{D}_0|}|\phi|_{\mathcal{B}_\alpha} \\ &\quad + \frac{M_p b(p)}{1-K_p|\mathcal{D}_0|} \sup_{0\le s\le p}|h(s)|_\alpha \\ &\le \frac{M_p a(p)}{1-K_p|\mathcal{D}_0|}|\phi|_{\mathcal{B}_\alpha} + \max\left\{\frac{K_p}{1-K_p|\mathcal{D}_0|}, \frac{M_p b(p)}{1-K_p|\mathcal{D}_0|}\right\} \sup_{0\le s\le p}|h(s)|_\alpha \\ &\quad + \max\left\{\frac{K_p}{1-K_p|\mathcal{D}_0|}, \frac{M_p b(p)}{1-K_p|\mathcal{D}_0|}\right\} \sup_{p\le s\le t}|h(s)|_\alpha \\ &\le \frac{M_p a(p)}{1-K_p|\mathcal{D}_0|}|\phi|_{\mathcal{B}_\alpha} + 2\max\left\{\frac{K_p}{1-K_p|\mathcal{D}_0|}, \frac{M_p b(p)}{1-K_p|\mathcal{D}_0|}\right\} \sup_{0\le s\le t}|h(s)|_\alpha. \end{aligned}$$

Thus, for all $t \in [p, 2p]$,

$$|x_t|_{\mathcal{B}_\alpha} \leq a_1(t)\,|\phi|_{\mathcal{B}_\alpha} + b_1(t) \sup_{0 \leq s \leq t} |h(s)|_\alpha ,$$

where $a_1$ can be seen as the extension of $a$ on $[0, 2p]$ and $b_1$ the extension of $b$ on $[0, 2p]$. It is exactly to say there exist $a, b \in L^\infty_{loc}([0, 2p]; \mathbb{R}^+)$ such that

$$|x_t|_{\mathcal{B}_\alpha} \leq a(t)|\phi|_{\mathcal{B}_\alpha} + b(t) \sup_{0 \leq s \leq t} |h(s)|_\alpha, \quad \text{for} \quad t \in [0, 2p].$$

Inductively, one can show the existence of an extension $u$ of $x$ on $[np, (n+1)p]$ and the extension $a_{np}$ of $a$, $b_{np}$ of $b$ on $[np, (n+1)p]$. Finally, the solution $x$ is unique and continuous defined on $\mathbb{R}^+$. Also, the functions $a \in L^\infty_{loc}(\mathbb{R}^+; \mathbb{R}^+)$ and $b \in L^\infty_{loc}(\mathbb{R}^+; \mathbb{R}^+)$ are well-defined. □

We have the following result.

**Theorem 8.3.2 ([14])** *Assume that* $(\mathbf{H}_1)$, $(\mathbf{H}_2)$, $(\mathbf{H}_3)$, *and* $(\mathbf{H}_4)$ *hold. Let $u$ and $v$ be two mild solutions of Eq.* (8.1) *on* $\mathbb{R}$, *respectively, associated to the initial data $\phi$ and $\psi$. Then, for any $a > 0$, there exists $l(a) > 0$ such that*

$$|u_t(\phi) - v_t(\psi)|_{\mathcal{B}_\alpha} \leq l(a)|\phi - \psi|_{\mathcal{B}_\alpha} \; \textit{for } t \in [0, a]. \tag{8.8}$$

For the regularity of the mild solution, we suppose that $\mathcal{B}$ satisfies the following axiom:

$(\mathbf{B}_1)$ If $(\phi_n)_{n \geq 0}$ is a Cauchy sequence in $\mathcal{B}$ and converges compactly to $\phi$ in $(-\infty, 0]$, then $\phi \in \mathcal{B}$ and $|\phi_n - \phi|_{\mathcal{B}} \to 0$ as $n \to +\infty$.

Now, we can claim the existence and uniqueness of strict solution for Eq. (8.1).

**Theorem 8.3.3** *Assume that* $(\mathbf{H}_1)$, $(\mathbf{H}_2)$, $(\mathbf{H}_3)$, *and* $(\mathbf{H}_4)$ *hold. Furthermore, assume that $\mathcal{B}$ satisfies axiom:* $(\mathbf{B}_1)$ *$f : \mathcal{B}_\alpha \to X$ is continuously differentiable with $f'$ locally Lipschitz continuous. Let $\phi \in \mathcal{B}_\alpha$ be such that*

$$\phi' \in \mathcal{B}_\alpha, \;\; \mathcal{D}(\phi) \in D(A) \;\; \textit{and} \;\; \mathcal{D}(\phi') = -A\mathcal{D}(\phi) + f(\phi).$$

*Then, the mild solution $u$ of the problem* (8.1) *is a strict solution of the problem* (8.1). □

***Proof*** Let $p > 0$ and $u$ be the mild solution of the problem (8.1) associated to $\phi$. We consider the following problem:

$$\begin{cases} \mathcal{D}(w_t) = T(t)\mathcal{D}(\phi') + \displaystyle\int_0^t T(t-s) f'(u_s) w_s ds, & \text{for } t \in [0, p] \\ w_0 = \phi' \end{cases} \tag{8.9}$$

and $z \in C((-\infty, p]; X_\alpha)$ defined by

$$z(t) = \begin{cases} \phi(0) + \displaystyle\int_0^t w(s)ds, & \text{for } t \in [0, p] \\ \phi(t) & \text{for } t \leq 0. \end{cases} \tag{8.10}$$

Then (8.9) has a unique mild and continuous solution $w$ on $(-\infty, p]$. Also, one can recall the following lemma that plays an important role in the proof of this current theorem.

**Lemma 8.3.2 ([7])** *The function $z$ defined above verifies*

$$z_t = \phi + \int_0^t w_s ds, \qquad \textit{for} \quad t \in [0, p]. \tag{8.11}$$

Note that our objective is to show that $u = z$ on $[0, p]$. Using (8.9), we get

$$\int_0^t \mathcal{D}(w_s)ds = \int_0^t T(t-s)\mathcal{D}(\phi')ds + \int_0^t \int_0^s T(s-\tau)f'(u_\tau)w_\tau d\tau ds. \tag{8.12}$$

For $t \in [0, p]$, we have

$$\frac{d}{dt}\int_0^t T(t-s)f(z_s)ds = T(t)f(\phi) + \int_0^t T(t-s)f'(z_s)w_s ds. \tag{8.13}$$

Consequently,

$$\int_0^t T(s)f(\phi)ds = \int_0^t T(t-s)f(z_s)ds - \int_0^t \int_0^s T(s-\tau)f'(z_\tau)w_\tau d\tau ds. \tag{8.14}$$

Using Eq. (8.11), it follows that

$$\begin{aligned} \mathcal{D}(z_t) &= \mathcal{D}(\phi) + \int_0^t T(t-s)\Big(-A\mathcal{D}(\phi) + f(\phi)\Big)ds \\ &\quad + \int_0^t \int_0^s T(s-\tau)f'(u_\tau)w_\tau d\tau ds \\ &= T(t)\mathcal{D}(\phi) + \int_0^t T(s)f(\phi)ds + \int_0^t \int_0^s T(s-\tau)f'(u_\tau)w_\tau d\tau ds. \end{aligned}$$

Using Eq. (8.14), we have

$$\begin{aligned} \mathcal{D}(z_t) &= T(t)\mathcal{D}(\phi) + \int_0^t T(t-s)f(z_s)ds \\ &\quad + \int_0^t \int_0^s T(s-\tau)\Big(f'(u_\tau) - f'(z_\tau)\Big)w_\tau d\tau ds. \end{aligned} \tag{8.15}$$

Therefore,

$$\mathcal{D}(u_t - z_t) = \int_0^t T(t-s)\Big(f(u_s) - f(z_s)\Big)ds$$
$$- \int_0^t \int_0^s T(s-\tau)\Big(f'(u_\tau) - f'(z_\tau)\Big)w_\tau d\tau ds. \tag{8.16}$$

By Fubini's theorem, we get that

$$\mathcal{D}(u_t - z_t) = \int_0^t T(t-s)\Big(f(u_s) - f(z_s)\Big)ds$$
$$- \int_0^t \Big(\int_0^{t-s} T(\tau)d\tau\Big)\Big(f'(u_s) - f'(z_s)\Big)w_s ds. \tag{8.17}$$

Then, we put for $t \in [0, p]$,

$$h(t) = \int_0^t T(t-s)\Big(f(u_s) - f(z_s)\Big)ds$$
$$- \int_0^t \Big(\int_0^{t-s} T(\tau)d\tau\Big)\Big(f'(u_s) - f'(z_s)\Big)w_s ds,$$

to obtain for some positive constants $k$ and $C_1$,

$$\begin{aligned}
|h(t)|_\alpha &= \Big|\int_0^t T(t-s)\Big(f(u_s) - f(z_s)\Big)ds \\
&\quad - \int_0^t \Big(\int_0^{t-s} T(\tau)d\tau\Big)\Big(f'(u_s) - f'(z_s)\Big)w_s ds\Big|_\alpha \\
&\le \int_0^t \Big|T(t-s)\Big(f(u_s) - f(z_s)\Big)\Big|_\alpha ds \\
&\quad + \int_0^t \int_0^{t-s} \Big|T(\tau)(f'(u_s) - f'(z_s))w_s\Big|_\alpha d\tau ds \\
&\le kM_\alpha \int_0^t \frac{e^{\omega(t-s)}}{(t-s)^\alpha}|u_s - z_s|_{\mathcal{B}_\alpha} ds \\
&\quad + C_1 M_\alpha \int_0^t \Big(\int_0^{t-s} \frac{e^{\omega\tau}}{\tau^\alpha} d\tau\Big)|u_s - z_s|_{\mathcal{B}_\alpha} ds.
\end{aligned}$$

One can write for $\omega > 0$

$$\begin{aligned}
\int_0^{t-s} \frac{e^{\omega\tau}}{\tau^\alpha} d\tau &\le e^{\omega(t-s)} \int_0^{t-s} \frac{1}{\tau^\alpha} d\tau \\
&\le e^{\omega(t-s)} \Big[\frac{1}{1-\alpha}\frac{1}{\tau^{\alpha-1}}\Big]_0^{t-s}
\end{aligned}$$

$$\leq e^{\omega(t-s)} \frac{t-s}{1-\alpha} \frac{1}{(t-s)^{\alpha}}$$
$$\leq e^{\omega(t-s)} \frac{p}{1-\alpha} \frac{1}{(t-s)^{\alpha}}.$$

Therefore,

$$|h(t)|_{\alpha} \leq kM_{\alpha} \int_0^t \frac{e^{\omega(t-s)}}{(t-s)^{\alpha}} |u_s - z_s|_{\mathcal{B}_{\alpha}} ds + \frac{C_1 p M_{\alpha}}{1-\alpha} \int_0^t \frac{e^{\omega(t-s)}}{(t-s)^{\alpha}} |u_s - z_s|_{\mathcal{B}_{\alpha}} ds.$$

Moreover, since for all $\theta \in (-\infty, 0]$, $u(\theta) = z(\theta)$, then one has for all $s \in [0, t]$,

$$|u_s - z_s|_{\mathcal{B}_{\alpha}} \leq \max_{0 \leq \tau \leq t} \left| u(\tau) - z(\tau) \right|_{\alpha}.$$

Thus,

$$|h(t)|_{\alpha} \leq \left( kM_{\alpha} + \frac{C_1 p M_{\alpha}}{1-\alpha} \right) \left( \int_0^p \frac{e^{\omega \tau}}{\tau^{\alpha}} d\tau \right) \max_{0 \leq \tau \leq p} \left| u(\tau) - z(\tau) \right|_{\alpha}.$$

Using Lemma 8.3.1, one obtains

$$|u_t - z_t|_{\mathcal{B}_{\alpha}} \leq \left( kM_{\alpha} + \frac{C_1 p M_{\alpha}}{1-\alpha} \right) \left( \int_0^p \frac{e^{\omega \tau}}{\tau^{\alpha}} d\tau \right) \max_{0 \leq \tau \leq p} \left| u(\tau) - z(\tau) \right|_{\alpha}.$$

One can choose $p > 0$ small enough such that

$$\left( kM_{\alpha} + \frac{C_1 p M_{\alpha}}{1-\alpha} \right) \left( \int_0^p \frac{e^{\omega \tau}}{\tau^{\alpha}} d\tau \right) < 1.$$

It follows that $u = z$ in $(-\infty, p]$ and that leads to $u$ continuously differentiable on $[0, p]$ with respect to the $\alpha$-norm. In order to extend the solution to $[p, 2p]$, we consider the following problems:

$$\begin{cases} \mathcal{D}(w_t) = T(t-p)\mathcal{D}(u'_p) + \int_p^t T(t-s) f'(u_s) w_s ds & \text{for } t \in [p, 2p] \\ w_p = u'_p, \end{cases}$$

and $\tilde{z} \in C((-\infty, 2p]; X_{\alpha})$ defined by

$$\tilde{z}(t) = \begin{cases} u_p(0) + \int_p^t w(s) ds & \text{for } t \in [p, 2p] \\ z(t) & \text{for } t \leq p. \end{cases}$$

Using the same technique, one obtains that $u = \tilde{z}$ on $(-\infty, 2p]$. Proceeding inductively, solution $u$ is uniquely extended to $[np, (n+1)p]$ for all $n \in \mathbb{N}^*$ with respect to the $\alpha$-norm. Since $X_\alpha \hookrightarrow X$, one obtains that $u \in C^1([0, +\infty); X)$. Finally, using Theorem 8.2.2 , $u$ is the strict solution defined on $\mathbb{R}$. □

## 8.4 Smoothness Results of the Operator Solution

Let $K : D(K) \subseteq Y \to Y$ be a closed linear operator with dense domain $D(K)$ in a Banach space $Y$. We denote by $\sigma(K)$ the spectrum of $K$.

**Definition 8.4.1** The essential spectrum $\sigma_{ess}(K)$ of $K$ is the set of all $\lambda \in \mathbb{C}$ such that at least one of the following relations holds:

*(i)* The range $Im(\lambda I - K)$ is not closed.

*(ii)* The generalized eigenspace $M_\lambda(K) = \bigcup_{n \geq 0} ker(\lambda I - K)^n$ of $\lambda$ is infinite-dimensional.

*(iii)* $\lambda$ is a limit of $\sigma(K)$, that is, $\lambda \in \overline{\sigma(K) - \{\lambda\}}$.

□

The essential radius denoted by $r_{ess}(K)$ is given by

$$r_{ess}(K) = \sup\{|\lambda| : \lambda \in \sigma_{ess}(K)\}.$$

**Definition 8.4.2** The spectral bound $s(A)$ of the linear operator $A$ is defined as

$$s(A) = \sup\{Re\lambda : \lambda \in \sigma(A)\}.$$

**Definition 8.4.3** The type of the linear operator $(T(t))_{t\geq 0}$ is defined by

$$\omega_0(T) = \inf\left\{\omega \in \mathbb{R} : \sup_{t\geq 0}\left\{e^{-\omega t}|T(t)| < \infty\right\}\right\}.$$

In the sequel, we recall the $\chi$ measure of noncompactness, which will be used in the next to analyse the spectral properties of semigroup solution. The $\chi$ measure of noncompactness for a bounded set $H$ of a Banach space $Y$ with the norm $|.|_Y$ is defined by

$$\chi(H) = \inf\{\epsilon > 0 : \ H \text{ has a finite cover of diameter} < \epsilon\}.$$

The following results are some basic properties of the $\chi$ measure of noncompactness.

**Lemma 8.4.1 ([17])** *Let $A_1$ and $A_2$ be bounded sets of a Banach space $Y$. Then:*

(i) $\chi(A_1) \leq dia(A_1)$, *where* $dia(A_1) = \sup_{x,y \in A_1} |x-y|$.
(ii) $\chi(A_1) = 0$ *if and only if $A_1$ is relatively compact in $Y$.*
(iii) $\chi(A_1 \bigcup A_2) = max\{\chi(A_1), \chi(A_2)\}$.
(iv) $\chi(\lambda A_1) = |\lambda|\chi(A_1)$, $\lambda \in \mathbb{R}$, *where* $\lambda A_1 = \{\lambda x : x \in A_1\}$.
(v) $\chi(A_1 + A_2) \leq \chi(A_1) + \chi(A_2)$, *where* $A_1 + A_2 = \{x + y : x \in A_1, y \in A_2\}$.
(vi) $\chi(A_1) \leq \chi(A_2)$ *if* $A_1 \subseteq A_2$.

**Definition 8.4.4** The essential norm of a bounded linear operator $K$ on $Y$ is defined by

$$|K|_{ess} = \inf\{M \geq 0 : \chi(K(B)) \leq M\chi(B) \text{ for any bounded set B in } Y\}.$$

Let $V = (V(t))_{t \geq 0}$ be a $c_0$-semigroup on a Banach space $Y$.

**Definition 8.4.5** The essential growth $\omega_{ess}(V)$ of $(V(t))_{t \geq 0}$ is defined by

$$\omega_{ess}(V) = \inf\left\{\omega \in \mathbb{R} : \sup_{t \geq 0} e^{-\omega t}|V(t)|_{ess} < \infty\right\}.$$

**Theorem 8.4.1 ([7])** *The essential growth bound of $(V(t))_{t \geq 0}$ is given by*

$$\omega_{ess}(V) = \lim_{t \to +\infty} \frac{1}{t} \log|V(t)|_{ess} = \inf_{t>0} \frac{1}{t} \log|V(t)|_{ess}. \tag{8.18}$$

*Moreover,*

$$r_{ess}(V(t)) = exp(t\omega_{ess}(V)), \text{ for } t \geq 0. \tag{8.19}$$

Assume now that:

**(H$_5$)** The semigroup $(T(t))_{t \geq 0}$ is compact for $t > 0$.

**Theorem 8.4.2** *Assume that* **(H$_1$)**, **(H$_2$)**, **(H$_3$)**, **(H$_4$)**, *and* **(H$_5$)** *hold. Then, the solution $u(., \phi)$ of Eq.* (8.1) *is decomposed as follows:*

$$u_t(., \phi) = \mathcal{U}(t)\phi + \mathcal{W}(t)\phi, \text{ for } t \geq 0,$$

*where $\mathcal{W}(t)$ is a compact operator on $\mathcal{B}_\alpha$, for each $t > 0$, and $\mathcal{U}(t)$ is the semigroup solution of the following equation:*

$$\begin{cases} \frac{d}{dt}\mathcal{D}(x_t) = -A\mathcal{D}(x_t) & \text{for } t \geq 0, \\ x_0 = \phi \in \mathcal{B}_\alpha. \end{cases} \tag{8.20}$$

***Proof*** Let $\mathcal{U}(t)$ be defined by

$$(\mathcal{U}(t)\phi)(\theta) = \begin{cases} \phi(t+\theta) & \text{for} \quad t+\theta \le 0 \\ \\ v(t+\theta) & \text{for} \quad t+\theta \ge 0, \end{cases} \tag{8.21}$$

where $v$ is a unique solution of the problem

$$\begin{cases} \mathcal{D}(v_t) = T(t)\mathcal{D}(\phi) & \text{for} \quad t \ge 0 \\ \\ v(t) = \phi(t) & \text{for} \quad t \le 0. \end{cases} \tag{8.22}$$

We can write $\mathcal{W}(t)\phi = w_t(.,\phi) = u_t(.,\phi) - \mathcal{U}(t)\phi = u_t(.,\phi) - v_t(.,\phi)$. Then,

$$\mathcal{D}(\mathcal{W}(t)\phi) = \mathcal{D}(u_t(.,\phi)) - \mathcal{D}(v_t(.,\phi)) = \int_0^t T(t-s)f(u_s)ds.$$

Consequently,

$$\begin{cases} \mathcal{D}(w_t) = h(t,\phi) = \displaystyle\int_0^t T(t-s)f(u_s)ds & \text{for} \quad t \ge 0, \\ \\ w_0 = 0 \quad \text{for} \quad t \le 0. \end{cases} \tag{8.23}$$

Let $\{\phi_k\}_{k\ge 0}$ be a bounded sequence in $\mathcal{B}_\alpha$. We will show that the family $\{h(.,\phi_k) : k \ge 0\}$ is equicontinuous and bounded on $C([0,\sigma]; X_\alpha)$, for any $\sigma > 0$ fixed. For all $0 < \alpha < \beta < 1$, there exists a positive constant $C$ such that

$$\begin{aligned} |A^\beta h(t,\phi_k)| &= |A^\beta \int_0^t T(t-s)f(u_s(.,\phi_k))ds| \\ &\le \int_0^t |A^\beta T(t-s)f(u_s(.,\phi_k))|ds \\ &\le M_\beta C \int_0^t \frac{e^{\omega s}}{s^\beta} ds, \end{aligned}$$

for every $k \ge 0$.

Using the compactness of the operator $A^{-\beta} : X \to X_\alpha$, we get that the set $\{h(t,\phi_k) : k \ge 0\}$ is relatively compact in $X_\alpha$ for each $t \ge 0$. Now, let us prove the equicontinuity of the family $\{h(.,\phi_k) : k \ge 0\}$ in the $\alpha$-norm. For this purpose, let $t > t_0 \ge 0$. Then,

$$\begin{aligned} A^\alpha h(t,\phi_k) - A^\alpha h(t_0,\phi_k) &= \int_0^t A^\alpha T(t-s) f(u_s) ds - \int_0^{t_0} A^\alpha T(t_0-s) f(u_s) ds \\ &= \int_0^{t_0} A^\alpha [T(t-s) - T(t_0-s)] f(u_s) ds \\ &\quad + \int_{t_0}^t A^\alpha T(t-s) f(u_s) ds \\ &= [T(t-t_0) - I] \int_0^{t_0} A^\alpha T(t_0-s) f(u_s) ds \\ &\quad + \int_{t_0}^t A^\alpha T(t-s) f(u_s) ds. \end{aligned}$$

We obtain that

$$\left| \int_{t_0}^t A^\alpha T(t-s) f(u_s) ds \right| \leq M_\alpha k \int_{t_0}^t \frac{e^{\omega s}}{s^\alpha} ds \to 0 \quad \text{as} \quad t \to t_0 \text{ uniformly in } \phi_k.$$

Moreover, since $\{A^\alpha \int_0^{t_0} T(t_0-s) f(u_s(.,\phi_k)) ds : \; k \geq 0\}$ is relatively compact in $X$, then there is a compact set $\Gamma$ in $X$ such that

$$\int_0^{t_0} A^\alpha T(t_0-s) f(u_s(.,\phi_k)) ds \subset \Gamma \quad \text{for all} \quad \phi_k.$$

It is well-known by the Banach–Steinhaus theorem that

$$\lim_{t \to t_0} \sup_{x \in \Gamma} |(T(t-t_0) - I)x| = 0.$$

Thus,

$$\lim_{t \to t_0} |h(t,\phi_k) - h(t_0,\phi_k)|_\alpha = 0 \quad \text{uniformly in} \quad \phi_k.$$

Using the same argument, we also obtain for $t_0 > t$,

$$\lim_{t \to t_0} |h(t,\phi_k) - h(t_0,\phi_k)|_\alpha = 0 \quad \text{uniformly in} \quad \phi_k.$$

Therefore, the family $\{h(.,\phi_k) : \; k \geq 0\}$ is relatively compact on $C([0,\sigma]; X_\alpha)$ for each $\sigma > 0$. Then, there exists a subsequence $\{\phi_k : \; k \geq 0\}$ such that $h(t,\phi_k)$ converges as $k \to +\infty$ uniformly on $[0,\sigma]$ to some function $h(t)$ with respect to the $\alpha$-norm. Let $w_t^k$ be the solution of problem (8.23) with the initial data $\phi = \phi_k$. Then,

$$\mathcal{D}(w_t^j - w_t^k) = h(t,\phi_j) - h(t,\phi_k).$$

Using Lemma 8.3.1, we obtain

$$|w_t^j - w_t^k|_{\mathcal{B}_\alpha} \leq b(t) \sup_{0 \leq s \leq t} |h(t, \phi_j) - h(t, \phi_k)|_\alpha,$$

which implies that $\{w_t^k\}_{k \geq 0} = \{w_t(., \phi_k)\}_{k \geq 0}$ is a Cauchy sequence in $\mathcal{B}_\alpha$. Therefore, $\mathcal{W}(t)$ is compact in $\mathcal{B}_\alpha$. □

**Definition 8.4.6** $\mathcal{D}$ is said to be stable if the zero solution of the difference system

$$\begin{cases} \mathcal{D}(x_t) = 0 \quad \text{for} \quad t \geq 0, \\ x_0(t) = \phi(t) \quad \text{for} \quad t \leq 0 \end{cases}$$

is exponentially stable. □

Now, we give the definitions of fading memory spaces that will be used later on. For $\phi \in \mathcal{B}$, $t \geq 0$ and $\theta \leq 0$, we define the following:

$$[S(t)\phi](\theta) = \begin{cases} \phi(0) \quad \text{if} \quad t + \theta \geq 0, \\ \phi(t + \theta) \quad \text{if} \quad t + \theta < 0. \end{cases} \tag{8.24}$$

Then, $\{S(t)\}_{t \geq 0}$ is a strongly continuous semigroup on $\mathcal{B}$. We set

$$S_0(t) = S(t)/\mathcal{B}_0, \quad \text{where} \quad \mathcal{B}_0 = \{\phi \in \mathcal{B} : \ \phi(0) = 0\}.$$

**Definition 8.4.7** [7] We say that $\mathcal{B}$ is a uniform fading memory space if the following conditions hold:

(i) If a uniformly bounded sequence $(\phi_n)_{n \in \mathbb{N}}$ in $C_{00}$ converges to a function $\phi$ compactly on $(-\infty, 0]$, then $\phi$ is in $\mathcal{B}$ and $|\phi_n - \phi|_{\mathcal{B}} \to 0$ as $n \to +\infty$.
(ii) $|S_0(t)|_{\mathcal{B}} \to 0$ as $t \to +\infty$. □

**Lemma 8.4.2 ([7])** *If $\mathcal{B}$ is a uniform fading memory space, then $K$ and $M$ can be chosen such that $K$ is bounded on $\mathbb{R}^+$ and $M(t) \to 0$ as $t \to +\infty$.* □

**Lemma 8.4.3** *If $\mathcal{B}$ is a uniform fading memory space, then $\mathcal{B}_\alpha$ is a uniform fading memory space.* □

***Proof*** Let $(\phi_n)_{n \in \mathbb{N}}$ in $C_{00}$ be a uniformly bounded sequence that converges to a function $\phi$ compactly on $(-\infty, 0]$. Then $\phi$ is in $\mathcal{B}$ and $|\phi_n - \phi|_{\mathcal{B}} \to 0$ as $n \to +\infty$ since $\mathcal{B}$ is a uniform fading memory space. Using $(\mathbf{H}_2)$, one can write $A^{-\alpha}\phi \in \mathcal{B}$ since $\phi \in \mathcal{B}$. $A^{-\alpha}\phi \in \mathcal{B}$ leads to the existence of $A^{-\alpha}\phi(\theta)$. We know that $R(A^{-\alpha}) = D(A^\alpha)$. For this reason, $\left|A^{-\alpha}\phi(\theta)\right|_\alpha$ is well-defined. The fact that $A^{-\alpha}$ is bounded linear operator implies $|\phi(\theta)|_\alpha$ exists. Therefore, $\phi(\theta) \in D(A^\alpha)$ for all $\theta \leq 0$. Also,

$$\left|A^{-\alpha} A^\alpha \phi\right|_{\mathcal{B}} = |\phi|_{\mathcal{B}} < \infty.$$

Using again the boundedness of $A^{-\alpha}$, one obtains the existence of $|A^{\alpha}\phi|_{\mathcal{B}}$. Thus, $A^{\alpha}\phi \in \mathcal{B}$. Hence, we establish that $\phi \in \mathcal{B}_{\alpha}$. Moreover,

$$|\phi_n - \phi|_{\mathcal{B}} = \left|A^{-\alpha}A^{\alpha}(\phi_n - \phi)\right|_{\mathcal{B}} \to 0 \quad \text{as} \quad n \to +\infty.$$

Since $A^{-\alpha}$ is a bounded linear operator, one obtains

$$\left|A^{\alpha}(\phi_n - \phi)\right|_{\mathcal{B}} = |\phi_n - \phi|_{\mathcal{B}_{\alpha}} \to 0 \quad \text{as} \quad n \to +\infty.$$

Consequently, the condition $(i)$ of Definition 8.4.7 is satisfied.

Now, we have to show that the condition $(ii)$ of Definition 8.4.7 is verified. In order to do this, we use the fact that $A^{-\alpha}$ is a bounded linear operator and $\mathcal{B}$ is a uniform fading memory space to write

$$|S_0(t)|_{\mathcal{B}} = \left|A^{-\alpha}A^{\alpha}S_0(t)\right|_{\mathcal{B}} \to 0 \quad \text{as} \quad t \to +\infty$$

and

$$|S_0(t)|_{\mathcal{B}_{\alpha}} \to 0 \quad \text{as} \quad t \to +\infty.$$

Hence, the condition $(ii)$ is satisfied. Finally, $\mathcal{B}_{\alpha}$ is a uniform fading memory space. □

Now, we have to prove that $\mathcal{U}(t)$ is exponentially stable. It is known that $\mathcal{U}(t)$ in Theorem 8.4.2 is defined by

$$(\mathcal{U}(t)\phi)(\theta) = \begin{cases} \phi(t+\theta) & \text{for} \quad t+\theta \leq 0 \\ v(t+\theta) & \text{for} \quad t+\theta \geq 0, \end{cases}$$

where $v$ is a unique solution for the same initial data $\phi$ of the following problem:

$$\begin{cases} \mathcal{D}(v_t) = T(t)\mathcal{D}(\phi) & \text{for} \quad t \geq 0 \\ v(t) = \phi(t) & \text{for} \quad t \leq 0. \end{cases}$$

Using the superposition principle of solutions of linear systems, we have

$$v(t) = x(t) + y(t) \ \text{ for } t \in \mathbb{R},$$

where

$$\begin{cases} \mathcal{D}(x_t) = 0 & \text{for} \quad t \geq 0, \\ x(t) = \phi(t) & \text{for} \quad t \leq 0 \end{cases} \tag{8.25}$$

and

$$\begin{cases} \mathcal{D}(y_t) = T(t)\mathcal{D}(\phi) \quad \text{for} \quad t \geq 0, \\ y(t) = 0 \quad \text{for} \quad t \leq 0. \end{cases} \tag{8.26}$$

Now, let $K_\infty = \sup_{s\geq 0} K(s)$. We have the following result.

**Theorem 8.4.3** *Assume that* $(\mathbf{H}_1)$, $(\mathbf{H}_2)$, *and* $(\mathbf{H}_3)$ *hold. Moreover, suppose that* $\mathcal{B}_\alpha$ *is a uniform fading memory space,* $\mathcal{D}$ *is stable, the semigroup* $\{T(t)\}_{t\geq 0}$ *is exponentially stable, and* $K_\infty|\mathcal{D}_0| < 1$. *Then, the semigroup solution* $\{\mathcal{U}(t)\}_{t\geq 0}$ *defined in Theorem 8.4.2 is exponentially stable.* □

***Proof*** Since $y$ verifies problem (8.26) and $\mathcal{B}_\alpha$ is a uniform fading memory space, then, using Axiom (**A**)-(iii), one can write for $t \geq s \geq \epsilon > 0$

$$\begin{aligned} |y_s|_{\mathcal{B}_\alpha} &\leq K(\epsilon) \sup_{s-\epsilon\leq\tau\leq s} |y(\tau)|_\alpha + M(\epsilon)|y_{s-\epsilon}|_{\mathcal{B}_\alpha} \\ &\leq K(\epsilon)|\mathcal{D}_0| \sup_{s-\epsilon\leq\tau\leq s} |y_\tau|_{\mathcal{B}_\alpha} + K(\epsilon) \sup_{s-\epsilon\leq\tau\leq s} |T(\tau)\mathcal{D}(\phi)|_\alpha + M(\epsilon)|y_{s-\epsilon}|_{\mathcal{B}_\alpha} \\ &\leq K(\epsilon)|\mathcal{D}_0| \sup_{s-\epsilon\leq\tau\leq s} |y_\tau|_{\mathcal{B}_\alpha} + K(\epsilon) \sup_{s-\epsilon\leq\tau\leq s} |T(\tau)\mathcal{D}(\phi)|_\alpha \\ &\quad + M(\epsilon) \sup_{s-\epsilon\leq\tau\leq s} |y_\tau|_{\mathcal{B}_\alpha}. \end{aligned}$$

Therefore, taking $\epsilon > 0$ such that $s - \epsilon \geq t - 2\epsilon \geq 0$, then

$$\begin{aligned} |y_s|_{\mathcal{B}_\alpha} \leq K(\epsilon)|\mathcal{D}_0| \sup_{t-2\epsilon\leq\tau\leq s} |y_\tau|_{\mathcal{B}_\alpha} + K(\epsilon) \sup_{t-2\epsilon\leq\tau\leq s} |T(\tau)\mathcal{D}(\phi)|_\alpha \\ + M(\epsilon) \sup_{t-2\epsilon\leq\tau\leq s} |y_\tau|_{\mathcal{B}_\alpha}. \end{aligned}$$

Now, one can write

$$\begin{aligned} \sup_{t-2\epsilon\leq s\leq t} |y_s|_{\mathcal{B}_\alpha} &\leq \sup_{t-2\epsilon\leq s\leq t} \Big\{ K_\infty|\mathcal{D}_0| \sup_{t-2\epsilon\leq\tau\leq s} |y_\tau|_{\mathcal{B}_\alpha} \\ &\quad + K_\infty \sup_{t-2\epsilon\leq\tau\leq s} |T(\tau)\mathcal{D}(\phi)|_\alpha + M(\epsilon) \sup_{t-2\epsilon\leq\tau\leq s} |y_\tau|_{\mathcal{B}_\alpha} \Big\} \\ &\leq K_\infty|\mathcal{D}_0| \sup_{t-2\epsilon\leq s\leq t} |y_s|_{\mathcal{B}_\alpha} + K_\infty \sup_{t-2\epsilon\leq s\leq t} |T(s)\mathcal{D}(\phi)|_\alpha \\ &\quad + M(\epsilon) \sup_{t-2\epsilon\leq s\leq t} |y_s|_{\mathcal{B}_\alpha}. \end{aligned}$$

Since $M(\epsilon) \to 0$ as $\epsilon \to +\infty$, then we choose $\epsilon$ big enough such that $0 < 1 - K_\infty|\mathcal{D}_0| - M(\epsilon)$. We obtain that

$$|y_t|_{\mathcal{B}_\alpha} \leq \frac{K_\infty}{\left(1 - K_\infty|\mathcal{D}_0| - M(\epsilon)\right)} \sup_{t-2\epsilon \leq s \leq t} |T(s)\mathcal{D}(\phi)|_\alpha.$$

Since $\{T(t)\}_{t\geq 0}$ is exponentially stable, then there exist positive constants $\alpha'$ and $\beta'$ such that $|y_t|_{\mathcal{B}_\alpha} \leq \beta' e^{-\alpha' t}$ for all $t \geq 0$.

Since $\mathcal{D}$ is stable, then $x_t(\phi) \to 0$ as $t \to +\infty$. On the other hand, we have

$$\mathcal{U}(t)\phi = x_t(\phi) + y_t(\phi).$$

Then, it follows that $\mathcal{U}(t) \to 0$ as $t \to 0$ and $\{\mathcal{U}(t)\}_{t\geq 0}$ is exponentially stable. □

In the sequel, we give the following.

**Theorem 8.4.4** *Assume that there exists $r > 0$ such that the elements $\phi \in \mathcal{B}_\alpha$ are continuous from $[-r, 0]$ to $X_\alpha$. If $\mathcal{D}(\phi) = \phi(0) - q\phi(-r)$ for all $\phi \in \mathcal{B}_\alpha$ with $0 < q < 1$ and $\mathcal{B}_\alpha$ a uniform fading memory space, then $\mathcal{D}$ is stable.* □

***Proof*** Since $\mathcal{D}(x_t) = 0$ and $x_0 = \phi$, then for all $t \in [0, r]$, we have $x(t) = qx(t - r)$. Therefore,

$$|x(t)|_\alpha \leq q|\phi(t - r)|_\alpha.$$

Also, for all $t \in [r, 2r]$,

$$|x(t, \phi)|_\alpha \leq q^2|\phi(t - 2r)|_\alpha.$$

Inductively, for all $t \in [(n - 1)r, nr]$, we have

$$|x(t, \phi)|_\alpha \leq q^n|\phi(t - nr)|_\alpha;$$

since $t \in [(n - 1)r, nr]$, then $t - nr \in [-r, 0]$. Furthermore, $\mathcal{B}_\alpha$ is assumed to be the space of functions from $(-\infty, 0]$ to $X_\alpha$ that are continuous on $[-r, 0]$. Thus, for all $t \in [(n - 1)r, nr]$,

$$|x(t, \phi)|_\alpha \leq q^n \sup_{-r\leq s\leq 0} |\phi(s)|_\alpha,$$

for all $\phi \in \mathcal{B}_\alpha$.

Thus, there exist $\alpha = -\frac{\ln(q)}{r} > 0$ and $C > 0$ such that

$$\begin{aligned}|x(t, \phi)|_\alpha &\leq q^n \sup_{-r\leq s\leq 0} |\phi(s)|_\alpha \\ &\leq Ce^{-\alpha t}.\end{aligned}$$

Hence, for all $\phi \in \mathcal{B}_\alpha$,

$$\lim_{t\to+\infty} x(t,\phi) = 0.$$

Now, let $\phi \in \mathcal{B}_\alpha$ such that $|\phi|_{\mathcal{B}_\alpha} \le 1$.

Using again Axiom (**A**)-(iii) and the fact that $\mathcal{B}_\alpha$ is a uniform fading memory space, we have for $t \ge s \ge \epsilon > 0$

$$\begin{aligned}|x_s(.,\phi)|_{\mathcal{B}_\alpha} &\le K(\epsilon) \sup_{s-\epsilon\le\tau\le s} |x(\tau,\phi)|_\alpha + M(\epsilon)|x_{s-\epsilon}(.,\phi)|_{\mathcal{B}_\alpha}\\ &\le K_\infty \sup_{s-\epsilon\le\tau\le s} |x(\tau,\phi)|_\alpha + M(\epsilon)|x_{s-\epsilon}(.,\phi)|_{\mathcal{B}_\alpha}.\end{aligned}$$

Choosing $\epsilon > 0$ such that $s-\epsilon \ge t-2\epsilon \ge 0$, we have

$$\begin{aligned}|x_s(.,\phi)|_{\mathcal{B}_\alpha} &\le K_\infty \sup_{s-\epsilon\le\tau\le s} |x(\tau,\phi)|_\alpha + M(\epsilon)|x_{s-\epsilon}(.,\phi)|_{\mathcal{B}_\alpha}\\ &\le K_\infty \sup_{t-2\epsilon\le\tau\le s} |x(\tau,\phi)|_\alpha + M(\epsilon)|x_{s-\epsilon}(.,\phi)|_{\mathcal{B}_\alpha}\\ &\le \sup_{t-2\epsilon\le s\le t}\left\{K_\infty \sup_{t-2\epsilon\le\tau\le s} |x(\tau,\phi)|_\alpha + M(\epsilon)|x_{s-\epsilon}(.,\phi)|_{\mathcal{B}_\alpha}\right\}\\ &\le K_\infty \sup_{t-2\epsilon\le s\le t} |x(s,\phi)|_\alpha + M(\epsilon) \sup_{t-2\epsilon\le s\le t} |x_s(.,\phi)|_{\mathcal{B}_\alpha}.\end{aligned}$$

Thus,

$$\sup_{t-2\epsilon\le s\le t} |x_s(.,\phi)|_{\mathcal{B}_\alpha} \le K_\infty \sup_{t-2\epsilon\le s\le t} |x(s,\phi)|_\alpha + M(\epsilon) \sup_{t-2\epsilon\le s\le t} |x_s(.,\phi)|_{\mathcal{B}_\alpha}.$$

Since $M(\epsilon) \to 0$ as $\epsilon \to +\infty$, then we can choose $\epsilon$ big enough such that $0 < 1 - M(\epsilon)$. Therefore,

$$|x_t(.,\phi)|_{\mathcal{B}_\alpha} \le \frac{K_\infty}{(1-M(\epsilon))} \sup_{t-2\epsilon\le s\le t} |x(s,\phi)|_\alpha, \quad \text{for all } \phi \in \mathcal{B}_\alpha \text{ with } |\phi|_{\mathcal{B}_\alpha} \le 1.$$

Thus, $x_t(.,\phi) \to 0$ as $t \to +\infty$ whenever $\phi \in \mathcal{B}_\alpha$ and $|\phi|_{\mathcal{B}_\alpha} \le 1$. Hence, $\mathcal{D}$ is stable. □

*Example 8.4.1* Let $\gamma$ be a real number, $1 \le p < +\infty$, and $r > 0$. We define the space $C_r \times L^p_\gamma$ that consists of measurable functions $\varphi : (-\infty, 0] \to X$ that are continuous on $[-r, 0]$ such that $e^{\gamma\theta}|\varphi(\theta)|^p$ is measurable on $(-\infty, -r]$. Let us provide the space $C_r \times L^p_\gamma$ with the following norm:

$$|\varphi|_{\mathcal{B}} = \sup_{-r\le\theta\le 0} |\varphi(\theta)| + \int_{-\infty}^{-r} e^{\gamma\theta}|\varphi(\theta)|^p d\theta.$$

$(C_r \times L^p_\gamma, |.|_{\mathcal{B}})$ is a normed linear space satisfying Axioms (**A**) and (**B**).

**Corollary 8.4.1** *Suppose that assumptions* **(H**$_1$**)**, **(H**$_2$**)**, *and* **(H**$_3$**)** *hold, and there exists a positive constant $r$ such that all $\phi \in \mathcal{B}_\alpha$ imply that $\phi$ is continuous on $[-r, 0]$ with values in $X_\alpha$. Moreover, suppose that $\mathcal{B}_\alpha$ is a uniform fading memory space, $\mathcal{D}(\phi) = \phi(0) - q\phi(-r)$ for all $\phi \in \mathcal{B}_\alpha$, the semigroup $\{T(t)\}_{t\geq 0}$ is exponentially stable, and $K_\infty|\mathcal{D}_0| < 1$. Then, the semigroup solution $\{\mathcal{U}(t)\}_{t\geq 0}$ defined in Theorem 8.4.2 is exponentially stable.* □

## 8.5 Linearized Stability of Solutions

Coming back to the operator $U(t)$ for $t \geq 0$ defined on $\mathcal{B}_\alpha$ by

$$U(t)(\phi) = u_t(., \phi),$$

where $u_t(., \phi)$ is the unique mild solution of the problem (8.1) for the initial condition $\phi \in \mathcal{B}_\alpha$, it is proved that the following result holds.

**Proposition 8.5.1 ([4])** *The family $(U(t))_{t\geq 0}$ is a nonlinear strongly continuous semigroup on $\mathcal{B}_\alpha$, that is:*

(i) $U(0) = I$.
(ii) $U(t + s) = U(t)U(s)$, *for* $t, s \geq 0$.
(iii) *For all $\phi \in \mathcal{B}_\alpha$, $U(t)(\phi)$ is a continuous function of $t \geq 0$ with values in $\mathcal{B}_\alpha$.*
(iv) *For $t \geq 0$, $U(t)$ is continuous from $\mathcal{B}_\alpha$ to $\mathcal{B}_\alpha$.*
(v) *$(U(t))_{t\geq 0}$ satisfies the following translation property, for $t \geq 0$ and $\theta \leq 0$:*

$$(U(t))(\theta) = \begin{cases} (U(t+\theta)(\phi))(0), & \text{if } t+\theta \geq 0, \\ \phi(t+\theta) & \text{if } t+\theta \leq 0. \end{cases} \tag{8.27}$$

It is now interesting to investigate the stability results of the equilibriums of the problem (8.1). Recalling that equilibrium means a constant solution $u^*$ of the problem (8.1). To preserve the generality, we can suppose that $u^* = 0$.

Now, let us assume that:

**(H**$_6$**)** $f : \mathcal{B}_\alpha \to X$ is differentiable at zero.

It is well-known that the linearized problem associated to problem (8.1) is given by

$$\begin{cases} \dfrac{d}{dt}\mathcal{D}(y_t) = -A\mathcal{D}(y_t) + L(y_t), & \text{for } t \geq 0, \\ y_0 = \phi \in \mathcal{B}_\alpha, \end{cases} \tag{8.28}$$

with $L = f'(0)$.

Let $(V(t))_{t\geq 0}$ be the semigroup solution on $\mathcal{B}_\alpha$ associated to the problem (8.28).

**Theorem 8.5.1** *Assume that* $(\mathbf{H}_1)$, $(\mathbf{H}_2)$, $(\mathbf{H}_3)$, $(\mathbf{H}_4)$, *and* $(\mathbf{H}_6)$ *hold. Then, for every* $t > 0$, *the derivative of* $U(t)$ *is* $V(t)$. □

***Proof*** Let $t \geq 0$ be fixed and $\phi \in \mathcal{B}_\alpha$. One has

$$\mathcal{D}\Big[U(t)\phi - V(t)\phi\Big] = \int_0^t T(t-s)\Big[f(U(s)(\phi)) - L(V(s)(\phi))\Big]ds.$$

Let us set

$$w_t = U(t)(\phi) - V(t)(\phi)$$

and

$$h(t) = \int_0^t T(t-s)\Big[f(U(s)(\phi)) - L(V(s)(\phi))\Big]ds.$$

Then, we can write

$$h(t) = \int_0^t T(t-s)\Big[f(U(s)(\phi)) - f(V(s)(\phi))\Big]ds + \int_0^t T(t-s)\Big[f(V(s)(\phi)) - L(V(s)(\phi))\Big]ds.$$

Using Lemma 8.3.1, we obtain

$$|w_t|_{\mathcal{B}_\alpha} \leq b(t) \sup_{0\leq s\leq t} |h(s)|_\alpha, \quad \text{for} \quad t \in [0, T].$$

Moreover,

$$|h(t)|_\alpha \leq kM_\alpha \int_0^t \frac{e^{\omega(t-s)}}{(t-s)^\alpha} |w_s|_{\mathcal{B}_\alpha} ds + M_\alpha \int_0^t \frac{e^{\omega(t-s)}}{(t-s)^\alpha} \Big|f(V(s)(\phi)) - L(V(s)(\phi))\Big| ds.$$

Using the fact that $f$ is differentiable at zero with differential $L$ at zero, we can state that for all $\epsilon > 0$, there exists $\eta > 0$ such that

$$M_\alpha \int_0^t \frac{e^{\omega(t-s)}}{(t-s)^\alpha} \Big|f(V(s)(\phi)) - L(V(s)(\phi))\Big| ds \leq \epsilon |\phi|_{\mathcal{B}_\alpha} \qquad \forall \phi \in \mathcal{B}_\alpha \text{ with } |\phi|_{\mathcal{B}_\alpha} < \eta.$$

Note that $w_0 = 0$, so we can write

$$|w_s|_{\mathcal{B}_\alpha} \le \sup_{0\le\tau\le t} |w(\tau)|_\alpha \le |w_t|_{\mathcal{B}_\alpha}, \quad \text{for } s \in [0, t].$$

Therefore, for $t \in [0, T]$,

$$\begin{aligned}\sup_{0\le s\le t} |h(s)|_\alpha &\le \epsilon|\phi|_{\mathcal{B}_\alpha} + kM_\alpha\Big(\int_0^t \frac{e^{\omega s}}{s^\alpha} ds\Big)|w_t|_{\mathcal{B}_\alpha} \\ &\le \epsilon|\phi|_{\mathcal{B}_\alpha} + kM_\alpha\Big(\int_0^T \frac{e^{\omega s}}{s^\alpha} ds\Big)|w_t|_{\mathcal{B}_\alpha}.\end{aligned}$$

We can choose $T > 0$ small enough such that $kM_\alpha b(T)\Big(\int_0^T \frac{e^{\omega s}}{s^\alpha} ds\Big) < 1$.

Consequently, for all $|\phi|_{\mathcal{B}_\alpha} < \eta$,

$$|w_t|_{\mathcal{B}_\alpha} \le \frac{b(T)}{1 - kM_\alpha b(T)\Big(\int_0^T \frac{e^{\omega s}}{s^\alpha} ds\Big)}\epsilon|\phi|_{\mathcal{B}_\alpha}.$$

Thus, $U(t)$ is differentiable at zero for all $t \in [0, T]$ with $d_\phi U(t)(0) = V(t)$. Proceeding by steps, one can prove that $d_\phi U(t)(0) = V(t)$, for all $t > 0$. □

**Theorem 8.5.2** *Assume that* $(\mathbf{H}_1)$, $(\mathbf{H}_2)$, $(\mathbf{H}_3)$, $(\mathbf{H}_4)$, *and* $(\mathbf{H}_6)$ *hold. If the zero equilibrium of* $(V(t))_{t\ge0}$ *is exponentially stable, then the zero equilibrium of* $(U(t))_{t\ge0}$ *is locally exponentially stable, which means that there exist* $\eta > 0$, $\beta > 0$, *and* $C \ge 1$ *such that for* $t \ge 0$,

$$|U(t)(\phi)|_{\mathcal{B}_\alpha} \le Ce^{-\beta t}|\phi|_{\mathcal{B}_\alpha} \quad \text{for all } \phi \in \mathcal{B}_\alpha \text{ with } |\phi|_{\mathcal{B}_\alpha} \le \eta.$$

*Moreover, if* $\mathcal{B}_\alpha$ *can be decomposed as* $\mathcal{B}_\alpha = \mathcal{B}_\alpha^1 \oplus \mathcal{B}_\alpha^2$, *where* $\mathcal{B}_\alpha^i$ *are* $V$*-invariant subspaces of* $\mathcal{B}_\alpha$ *and* $\mathcal{B}_\alpha^1$ *a finite-dimensional with*

$$\omega_0 = \lim_{h\to+\infty} \frac{1}{h} \log \Big|V(h)/\mathcal{B}_\alpha^2\Big|_\alpha$$

*and*

$$\inf\{|\lambda| : \ \lambda \in \sigma(V(t)/\mathcal{B}_\alpha^1)\} > e^{\omega_0 t},$$

*then the zero equilibrium of* $(U(t))_{t\ge0}$ *is not stable, in the sense that there exist* $\epsilon > 0$, *a sequence* $(\phi_n)_{n\in\mathbb{N}}$ *converging to* 0, *and a sequence* $(t_n)_{n\in\mathbb{N}}$ *of positive real numbers such that* $|U(t_n)\phi_n|_\alpha > \epsilon$. □

The proof of this theorem is based on the Theorem 8.5.1 and the following theorem.

**Theorem 8.5.3 ([12])** *Let $(W(t))_{t\geq 0}$ be a nonlinear strongly continuous semigroup on the subset $\Omega$ of a Banach space $(X; ||.||)$. Assume that $x_0 \in \Omega$ is an equilibrium of $(W(t))_{t\geq 0}$ such that $W(t)$ is differentiable at $x_0$ for each $t \geq 0$ with $Z(t)$ the derivative of $W(t)$ at $x_0$. Then, $(Z(t))_{t\geq 0}$ is a strongly nonlinear continuous semigroup of bounded linear operators on $X$, and if the zero equilibrium of $(Z(t))_{t\geq 0}$ is exponentially stable, then the equilibrium $x_0$ of $(W(t))_{t\geq 0}$ is locally exponentially stable. Moreover, if $X$ can be decomposed as $X = X_1 \oplus X_2$, where $X_i$ are $Z$-invariant subspaces of $X$, $X_1$ a finite-dimensional with*

$$\omega = \lim_{h\to+\infty} \frac{1}{h} \log \left| |Z(h)/X_2| \right|$$

*and*

$$\inf\{|\lambda| : \quad \lambda \in \sigma(Z(t)/X_1)\} > e^{\omega t},$$

*then the zero equilibrium of $(W(t))_{t\geq 0}$ is not stable, in the sense that there exist $\epsilon > 0$, a sequence $(\phi_n)_{n\in\mathbb{N}}$ converging to 0, and a sequence $(t_n)_{n\in\mathbb{N}}$ of positive real numbers such that*

$$|W(t_n)\phi_n|_\alpha > \epsilon.$$

**Lemma 8.5.1 ([19], Corollary 1.2, page 43)** *Let $\Theta$ be a continuous and right differentiable function on $[a, b)$. If the right derivative function $d^+\Theta$ is continuous on $[a, b)$, then $\Theta$ is continuously differentiable on $[a, b)$.* □

Now, we make some sufficient conditions on $\mathcal{B}$ in order to determine $(A_V, D(A_V))$, the generator of the semigroup $(V(t))_{t\geq 0}$. So, we assume the following axiom:

**(C)**: Let $(\phi_n)_{n\geq 0}$ be a sequence in $\mathcal{B}$ such that $\phi_n \to 0$ as $n \to +\infty$ in $\mathcal{B}$; then, $\phi_n(\theta) \to 0$ as $n \to +\infty$ for all $\theta \leq 0$.

We can state the following result.

**Theorem 8.5.4** *Assume that* $(\mathbf{H}_1)$, $(\mathbf{H}_2)$, $(\mathbf{H}_3)$, $(\mathbf{H}_4)$, *and* $(\mathbf{H}_6)$ *hold. Moreover, suppose that $\mathcal{B}$ satisfies axioms* **(A)**, **(B)**, *and* **(C)**. *If $\mathcal{B}$ is a subspace of the space of continuous functions from $(-\infty, 0]$ into $X$, then $(A_V, D(A_V))$ is given by*

$$\begin{cases} D(A_V) = \Big\{\phi \in \mathcal{B}_\alpha : \ \phi' \in \mathcal{B}_\alpha, \ \mathcal{D}(\phi) \in D(A) \quad \textit{and} \\ \ \mathcal{D}(\phi') = -A\mathcal{D}(\phi) + L(\phi)\Big\}, \\ \\ A_V\phi = \phi' \quad \textit{for } \phi \in D(A_V). \end{cases}$$

***Proof*** Let $B$ be the infinitesimal generator of the semigroup $(V(t))_{t\geq 0}$ on $\mathcal{B}_\alpha$ and $\phi \in D(B)$. Then, one can write

$$\begin{cases} \lim\limits_{t\to 0^+} \dfrac{1}{t}(V(t)\phi - \phi) = \psi \text{ exists in } \mathcal{B}_\alpha, \\ \\ B\phi = \psi. \end{cases}$$

Using axiom (**C**), one obtains

$$\lim_{t\to 0^+} \frac{1}{t}(\phi(t+\theta) - \phi(\theta)) = \psi(\theta) \quad \text{ for } \theta \in (-\infty, 0).$$

It follows that the right derivative $d^+\phi$ exists on $(-\infty, 0)$ and is equal to $\psi$. The fact that each function in $\mathcal{B}_\alpha$ is continuous on $(-\infty, 0]$ leads to $d^+\phi$ continuous on $(-\infty, 0)$.

Using Lemma 8.5.1, we deduce that the function $\phi$ is continuously differentiable and $\phi' = \psi$ on $(-\infty, 0)$. Moreover,

$$\lim_{\theta\to 0} d^+\phi(\theta) = \psi(0),$$

which implies that the function $\phi$ is continuously differentiable from $(-\infty, 0]$ to $X_\alpha$ and $\phi' = \psi$ on $(-\infty, 0]$.

We have

$$\frac{1}{t}(T(t)\mathcal{D}(\phi) - \Leftarrow\phi)) = \frac{1}{t}\mathcal{D}(V(t)\phi - \phi) - \frac{1}{t}\int_0^t T(t-s)L(V(s)\phi)ds.$$

It is well-known that

$$\lim_{t\to 0} \frac{1}{t}\int_0^t T(t-s)L(V(s)\phi)ds = L(\phi)$$

in $X$-norm and

$$\lim_{t\to 0} \frac{1}{t}\mathcal{D}(V(t)\phi - \phi) = \mathcal{D}(\phi')$$

in $\alpha$-norm. The fact that $X_\alpha \hookrightarrow X$ implies

$$\lim_{t\to 0} \frac{1}{t}\mathcal{D}(V(t)\phi - \phi) = \mathcal{D}(\phi')$$

in $X$-norm. Consequently,

$$\mathcal{D}(\phi) \in D(A) \text{ and } \lim_{t\to 0} \frac{1}{t}(T(t)\mathcal{D}(\phi) - \Leftarrow\phi)) = A\mathcal{D}(\phi)$$

in $X$-norm. It follows that

$$\begin{cases} D(B) \subseteq \left\{\phi \in \mathcal{B}_\alpha : \ \phi' \in \mathcal{B}_\alpha, \ \mathcal{D}(\phi) \in D(A) \text{ and } \mathcal{D}(\phi') = -A\mathcal{D}(\phi) + L(\phi)\right\}, \\ B(\phi) = \phi'. \end{cases}$$

Conversely, let $\phi \in \mathcal{B}_\alpha$ be such that

$$\phi' \in \mathcal{B}_\alpha, \ \mathcal{D}(\phi) \in D(A) \quad \text{and } \mathcal{D}(\phi') = -A\mathcal{D}(\phi) + L(\phi).$$

Since $t \to T(t)\phi$ is continuously differentiable from $\mathbb{R}^+$ to $X_\alpha$, then $\phi \in D(B)$.

□

Now, let us study the spectral of the linear equation. We assume that $\mathcal{B}_\alpha$ satisfies the following axiom:

(**D**) There exists a constant $\nu \in \mathbb{R}$ such that for every $x \in X$ and $\lambda \in \mathbb{C}$ with $\Re(\lambda) > \nu$, one has

$$\epsilon_\lambda \otimes x \in \mathcal{B}_\alpha \quad \text{and} \quad \sup_{|x|\le 1} |\epsilon_\lambda \otimes x| < \infty,$$

where $(\epsilon_\lambda \otimes x)(\theta) = e^{\lambda\theta}x$ for $\theta \le 0$.

For $\lambda \in \mathbb{C}$ such that $\Re(\lambda) > \nu$, we define the linear operator $\Delta(\lambda)$ by

$$\begin{cases} D(\Delta(\lambda)) = \left\{x \in X_\alpha : \ \mathcal{D}(e^{\lambda\cdot}x) \in D(A) \quad \text{and } A\mathcal{D}(e^{\lambda\cdot}x) - L(e^{\lambda\cdot}x) \in X_\alpha\right\}, \\ \Delta(\lambda) = \lambda\mathcal{D}(e^{\lambda\cdot}I) + A\mathcal{D}(e^{\lambda\cdot}I) - L(e^{\lambda\cdot}I). \end{cases}$$

Let $(A_V, D(A_V))$ be the infinitesimal generator of the semigroup $(V(t))_{t\ge 0}$ and $\sigma_p(A_V)$ be the point spectrum of $A_V$.

**Theorem 8.5.5** *Assume that* (**H**$_1$), (**H**$_2$), (**H**$_3$), (**H**$_4$), *and* (**H**$_6$) *hold. Assume furthermore that the axioms* (**A**), (**B**), (**C**), *and* (**D**) *are satisfied. Let* $\lambda \in \mathbb{C}$ *with* $\Re(\lambda) > \nu$. *If* $\mathcal{B}_\alpha$ *is a uniform fading memory space and* $\mathcal{D}$ *is stable, then the following are equivalent:*

(i) $\lambda \in \sigma_p(A_V)$.
(ii) $ker\,\Delta(\lambda) \neq \{0\}$.

□

***Proof*** Let $\lambda \in \sigma_p(A_V)$ with $\Re(\lambda) > \nu$. Then, there exists $\phi \in D(A_V)$, $\phi \neq 0$, with $A_V\phi = \lambda\phi$. That leads to

$$\lim_{t\to 0} \frac{1}{t}(V(t)\phi - \phi) = \lambda\phi$$

and

$$\lim_{t\to 0}\frac{1}{t}\mathcal{D}(V(t)\phi-\phi)(0)=\lambda\mathcal{D}(\phi(0)).$$

Since for all $t>0$,

$$\frac{1}{t}(T(t)\mathcal{D}(\phi(0))-\Leftarrow\phi(0)))=\frac{1}{t}\mathcal{D}(V(t)\phi-\phi)(0)-\frac{1}{t}\int_0^t T(t-s)L(V(s)\phi)ds,$$

then letting $t$ goes to 0, and one obtains

$$\mathcal{D}(\phi(0))\in D(A)\quad\text{and}\quad -A\mathcal{D}(\phi(0))=\lambda\mathcal{D}(\phi(0))-l(\phi). \tag{8.29}$$

Moreover, using the spectral mapping (Theorem 2.4 in [18]), we have

$$e^{\lambda t}\in\sigma_p(V(t))\quad\text{and}\quad V(t)\phi=e^{\lambda t}\phi\quad\text{for all}\quad t>0.$$

Letting $t>0$ and $\theta\leq 0$ such that $t+\theta\geq 0$, the translation property of the semigroup solution leads to

$$(V(t)\phi)(\theta)=(V(t+\theta)\phi)(0)=e^{\lambda t}\phi(\theta)=e^{\lambda(t+\theta)}\phi(0).$$

Thus, $\phi(\theta)=e^{\lambda\theta}\phi(0)$ for $\theta\geq 0$. Since $\phi\neq 0$, using (8.29), it follows that $\mathcal{D}(\phi(0))\in ker\Delta(\lambda)$.

Conversely, if $\phi$ verifies all conditions of Theorem 8.3.3, then $A_V\phi=\phi'$. Taking $x\in D(A)$ such that $x\neq 0$ and $\Delta(\lambda)x=0$, then the function $\epsilon_\lambda\otimes x$ satisfies all conditions of Theorem 8.3.3, and we deduce that

$$A_V(\epsilon_\lambda\otimes x)=\lambda(\epsilon_\lambda\otimes x).$$

□

Now, let

$$\nu_0=\inf\{\nu\in\mathbb{R}:\quad\text{such that}\quad\textbf{(D)}\quad\text{is satisfied}\}.$$

**Lemma 8.5.2 ([18])** *If $\mathcal{B}$ is a uniform fading memory space, then $\nu_0<0$.* □

**Definition 8.5.1** $\lambda\in\mathbb{C}$ is a characteristic value of Eq. (8.28) if

$$\Re(\lambda)>\nu_0\quad\text{and}\quad ker\Delta(\lambda)\neq\{0\}.$$

Let

$$s'(A_V)=\sup\{\Re(\lambda):\quad\lambda\in\sigma(A_V)-\sigma_{ess}(A_V)\}.$$

It is well-known that $\sigma(A_V) - \sigma_{ess}(A_V)$ contains a finite number of eigenvalues of $A_V$. Consequently, the stability of $(V(t))_{t\geq 0}$ is completely determined by $s'(A_V)$.

**Theorem 8.5.6** *Assume that* $(\mathbf{H}_1)$, $(\mathbf{H}_2)$, $(\mathbf{H}_3)$, $(\mathbf{H}_4)$, $(\mathbf{H}_5)$, *and* $(\mathbf{H}_6)$ *hold. Furthermore, assume that the axioms* **(A)**, **(B)**, **(C)**, *and* **(D)** *are satisfied. If* $\mathcal{B}$ *is a uniform fading memory space and* $\mathcal{D}$ *is stable, then the following holds:*

(i) *If* $s'(A_V) < 0$, *then* $(V(t))_{t\geq 0}$ *is exponentially stable.*
(ii) *If* $s'(A_V) = 0$, *then there exists* $\phi \in \mathcal{B}_\alpha$ *such that* $|V(t)\phi|_{\mathcal{B}_\alpha} = |\phi|_{\mathcal{B}_\alpha}$.
(iii) *If* $s'(A_V) > 0$, *then there exists* $\phi \in \mathcal{B}_\alpha$ *such that* $\lim_{t\to+\infty} |V(t)\phi|_{\mathcal{B}_\alpha} = +\infty$.

□

We deduce the following stability result in the nonlinear case, from Theorem 8.5.2.

**Theorem 8.5.7** *Assume that* $(\mathbf{H}_1)$, $(\mathbf{H}_2)$, $(\mathbf{H}_3)$, $(\mathbf{H}_4)$, $(\mathbf{H}_5)$, *and* $(\mathbf{H}_6)$ *hold. Furthermore, assume that the axioms* **(A)**, **(B)**, **(C)**, *and* **(D)** *are satisfied. If* $\mathcal{B}$ *is a uniform fading memory space and* $\mathcal{D}$ *is stable, then the following holds:*

(i) *If* $s'(A_V) < 0$, *then the zero equilibrium of* $(U(t))_{t\geq 0}$ *is locally exponentially stable.*
(ii) *If* $s'(A_V) > 0$, *then the zero equilibrium of* $(U(t))_{t\geq 0}$ *is unstable.*

□

## 8.6 Application

To apply the theoretical results of this chapter, we consider the following nonlinear system with infinite delay:

$$\begin{cases} \dfrac{\partial}{\partial t}\Big[v(t,\xi) - qv(t-r,\xi)\Big] = \dfrac{\partial^2}{\partial \xi^2}\Big[v(t,\xi) - qv(t-r,\xi)\Big] \\ +b\frac{\partial}{\partial \xi}\Big[v(t,\xi) - qv(t-r,\xi)\Big] \\ +c\displaystyle\int_{-\infty}^{0} g(\theta, v(t+\theta,\xi))d\theta \quad \text{for } t \geq 0 \text{ and } \xi \in [0,\pi] \\ v(t,0) - qv(t-r,0) = v(t,\pi) - qv(t-r,\pi) = 0 \text{ for } t \geq 0 \\ v(\theta,\xi) = \psi(\theta,\xi) \text{ for } \theta \in (-\infty,0] \text{ and } \xi \in [0,\pi], \end{cases} \tag{8.30}$$

where $g : (-\infty, 0] \times \mathbb{R} \to \mathbb{R}$ is a function and $c \in \mathbb{R}_+^*$, $b \in \mathbb{R}$. $q$ is a positive constant such that $|q| < 1$. $H : \mathbb{R}^2 \to \mathbb{R}$ is a Lipschitz continuous with $H(0,0) = 0$. The initial data $\psi$ will be precised in the next.

In order to write system (8.30) in an abstract form, we introduce the space $X = L^2((0,\pi);\mathbb{R})$. Let $A$ be the operator defined on $X$ by

$$\begin{cases} D(A) = H^2((0,\pi);\mathbb{R}) \cap H_0^1((0,\pi);\mathbb{R}), \\ Ay = -y'' \text{ for } y \in D(A). \end{cases}$$

Then, $(-A)$ generates an analytic semigroup $(T(t))_{t\geq 0}$ on $X$. Moreover, $T(t)$ is compact on $X$ for every $t > 0$. The spectrum $\sigma(-A)$ is equal to the point spectrum $P\sigma(-A)$ and is given by $\sigma(-A) = \{-n^2 : n \geq 1\}$, and the associated eigenfunctions $(\phi_n)_{n\geq 1}$ are given by $\phi_n = \sqrt{\frac{2}{\pi}}\sin(nx)$ for $x \in [0,\pi]$; the associated analytic semigroup is explicitly given by

$$T(t)y = \sum_{n=1}^{\infty} e^{-n^2 t}(y,\phi_n)\phi_n \quad \text{for } t \geq 0 \text{ and } y \in X,$$

where $(.,.)$ is an inner product on $X$.

**Lemma 8.6.1 ([21])** *If* $\alpha = \frac{1}{2}$, *then*

$$Ay = \sum_{n=1}^{+\infty} n^2 (y,\phi_n)\phi_n \ \textit{for}\ y \in D(A),$$

$$A^{\frac{1}{2}}y = \sum_{n=1}^{+\infty} n(y,\phi_n)\phi_n \ \textit{for}\ y \in X,$$

$$A^{\frac{1}{2}}T(t)y = \sum_{n=1}^{+\infty} ne^{-n^2 t}(y,\phi_n)\phi_n \ \textit{for}\ y \in X,$$

$$A^{-\frac{1}{2}}y = \sum_{n=1}^{+\infty}\left(\frac{1}{n}\right)(y,\phi_n)\phi_n \ \textit{for}\ y \in X,$$

*and*

$$A^{-\frac{1}{2}}T(t)y = \sum_{n=1}^{+\infty}\left(\frac{1}{n}\right)e^{-n^2 t}(y,\phi_n)\phi_n \ \textit{for}\ y \in X.$$

There exists $M \geq 1$ (see [21]) such that for $t \geq 0$, $|T(t)| \leq Me^{\omega t}$ for some $-1 < \omega < 0$.

Then, the semigroup $\{T(t)\}_{t\geq 0}$ is exponentially stable.

Note also that (see [21]) there exists $M_{\frac{1}{2}} \geq 0$ such that

$$|A^{\frac{1}{2}}T(t)| \leq M_{\frac{1}{2}} t^{-\frac{1}{2}} e^{\omega t} \text{ for each } t > 0.$$

Therefore, hypotheses ($\mathbf{H}_1$) and ($\mathbf{H}_5$) are satisfied.

**Lemma 8.6.2 ([7])** *If $m \in D(A^{\frac{1}{2}})$, then $m$ is absolutely continuous, $\frac{\partial}{\partial x} m \in X$. Moreover, there exist positive constants $N_0$ and $M_0$ such that*

$$N_0 |A^{\frac{1}{2}} m|_X \leq |\frac{\partial}{\partial x} m|_X \leq M_0 |A^{\frac{1}{2}} m|_X.$$

Let $\gamma > 0$. We consider the following phase space

$$\mathcal{B} = C_\gamma = \left\{ \phi \in C((-\infty, 0]; X) : \lim_{\theta \to -\infty} e^{\gamma\theta} |\phi(\theta)| \text{ exists in } X \right\}$$

provided with the following norm:

$$|\phi|_{C_\gamma} = \sup_{\theta \leq 0} e^{\gamma\theta} |\phi(\theta)|_X \text{ for } \phi \in C_\gamma.$$

According to [7], $\mathcal{B}$ satisfies Axioms (**A**), (**B**) and is a uniform fading memory space. Moreover, it is well-known that $K(t) = 1$ for every $t \in \mathbb{R}^+$ and $M(t) = e^{-\gamma t}$ for $t \in \mathbb{R}^+$. Therefore, the norm in $\mathcal{B}_{\frac{1}{2}}$ is given (see [7]) by

$$|\phi|_{\mathcal{B}_{\frac{1}{2}}} = \sup_{\theta \leq 0} e^{\gamma\theta} |A^{\frac{1}{2}} \phi(\theta)|_X.$$

One can write ( see, [21], p.144)

$$\int_0^\pi \left(\phi(\theta)(\xi)\right)^2 d\xi \leq |A^{\frac{1}{2}}\phi(\theta)|_X^2 = \int_0^\pi \left(\frac{\partial}{\partial \xi}\phi(\theta)(\xi)\right)^2 d\xi. \tag{8.31}$$

Next, we assume the following.

($\mathbf{H}_7$) For $\theta \leq 0$ and $\zeta_1, \zeta_2 \in \mathbb{R}$, $|g(\theta, \zeta_1) - g(\theta, \zeta_2)| \leq s(\theta)|\zeta_1 - \zeta_2|$, $g(\theta, 0) = 0$, $\frac{\partial}{\partial \zeta} g(\theta, 0) \neq 0$, where $s$ is some nonnegative function that verifies

$$\int_{-\infty}^0 e^{-2\gamma\theta} s(\theta) < \infty.$$

Let $f_1$, $f_2$, and $f$ be defined on $\mathcal{B}_{\frac{1}{2}}$ by

$$f_1(\phi)(\xi) = c\int_{-\infty}^{0} g(\theta, \phi(\theta)(\xi))d\theta \text{ for } \xi \in [0, \pi],$$

$$f_2(\phi)(\xi) = b\frac{\partial}{\partial \xi}\Big[\phi(0)(\xi) - q\phi(-r)(\xi)\Big] \text{ for } \xi \in [0, \pi],$$

and

$$f(\phi)(\xi) = f_1(\phi)(\xi) + f_2(\phi)(\xi) \text{ for } \xi \in [0, \pi].$$

**Proposition 8.6.1** *For each $\phi \in \mathcal{B}_{\frac{1}{2}}$, $f(\phi) \in L^2((0,\pi);\mathbb{R})$, and $f$ is continuous on $\mathcal{B}_{\frac{1}{2}}$.* □

***Proof*** Let $\phi \in \mathcal{B}_{\frac{1}{2}}$. Since for all $\xi \in [0, \pi]$ and for all $\theta \in (-\infty, 0]$, we have

$$\begin{aligned}|g(\theta, \xi)| &\le s(\theta)|\xi| + |g(\theta, 0)|\\ &= s(\theta)|\xi|,\end{aligned}$$

then for all $\xi \in [0, \pi]$,

$$|f_1(\phi)(\xi)| \le c\int_{-\infty}^{0} |s(\theta)|\,|\phi(\theta)(\xi)|d\theta.$$

Let us set

$$B(\xi) = \int_{-\infty}^{0} |s(\theta)||\phi(\theta)(\xi)|d\theta \text{ for } \xi \in [0, \pi].$$

Using Hölder inequality, one can write

$$\begin{aligned}B(\xi) &= \int_{-\infty}^{0} e^{-2\gamma\theta}|s(\theta)||\phi(\theta)(\xi)|e^{2\gamma\theta}d\theta\\ &\le \left(\int_{-\infty}^{0} |e^{-2\gamma\theta}s(\theta)|^2 d\theta\right)^{\frac{1}{2}} \left(\int_{-\infty}^{0} |\phi(\theta)(\xi)e^{2\gamma\theta}|^2 d\theta\right)^{\frac{1}{2}}.\end{aligned}$$

Then, using the above inequality and the inequality (8.31),

$$\begin{aligned}\int_0^{\pi} |B(\xi)|^2 d\xi &\le \int_0^{\pi} \left(\left(\int_{-\infty}^{0} |e^{-2\gamma\theta}s(\theta)|^2 d\theta\right)\left(\int_{-\infty}^{0} |\phi(\theta)(\xi)e^{2\gamma\theta}|^2 d\theta\right)\right) d\xi\\ &= \int_0^{\pi} \left(|e^{-2\gamma\cdot}s|^2_{L^2(\mathbb{R}^-)} \int_{-\infty}^{0} |\phi(\theta)(\xi)e^{2\gamma\theta}|^2 d\theta\right) d\xi\end{aligned}$$

$$
\begin{aligned}
&\le |e^{-2\gamma\cdot}s|^2_{L^2(\mathbb{R}^-)}\left(\int_{-\infty}^0 e^{2\gamma\theta}\left(e^{2\gamma\theta}\int_0^\pi |\phi(\theta)(\xi)|^2 d\xi\right)d\theta\right)\\
&\le |e^{-2\gamma\cdot}s|^2_{L^2(\mathbb{R}^-)}\left(\int_{-\infty}^0 e^{2\gamma\theta}\left(e^{2\gamma\theta}\int_0^\pi |\frac{\partial}{\partial\xi}\phi(\theta)(\xi)|^2 d\xi\right)d\theta\right)\\
&= |e^{-2\gamma\cdot}s|^2_{L^2(\mathbb{R}^-)}\left(\int_{-\infty}^0 e^{2\gamma\theta}\left(e^{2\gamma\theta}|\frac{\partial}{\partial\xi}\phi(\theta)|^2_{L^2([0,\pi];\mathbb{R})}\right)d\theta\right)\\
&\le |e^{-2\gamma\cdot}s|^2_{L^2(\mathbb{R}^-)}\left(\int_{-\infty}^0 e^{2\gamma\theta}\left(\sup_{\theta\le 0} e^{2\gamma\theta}|\frac{\partial}{\partial\xi}\phi(\theta)|^2_{L^2([0,\pi];\mathbb{R})}\right)d\theta\right)\\
&\le |e^{-2\gamma\cdot}s|^2_{L^2(\mathbb{R}^-)}\left(\int_{-\infty}^0 e^{2\gamma\theta}\left(\sup_{\theta\le 0} e^{2\gamma\theta}|A^{\frac{1}{2}}\phi(\theta)|^2_{L^2([0,\pi];\mathbb{R})}\right)d\theta\right)\\
&\le |e^{-2\gamma\cdot}s|^2_{L^2(\mathbb{R}^-)}\left(\int_{-\infty}^0 e^{2\gamma\theta}|\phi|^2_{\mathcal{B}_{\frac{1}{2}}}d\theta\right)\\
&\le |e^{-2\gamma\cdot}s|^2_{L^2(\mathbb{R}^-)}|\phi|^2_{\mathcal{B}_{\frac{1}{2}}}\int_{-\infty}^0 e^{2\gamma\theta}d\theta\\
&< \infty.
\end{aligned}
$$

Also, we refer to Minkowski inequality to obtain

$$
\begin{aligned}
\int_0^\pi |f_2(\phi)(\xi)|^2 d\xi &= \int_0^\pi \left|\frac{\partial}{\partial\xi}\Big[\phi(0)(\xi) - q\phi(-r)(\xi)\Big]\right|^2 d\xi\\
&\le \int_0^\pi \left|\frac{\partial}{\partial\xi}\phi(0)(\xi)\right|^2 d\xi + \int_0^\pi \left|q\frac{\partial}{\partial\xi}\phi(-r)(\xi)\right|^2 d\xi\\
&\quad +2\Big(\int_0^\pi \left|\frac{\partial}{\partial\xi}\phi(0)(\xi)\right|^2 d\xi\Big)^{\frac{1}{2}}\Big(\int_0^\pi \left|q\frac{\partial}{\partial\xi}\phi(-r)(\xi)\right|^2 d\xi\Big)^{\frac{1}{2}}\\
&\le \left|A^{\frac{1}{2}}\phi(0)\right|^2_{L^2([0,\pi];\mathbb{R})} + q^2\left|A^{\frac{1}{2}}\phi(-r)\right|^2_{L^2([0,\pi];\mathbb{R})}\\
&\quad +2q\left|A^{\frac{1}{2}}\phi(0)\right|_{L^2([0,\pi];\mathbb{R})}\left|A^{\frac{1}{2}}\phi(-r)\right|_{L^2([0,\pi];\mathbb{R})}\\
&\le \sup_{\theta\le 0} e^{2\gamma\theta}\left|A^{\frac{1}{2}}\phi(\theta)\right|^2_{L^2([0,\pi];\mathbb{R})}\\
&\quad +q^2 e^{2\gamma r}\sup_{\theta\le 0} e^{2\gamma\theta}\left|A^{\frac{1}{2}}\phi(\theta)\right|^2_{L^2([0,\pi];\mathbb{R})}\\
&\quad +2q\sup_{\theta\le 0} e^{2\gamma\theta}\left|A^{\frac{1}{2}}\phi(\theta)\right|_{L^2([0,\pi];\mathbb{R})} e^{2\gamma r}\\
&\quad \times\sup_{\theta\le 0} e^{2\gamma\theta}\left|A^{\frac{1}{2}}\phi(\theta)\right|_{L^2([0,\pi];\mathbb{R})}
\end{aligned}
$$

$$\leq \sup_{\theta\leq 0} e^{2\gamma\theta}\left|A^{\frac{1}{2}}\phi(\theta)\right|^2_{L^2([0,\pi];\mathbb{R})}$$
$$+q^2e^{2\gamma r}\sup_{\theta\leq 0} e^{2\gamma\theta}\left|A^{\frac{1}{2}}\phi(\theta)\right|^2_{L^2([0,\pi];\mathbb{R})}$$
$$+2qe^{2\gamma r}\sup_{\theta\leq 0} e^{2\gamma\theta}\left|A^{\frac{1}{2}}\phi(\theta)\right|^2_{L^2([0,\pi];\mathbb{R})}$$
$$< \infty.$$

We conclude that $f(\phi) = (f_1 + f_2)(\phi) \in L^2([0,\pi];\mathbb{R})$ for all $\phi \in \mathcal{B}_{\frac{1}{2}}$.

Let us show that $f$ is continuous. For this purpose, let $(\phi_n)_{n\in\mathbb{N}}$ be a sequence in $\mathcal{B}_{\frac{1}{2}}$ and $\phi \in \mathcal{B}_{\frac{1}{2}}$ such that $\phi_n \to \phi$ in $\mathcal{B}_{\frac{1}{2}}$ as $n \to +\infty$. Then

$$\begin{aligned}\Big(f_1(\phi_n) - f_1(\phi)\Big)(\xi) &= c\int_{-\infty}^0 g(\theta,\phi_n(\theta)(\xi))d\theta - c\int_{-\infty}^0 g(\theta,\phi(\theta)(\xi))d\theta\\ &= c\int_{-\infty}^0 \Big[g(\theta,\phi_n(\theta)(\xi)) - g(\theta,\phi(\theta)(\xi))\Big]d\theta,\end{aligned}$$

and we obtain that

$$|(f_1(\phi_n) - f_1(\phi))(\xi)| \leq c\int_{-\infty}^0 |s(\theta)|\,|\phi_n(\theta)(\xi) - \phi(\theta)(\xi))|d\theta.$$

Let us set for all $\xi \in [0,\pi]$,

$$J_n(\xi) = c\int_{-\infty}^0 |s(\theta)|\Big|\phi_n(\theta)(\xi) - \phi(\theta)(\xi)\Big|d\theta.$$

Then

$$\begin{aligned}|J_n(\xi)| &\leq c\int_{-\infty}^0 e^{-2\gamma\theta}|s(\theta)|\Big|\phi_n(\theta)(\xi) - \phi(\theta)(\xi)\Big|e^{2\gamma\theta}d\theta\\ &\leq c\left(\int_{-\infty}^0 \Big|e^{-2\gamma\theta}s(\theta)\Big|^2d\theta\right)^{\frac{1}{2}}\left(\int_{-\infty}^0 \Big|\Big(\phi_n(\theta)(\xi) - \phi(\theta)(\xi)\Big)e^{2\gamma\theta}\Big|^2d\theta\right)^{\frac{1}{2}},\end{aligned}$$

which leads to

$$\begin{aligned}\int_0^\pi |J_n(\xi)|^2d\xi \leq |c|^2|e^{-2\gamma\cdot}s|^2_{L^2(\mathbb{R}^-)}\int_{-\infty}^0 \Bigg(e^{2\gamma\theta}e^{2\gamma\theta}\int_0^\pi \Big|\phi_n(\theta)(\xi)\\ - \phi(\theta)(\xi)\Big|^2d\xi\Bigg)\,d\theta\end{aligned}$$

$$
\begin{aligned}
&\leq |c|^2 |e^{-2\gamma \cdot}s|^2_{L^2(\mathbb{R}^-)} \int_{-\infty}^{0} \left( e^{2\gamma\theta} e^{2\gamma\theta} \int_0^\pi \left| \frac{\partial}{\partial \xi}\phi_n(\theta)(\xi) \right.\right. \\
&\quad \left.\left. - \frac{\partial}{\partial \xi}\phi(\theta)(\xi) \right|^2 d\xi \right) d\theta \\
&\leq |c|^2 |e^{-2\gamma \cdot}s|^2_{L^2(\mathbb{R}^-)} \int_{-\infty}^{0} e^{2\gamma\theta} \left( \sup_{\theta\leq 0} e^{2\gamma\theta} \int_0^\pi \left| \frac{\partial}{\partial \xi}\phi_n(\theta)(\xi) \right.\right. \\
&\quad \left.\left. - \frac{\partial}{\partial \xi}\phi(\theta)(\xi) \right|^2 d\xi \right) d\theta \\
&\leq |c|^2 |e^{-2\gamma \cdot}s|^2_{L^2(\mathbb{R}^-)} \int_{-\infty}^{0} e^{2\gamma\theta} \left( \sup_{\theta\leq 0} e^{2\gamma\theta} \left| A^{\frac{1}{2}}(\phi_n(\theta) \right.\right. \\
&\quad \left.\left. -\phi(\theta)) \right|^2_{L^2([0,\pi];\mathbb{R})} \right) d\theta \\
&\leq |c|^2 \left| e^{-2\gamma \cdot}s \right|^2_{L^2(\mathbb{R}^-)} \left| \phi_n - \phi \right|^2_{\mathcal{B}_{\frac{1}{2}}} \int_{-\infty}^{0} e^{2\gamma\theta} d\theta.
\end{aligned}
$$

Since $\phi_n \to \phi$ in $\mathcal{B}_{\frac{1}{2}}$, then $\displaystyle\int_0^\pi |J_n(\xi)|^2 d\xi \to 0$ as $n \to +\infty$. Therefore, $f_1$ is continuous. Moreover,

$$
\begin{aligned}
\int_0^\pi \left| f_2\Big(\phi_n(\xi) - \phi(\xi)\Big) \right|^2 d\xi &= \int_0^\pi \left| \frac{\partial}{\partial \xi} \Big[ \Big(\phi_n(0) - \phi(0)\Big)(\xi) \right. \\
&\quad \left. -q\Big(\phi_n(-r) - \phi(-r)\Big)(\xi) \Big] \right|^2 d\xi \\
&\leq \int_0^\pi \left| \frac{\partial}{\partial \xi} \Big(\phi_n(0)(\xi) - \phi(0)(\xi)\Big) \right|^2 d\xi \\
&\quad + \int_0^\pi \left| q\frac{\partial}{\partial \xi} \Big(\phi_n(-r)(\xi) - \phi(-r)(\xi)\Big) \right|^2 d\xi \\
&\quad +2\left( \int_0^\pi \left| \frac{\partial}{\partial \xi} \Big(\phi_n(0)(\xi) - \phi(0)(\xi)\Big) \right|^2 d\xi \right)^{\frac{1}{2}} \\
&\quad \times \left( \int_0^\pi \left| q\frac{\partial}{\partial \xi} \Big(\phi_n(-r)(\xi) - \phi(-r)(\xi)\Big) \right|^2 d\xi \right)^{\frac{1}{2}} \\
&\leq \left| A^{\frac{1}{2}} \Big(\phi_n(0) - \phi(0)\Big) \right|^2_{L^2([0,\pi];\mathbb{R})} \\
&\quad +q^2 \left| A^{\frac{1}{2}} \Big(\phi_n(-r) - \phi(-r)\Big) \right|^2_{L^2([0,\pi];\mathbb{R})}
\end{aligned}
$$

$$
\begin{aligned}
&+2q\left|A^{\frac{1}{2}}\Big(\phi_n(0)-\phi(0)\Big)\right|_{L^2([0,\pi];\mathbb{R})}\Big|A^{\frac{1}{2}}\\
&\times\Big(\phi_n(-r)-\phi(-r)\Big)\Big|_{L^2([0,\pi];\mathbb{R})}\\
\le\ & \sup_{\theta\le 0} e^{2\gamma\theta}\left|A^{\frac{1}{2}}\Big(\phi_n(\theta)-\phi(\theta)\Big)\right|^2_{L^2([0,\pi];\mathbb{R})}\\
&+q^2e^{2\gamma r}\sup_{\theta\le 0} e^{2\gamma\theta}\left|A^{\frac{1}{2}}\Big(\phi_n(\theta)-\phi(\theta)\Big)\right|^2_{L^2([0,\pi];\mathbb{R})}\\
&+2q\sup_{\theta\le 0} e^{2\gamma\theta}\left|A^{\frac{1}{2}}\Big(\phi_n(\theta)-\phi(\theta)\Big)\right|_{L^2([0,\pi];\mathbb{R})}\\
&\times e^{2\gamma r}\sup_{\theta\le 0} e^{2\gamma\theta}\left|A^{\frac{1}{2}}\Big(\phi_n(\theta)-\phi(\theta)\Big)\right|_{L^2([0,\pi];\mathbb{R})}\\
\le\ & \left|\phi_n-\phi\right|^2_{\mathcal{B}_{\frac{1}{2}}}+q^2e^{2\gamma r}\left|\phi_n-\phi\right|^2_{\mathcal{B}_{\frac{1}{2}}}\\
&+2qe^{2\gamma r}\left|\phi_n-\phi\right|^2_{\mathcal{B}_{\frac{1}{2}}}.
\end{aligned}
$$

Using the fact that $\phi_n \to \phi$ in $\mathcal{B}_{\frac{1}{2}}$ as $n \to +\infty$, we obtain that $\int_0^{\pi} \left|f_2\Big(\phi_n(\xi) - \phi(\xi)\Big)\right|^2 d\xi \to 0$ when $n \to +\infty$.

Hence, $f(\phi_n) \to f(\phi)$ in $L^2([0,\pi];\mathbb{R})$ as $n \to +\infty$ and the proof is complete. □

Let

$$
\begin{cases}
u(t)(x) = v(t,x) & \text{for } t \ge 0 \ \text{ and } \ x \in [0,\pi],\\
u_0(\theta)(x) = \psi(\theta,x) & \text{for } \theta \in (-\infty,0] \ \text{ and } \ x \in [0,\pi].
\end{cases}
$$

We need the following result to prove that $(\mathbf{H}_3)$ is satisfied.

**Proposition 8.6.2** *Assume that* $(\mathbf{H}_7)$ *holds. Then,* $f$ *is Lipschitzian.*

***Proof*** We have to show that $f_1$ and $f_2$ are Lipschitz functions. So, let $\phi$ and $\psi$ be in $\mathcal{B}_{\frac{1}{2}}$. Then, for $\xi \in [0,\pi]$, one has

$$
(f_1(\phi) - f_1(\psi))(\xi) = c\int_{-\infty}^{0}\Big[g(\theta,\phi(\theta)(\xi)) - g(\theta,\psi(\theta)(\xi))\Big]d\theta \ \text{ for } \xi \in [0,\pi].
$$

Note that using Hölder inequality, one can write

$$
\begin{aligned}
|(f_1(\phi) - f_1(\psi))(\xi)| &\leq |c| \int_{-\infty}^{0} \Big| g(\theta, \phi(\theta)(\xi)) - g(\theta, \psi(\theta)(\xi)) \Big| d\theta \\
&\leq c \int_{-\infty}^{0} |s(\theta)| \Big| \phi(\theta)(\xi) - \psi(\theta)(\xi) \Big| d\theta \\
&= c \int_{-\infty}^{0} e^{-2\gamma\theta} |s(\theta)|\, e^{2\gamma\theta} \Big| \phi(\theta)(\xi) - \psi(\theta)(\xi) \Big| d\theta \\
&\leq c \left( \int_{-\infty}^{0} |e^{-2\gamma\theta} s(\theta)|^2 d\theta \right)^{\frac{1}{2}} \left( \int_{-\infty}^{0} e^{4\gamma\theta} \Big| \phi(\theta)(\xi) \right. \\
&\qquad \left. - \psi(\theta)(\xi) \Big|^2 d\theta \right)^{\frac{1}{2}}.
\end{aligned}
$$

Therefore,

$$
\begin{aligned}
|f_1(\phi)(\xi) - f_1(\psi)(\xi)|^2 \leq |c|^2 \Big( \int_{-\infty}^{0} |e^{-2\gamma\theta} s(\theta)|^2 d\theta \Big) \Big( \int_{-\infty}^{0} e^{4\gamma\theta} \Big| \phi(\theta)(\xi) \\
- \psi(\theta)(\xi) \Big|^2 d\theta \Big),
\end{aligned}
$$

for which we deduce that

$$
\begin{aligned}
&\int_0^{\pi} |f_1(\phi)(\xi) - f_1(\psi)(\xi)|^2 d\xi \\
&\quad \leq |c|^2 \Big( \int_{-\infty}^{0} |e^{-2\gamma\theta} s(\theta)|^2 d\theta \Big) \times \int_{-\infty}^{0} e^{4\gamma\theta} \left( \int_0^{\pi} |\phi(\theta)(\xi) - \psi(\theta)(\xi)|^2 \, d\xi \right) d\theta \\
&\quad \leq |c|^2 \Big( \int_{-\infty}^{0} |e^{-2\gamma\theta} s(\theta)|^2 d\theta \Big) \\
&\qquad \times \int_{-\infty}^{0} e^{2\gamma\theta} \left( \sup_{\theta \leq 0} e^{2\gamma\theta} \int_0^{\pi} \left| \frac{\partial}{\partial \xi} \phi(\theta)(\xi) - \frac{\partial}{\partial \xi} \psi(\theta)(\xi) \right|^2 d\xi \right) d\theta \\
&\quad \leq |c|^2 \Big( \int_{-\infty}^{0} |e^{-2\gamma\theta} s(\theta)|^2 d\theta \Big) \\
&\qquad \times \int_{-\infty}^{0} e^{2\gamma\theta} \left( \sup_{\theta \leq 0} e^{\gamma\theta} \sqrt{ \int_0^{\pi} \left| \frac{\partial}{\partial \xi} \phi(\theta)(\xi) - \frac{\partial}{\partial \xi} \psi(\theta)(\xi) \right|^2 d\xi } \right)^2 d\theta \\
&\quad \leq |c|^2 \Big( \int_{-\infty}^{0} |e^{-2\gamma\theta} s(\theta)|^2 d\theta \Big)
\end{aligned}
$$

$$\times \int_{-\infty}^{0} e^{2\gamma\theta} \left( \sup_{\theta \le 0} e^{\gamma\theta} |A^{\frac{1}{2}}(\phi(\theta) - \psi(\theta))|_{L^2([0,\pi];\mathbb{R})} \right)^2 d\theta$$

$$\le |c|^2 \Big( \int_{-\infty}^{0} |e^{-2\gamma\theta} s(\theta)|^2 d\theta \Big) \int_{-\infty}^{0} e^{2\gamma\theta} \, |\phi - \psi|^2_{\mathcal{B}_{\frac{1}{2}}} \, d\theta$$

$$\le \frac{|c|^2}{2\gamma} \Big| e^{-2\gamma \cdot} s \Big|^2_{L^2(\mathbb{R}^-)} \Big| \phi - \psi \Big|^2_{\mathcal{B}_{\frac{1}{2}}}.$$

Finally, we obtain that

$$|f_1(\phi) - f_2(\psi)|_{L^2([0,\pi];\mathbb{R})} \le k' \, |\phi - \psi|_{\mathcal{B}_{\frac{1}{2}}} \quad \text{for } \phi, \psi \in \mathcal{B}_{\frac{1}{2}},$$

where

$$k' = \frac{|c|}{\sqrt{2\gamma}} \left( \int_{-\infty}^{0} |e^{-2\gamma\theta} s(\theta)|^2 d\theta \right)^{\frac{1}{2}}.$$

Moreover,

$$\Big| f_1(\phi) - f_2(\psi) \Big|^2_{L^2([0,\pi];\mathbb{R})} \le \Big| \phi - \psi \Big|^2_{\mathcal{B}_{\frac{1}{2}}} + q^2 e^{2\gamma r} \Big| \phi - \psi \Big|^2_{\mathcal{B}_{\frac{1}{2}}} + 2q e^{2\gamma r} \Big| \phi - \psi \Big|^2_{\mathcal{B}_{\frac{1}{2}}}$$
$$\le k'' \Big| \phi - \psi \Big|^2_{\mathcal{B}_{\frac{1}{2}}}.$$

Therefore, $f$ is Lipschitzian and $(\mathbf{H}_4)$ is satisfied. □

Let us define the operators $\mathcal{D}$ and $\mathcal{D}_0$ on $\mathcal{B}_{\frac{1}{2}}$ by

$$(\mathcal{D}(\phi)(\xi)) = \phi(0)(\xi) - q\phi(-r)(\xi) \text{ for all } \quad \xi \in [0, \pi]$$

and

$$(\mathcal{D}_0(\phi))(\xi) = q\phi(-r)(\xi) \quad \text{for all} \quad \xi \in [0, \pi].$$

Then, $\mathcal{D}(\phi) = \phi(0) - \mathcal{D}_0(\phi)$.

**Proposition 8.6.3** $\mathcal{D} \in \mathcal{L}(\mathcal{B}_{\frac{1}{2}}; X)$. □

***Proof*** Let $\phi \in \mathcal{B}_{\frac{1}{2}}$. Then, $\mathcal{D}_0(\phi)(\xi) = q\phi(-r)(\xi)$ for all $\xi \in [0, \pi]$. We can write

$$\int_0^{\pi} |\mathcal{D}_0(\phi)(\xi)|^2 d\xi = \int_0^{\pi} q^2 |\phi(-r)(\xi)|^2 d\xi$$

$$= q^2 e^{2\gamma r} e^{-2\gamma r} \int_0^\pi |\phi(-r)(\xi)|^2$$

$$= q^2 e^{2\gamma r} e^{-2\gamma r} \int_0^\pi |\frac{\partial}{\partial \xi}\phi(-r)(\xi)|^2$$

$$\leq q^2 e^{2\gamma r} \sup_{\theta \leq 0} e^{2\gamma\theta} |A^{\frac{1}{2}}\phi(\theta)|^2_{L^2([0,\pi];\mathbb{R})}$$

$$= q^2 e^{2\gamma r} |\phi|^2_{\mathcal{B}_{\frac{1}{2}}}.$$

Hence, $\mathcal{D}_0 \in \mathcal{L}(\mathcal{B}_{\frac{1}{2}}; X)$. It is obvious that $\phi(0) \in \mathcal{L}(\mathcal{B}_{\frac{1}{2}}; X)$. Therefore, we can conclude that $\mathcal{D} \in \mathcal{L}(\mathcal{B}_{\frac{1}{2}}; X)$ and the proof is complete. □

Since $0 < q < 1$, then $\mathcal{D}$ is stable and $|\mathcal{D}_0| < 1$. Thus, hypothesis $(\mathbf{H}_3)$ is satisfied.

Now, let $\varphi$ be defined by $\varphi(\theta)(\xi) = \psi(\theta, \xi)$ for all $\theta \in (-\infty, 0]$ and $\xi \in [0, \pi]$. We make the following additional assumption.

$(\mathbf{H}_8)$ $\varphi(\theta) \in D(A^{\frac{1}{2}})$ for all $\theta \leq 0$, with

$$\sup_{\theta \leq 0} e^{\gamma\theta} \sqrt{\int_0^\pi \left(\frac{\partial}{\partial \xi}\psi(\theta, \xi)\right)^2 d\xi} < \infty$$

and

$$\lim_{\theta \to \theta_0} \int_0^\pi \left(\frac{\partial}{\partial \xi}\psi(\theta, \xi) - \frac{\partial}{\partial \xi}\psi(\theta_0, \xi)\right)^2 d\xi = 0 \text{ for all } \theta_0 \leq 0.$$

Remark that $(\mathbf{H}_8)$ implies $\varphi \in \mathcal{B}_{\frac{1}{2}}$. Then, Eq. (8.30) can be written as follows:

$$\begin{cases} \dfrac{d}{dt}\mathcal{D}(u_t) = -A\mathcal{D}(u_t) + f(u_t) \text{ for } t \geq 0, \\ \\ u_0 = \varphi. \end{cases} \tag{8.32}$$

Consequently, we obtain the existence and uniqueness of a mild solution of problem (8.32). Furthermore, it is clear that $f_1$ and $f_2$ are continuously differentiable and their differential functions are given for $\phi, \psi \in \mathcal{B}_{\frac{1}{2}}$ and $\xi \in [0, \pi]$ by

$$f_1'(\phi)(\psi)(\xi) = c \int_{-\infty}^0 \frac{\partial}{\partial \zeta} g(\theta, \phi(\theta)(\xi))\psi(\theta)(\xi) d\theta$$

and

$$f_2'(\phi)(\psi)(\xi) = b \frac{\partial}{\partial \xi}\Big[\psi(0)(\xi) - q\psi(-r)(\xi)\Big] \text{ for } \xi \in [0, \pi].$$

Let $v_0 = \psi \in \mathcal{B}_{\frac{1}{2}}$ such that:

**(a)** $v_0(0,.) - qv_0(-r,.) \in H^2(0,\pi) \cap H_0^1(0,\pi)$ and $\frac{\partial v_0}{\partial \theta} \in \mathcal{B}_{\frac{1}{2}}$.

**(b)** $\dfrac{\partial v_0(0,\xi)}{\partial \theta} - q\dfrac{\partial v_0(-r,\xi)}{\partial \theta} = \dfrac{\partial^2}{\partial \xi^2}\Big[v_0(0,\xi) - qv_0(-r,\xi)\Big] + b\dfrac{\partial}{\partial \xi}\Big[v_0(0,\xi) - qv_0(-r,\xi)\Big]$

$+ c\displaystyle\int_{-\infty}^{0} g(\theta, v_0(\theta,\xi))d\theta$ for and $\xi \in [0,\pi]$.

We deduce that

$$\psi \in \mathcal{B}_{\frac{1}{2}}, \psi' \in \mathcal{B}_{\frac{1}{2}}, \quad \mathcal{D}(\psi) \in D(A) \quad \text{, and} \quad \mathcal{D}(\psi') = -A\mathcal{D}(\psi) + f(\psi).$$

Then, problem (8.32) has a unique strict solution for every $\phi \in \mathcal{B}_{\frac{1}{2}}$.

Now, we can see that $f = f_1 + f_2$ is continuously differentiable, and zero is a solution of (8.30), ,i.e., $f(0) = 0$. The differential of $f$ in 0 is given for $\phi, \psi \in \mathcal{B}_{\frac{1}{2}}$ and $\xi \in [0,\pi]$ by

$$L(\psi)(\xi) = f'(0)(\psi)(\xi) = c\int_{-\infty}^{0} \frac{\partial}{\partial \zeta} g(\theta,0)\psi(\theta)(\xi)d\theta + b\frac{\partial}{\partial \xi}\Big[\psi(0)(\xi) - q\psi(-r)(\xi)\Big].$$

Consequently, the linearized equation of (8.30) can be written as follows:

$$\begin{cases} \dfrac{\partial}{\partial t}\Big[v(t,\xi) - qv(t-r,\xi)\Big] = \dfrac{\partial^2}{\partial \xi^2}\Big[v(t,\xi) - qv(t-r,\xi)\Big] \\ +b\frac{\partial}{\partial \xi}\Big[v(t,\xi) - qv(t-r,\xi)\Big] + c\displaystyle\int_{-\infty}^{0} p(\theta)v(t+\theta,\xi)d\theta \quad \text{for } t \geq 0 \text{ and } \xi \in [0,\pi] \\ v(t,0) - qv(t-r,0) = v(t,\pi) - v(t-r,\pi) = 0 \text{ for } t \geq 0 \\ v(\theta,\xi) = \psi(\theta,\xi) \text{ for } \theta \in (-\infty,0] \text{ and } \xi \in [0,\pi], \end{cases} \tag{8.33}$$

where $p = \frac{\partial}{\partial \zeta} g(.,0) : (-\infty,0] \to \mathbb{R}$ is a continuous and measurable function.

We state the main result of the stability of the solutions.

**Theorem 8.6.1** *Assume that* ($\mathbf{H}_7$) *and* ($\mathbf{H}_8$) *hold. Furthermore, suppose that*

$$0 < c\int_{-\infty}^{0} |p(\theta)|d\theta < \left(1 + \frac{b^2}{4}\right)(1-q). \tag{8.34}$$

*Then, the semigroup solution of* (8.33) *is exponentially stable.* □

The proof of Theorem 8.6.1 makes use of this following lemma.

**Lemma 8.6.3 ([4])** *The spectrum* $\sigma(\tilde{A})$ *of the operator* $\tilde{A} = \frac{\partial^2}{\partial \xi^2} + b\frac{\partial}{\partial \xi}$ *is equal to the point spectrum* $P\sigma(\tilde{A}) = \{-n^2 - \frac{b^2}{4} : \; n \in \mathbb{N}^*\}$. □

***Proof of Theorem 8.6.1*** The exponential stability of (8.33) is obtained when $s'(\tilde{A}) < 0$, which is true only if

$$\sup\left\{\Re(\lambda) : \;\; \lambda \in \sigma(\tilde{A}) - \sigma_{ess}(\tilde{A}) \quad \text{and} \quad \Re(\lambda) > -\gamma\right\} < 0.$$

Moreover, the characteristic equation is given by

$$\begin{cases} \Re(\lambda) > -\gamma, \quad f \in D(A), \quad f \neq 0 \\ \\ \lambda(1 - qe^{-\lambda r})f - (1 - qe^{-\lambda r})(f'' + bf') - c\Big(\int_{-\infty}^{0} p(\theta)e^{\lambda\theta}d\theta\Big)f = 0, \end{cases} \tag{8.35}$$

which leads to

$$\lambda - \frac{c}{1 - qe^{-\lambda r}} \int_{-\infty}^{0} p(\theta)e^{\lambda\theta}d\theta \in \sigma_p\left(\frac{\partial^2}{\partial \xi^2} + b\frac{\partial}{\partial \xi}\right).$$

Since

$$\sigma_p\left(\frac{\partial^2}{\partial \xi^2} + b\frac{\partial}{\partial \xi}\right) = P\sigma(\tilde{A}) = \{-n^2 - \frac{b^2}{4} : \;\; n \in \mathbb{N}^*\},$$

then the characteristic equation (8.35) becomes

$$\begin{cases} \Re(\lambda_n) > -\gamma, \\ \\ \lambda_n = \dfrac{c}{1 - qe^{-\lambda_n r}} \displaystyle\int_{-\infty}^{0} p(\theta)e^{\lambda_n\theta}d\theta - n^2 - \frac{b^2}{4} \quad \text{for some} \quad n \in \mathbb{N}^*. \end{cases} \tag{8.36}$$

Let $k_n = n + \frac{b^2}{4}$. Then, using (8.36), we obtain that

$$(\lambda_n + k_n)(1 - qe^{-\lambda_n r}) = c\int_{-\infty}^{0} p(\theta)e^{\lambda_n\theta}d\theta.$$

Therefore,

$$|\lambda_n + k_n||1 - qe^{-\lambda_n r}| = |c\int_{-\infty}^{0} p(\theta)e^{\lambda_n \theta}d\theta|$$
$$\leq c\int_{-\infty}^{0} |p(\theta)|e^{\Re(\lambda_n \theta)}d\theta.$$

We have also

$$|\lambda_n + k_n| \geq \sqrt{(\Re(\lambda_n) + k_n)^2}$$

and

$$\left|1 - qe^{-\lambda_n r}\right| \geq \left||1| - |qe^{-\lambda_n r}|\right| = |1 - qe^{-\Re(\lambda_n r)}|.$$

It follows that

$$\sqrt{(\Re(\lambda_n) + k_n)^2}\left|1 - qe^{-\Re(\lambda_n r)}\right| \leq c\int_{-\infty}^{0} |p(\theta)|e^{\Re(\lambda_n \theta)}d\theta.$$

Now, assume that $\Re(\lambda_n) \geq 0$. Then,

$$\left|1 - qe^{-\Re(\lambda_n r)}\right| \geq (1 - q).$$

Consequently,

$$(1 - q)\Big[\Re(\lambda_n) + k_n\Big] \leq c\int_{-\infty}^{0} |p(\theta)|d\theta.$$

Finally, since $(1 - q)\Re(\lambda_n)$, we obtain

$$(1 - q)k_n \leq c\int_{-\infty}^{0} |p(\theta)|d\theta.$$

Taking $n = 1$, we obtain a contraction with condition (8.34). That leads to $\Re(\lambda) < 0$. □

## References

1. M. Adimy, K. Ezzinbi, A class of linear partial neutral functional differential equations with nondense domains. J. Differ. Equ., **147**, 285–332 (1998)
2. M. Adimy, K. Ezzinbi, Existence and linearized stability for partial neutral functional differential equations with nondense domains. Differ. Equ. Dyn. Syst. **7**, 371–417 (1999)

3. M. Adimy, K. Ezzinbi, Strict solutions of nonlinear hyperbolic neutral differential equations. Appl. Math. Lett. **12**, 107–112 (1999)
4. M. Adimy, K. Ezzinbi, Existence and stability in the $\alpha$-norm for some partial functional differential equations of neutral type. Ann. Mate. Pura ed Appl. **185**(3), 437–460 (2006)
5. M. Adimy, H. Bouzahir, K. Ezzinbi, Local existence for a class of partial neutral functional differential equations with infinite delay. Differ. Equ. Dyn. Syst. **12**(3–4), 353–370 (2004)
6. C. Avramescu, A fixed point theorem for multivalued mappings. Electron. J. Qual. Theory Differ. Equ. **2004**(17), 1–10 (2004)
7. R. Benkhalti, K. Ezzinbi, Existence and stability in the $\alpha$-norm for some partial functional differential equations with infinite delay. Differ. Integral Equ. **19**(5), 545–572 (2006)
8. H. Bouzahir, On neutral functional differential equations with infinite delay. Fixed Point Theory **6**(1), 11–24 (2005)
9. T. Burton, *Stability and Periodic Solutions of Ordinary Differential Equations and Functional Differential Equations* (Academic Press, San Diego, 1985), pp. 197–308
10. M.A. Cruz, J.K. Hale, Stability of functional differential equations of neutral type. J. Differ. Equ. **7**, 334–355 (1970)
11. R. Datko, Linear autonomous neutral differential equations in a Banach space. J. Differ. Equ. **25**, 258–274 (1977)
12. W. Desch, W. Schappacher, Linearized stability for nonlinear semigroup, in *Differential Equations in Banach Spaces*, ed. by A. Favini, E. Obrecht. Lecture Notes, vol. 1223 (Springer, Berlin, 1986)
13. K. Engel, R. Nagel, *One-Parameter Semigroups for Linear Evolution Equations*. Graduate Texts in Mathematics, vol. 194 (Springer, Berlin, 2001)
14. K. Ezzinbi, B.A. Kyelem, S. Ouaro, Periodicity in $\alpha$-Norm for partial functional differential equations in fading memory spaces. Nonlinear Anal. Theory Methods Appl. **97**, 30–54 (2014)
15. J.K. Hale, *Asymptotic Behavior of Dissipative Systems* (American Mathematical Society, Providence, 1988)
16. J. Hale, J. Kato, Phase space for retarded equations with infinite delay. Funkcial. Ekvac. **21**, 11–41 (1978)
17. V. Lakshmikantham, S. Leela, *Differential and Integral Inequalities*, vol. 1 (Academic, New York, 1969)
18. T. Naito, J.S. Shin, S. Murakami, The generator of the solution semigroup for the general linear functional differential equation. Bull. Univ. Electro-Commun. **11**(1), 29–36 (1998)
19. A. Pazy, *Semigroups of Linear Operators and Applications to Partial Differential Equations* (Springer, New-York, 1993)
20. C.C. Travis, G.F. Webb, Partial differential equations with deviating arguments in the time variable. J. Math. Anal. Appl. **56**, 397–409 (1976)
21. C.C. Travis, G.F. Webb, Existence, stability and compactness in $\alpha$-norm for partial functional differential equations. Trans. Am. Math. Soc. **240**, 129–143 (1978)
22. J. Wu, H. Xia, Self-sustained oscillations in a ring array of coupled lossless transmission lines. J. Differ. Equ. **124**, 247–278 (1996)

# Chapter 9
# Pseudo-almost Periodic Solutions of Class $r$ in the $\alpha$-Norm Under the Light of Measure Theory

**Issa Zabsonre, Abdel Hamid Gamal Nsangou, Moussa El-KhalilL Kpoumiè, and Salifou Mboutngam**

**Abstract** We consider the existence of weak solutions for discrete nonlinear problems. The proof of the main result is based on a minimization method.

**Keywords** Discrete nonlinear problems · Minimization method · Anti-periodic

## 9.1 Introduction

In this chapter, we present a new approach dealing with weighted pseudo-almost periodic functions and their applications in evolution equations and partial functional differential equations. Here we use the measure theory to define an ergodic function, and we investigate many interesting properties of such functions. Weighted pseudo-almost periodic functions started recently and becomes an interesting field in dynamical systems. We can refer to [2–4] and the bibliography therein.

In this chapter, we study the existence and uniqueness of $\alpha - (\mu, \nu)$-pseudo-almost periodic solutions of class $r$ for the following partial functional differential equation

$$u'(t) = -Au(t) + L(u_t) + f(t) \text{ for } t \in \mathbb{R}, \tag{9.1}$$

I. Zabsonre (✉)
Université Joseph KI-ZERBO, Unité de Recherche et de Formation en Sciences Exactes et Appliquées, Département de Mathématiques, Ouagadougou, Burkina Faso

A. H. G. Nsangou · S. Mboutngam
Université de Maroua, Département de Mathématiques, Ecole Normale Supérieure, Maroua, Cameroun

M. E.-K. Kpoumiè
Université de N'gaoundéré, Département de Mathématiques et Informatique, Ecole de Géologie et d'exploitation minière, Meiganga, Cameroun



T. Diagana et al. (eds.), *Partial Differential Equations and Applications*,
Springer Proceedings in Mathematics & Statistics 420,
https://doi.org/10.1007/978-3-031-27661-3_9

where $-A : D(A) \to X$ is the infinitesimal generator of a compact analytic semigroup of uniformly bounded linear operators on a Banach space $X$, $C_\alpha = C([-r, 0], D(A^\alpha))$, $0 < \alpha < 1$, denotes the space of continuous functions from $[-r, 0]$ into $D(A^\alpha)$, and $A^\alpha$ is the fractional $\alpha$-power of A. This operator $(A^\alpha, D(A^\alpha))$ will be described later, and

$$\|\varphi\|_{C_\alpha} = \|A^\alpha \varphi\|_{C([-r,0],X)}.$$

For $t \geq 0$, and $u \in C([-r, a], D(A^\alpha))$, $r > 0$ and $u_t$ denotes the history function of $C_\alpha$ defined by

$$u_t(\theta) = u(t + \theta) \text{ for } \quad -r \leq \theta \leq 0.$$

$L$ is a bounded linear operator from $C_\alpha$ into $X$, and $f : \mathbb{R} \to X$ is a continuous function.

Some recent contributions concerning pseudo-almost periodic solutions for abstract differential equations similar to Eq. (9.1) have been made. For example in [2], the authors have shown if the inhomogeneous term $f$ depends only on variable $t$ and it is a pseudo-almost periodic function, then Eq. (9.1) has a unique pseudo-almost periodic solution. In [4], the authors have proven if $f : \mathbb{R} \times X_0 \to X$ is a suitable continuous function, where $X_0 = \overline{D(A)}$, the problem

$$x'(t) = Ax(t) + f(t, x(t)), t \in \mathbb{R}$$

has a unique pseudo-almost periodic solution, while in [3] the authors have treated the existence of almost periodic solutions for a class of partial neutral functional differential equations defined by a linear operator of Hille–Yosida type with non-dense domain. In [1], the authors studied the existence and uniqueness of pseudo-almost periodic solutions for a first-order abstract functional differential equation with a linear part dominated by a Hille–Yosida type operator with a non-dense domain.

In [7], the authors introduce some new classes of functions called weighted pseudo-almost periodic functions, which implement in a natural fashion the classical pseudo-almost periodic functions due to Zhang [13–15]. Properties of these weighted pseudo-almost periodic functions are discussed, including a composition result for weighted pseudo-almost periodic functions. The results obtained are subsequently utilized to study the existence and uniqueness of a weighted pseudo-almost periodic solution to the heat equation with Dirichlet conditions.

In [6], the authors present a new approach to study weighted pseudo-almost periodic functions using the measure theory. They present a new concept of weighted ergodic functions that is more general than the classical one. Then they establish many interesting results on the functional space of such functions like completeness and composition theorems. The theory of their work generalizes the classical results on weighted pseudo-almost periodic functions.

The aim of this chapter is to prove the existence of $(\mu, \nu)$-pseudo-almost periodic solutions of Eq. (9.1) when the delay is distributed on $[-r, 0]$. Our approach is based on the spectral decomposition of the phase space developed in [4] and a new approach developed in [6].

This chapter is organized as follows: in Sect. 9.2, we recall some preliminary results about analytic semigroups, and fractional power associated to its generator will be used throughout this chapter. In Sect. 9.3, we recall some preliminary results on spectral decomposition. In Sect. 9.4, we recall some preliminary results on $(\mu, \nu)$-pseudo-almost periodic functions and neutral partial functional differential equations that will be used in this chapter. In Sect. 9.5, we give some properties of $(\mu, \nu)$-pseudo-almost periodic functions of class $r$. In Sect. 9.6, we discuss the main result of this chapter. Using the strict contraction principle, we show the existence and uniqueness of $(\mu, \nu)$-pseudo-almost periodic solution of class $r$ for Eq. (9.1). Last section is devoted to some applications arising in population dynamics.

## 9.2 Analytic Semigroup

Let $(X, \|.\|)$ be a Banach space and $\alpha$ be a constant such that $0 < \alpha < 1$ and $-A$ be the infinitesimal generator of a bounded analytic semigroup of linear operator $(T(t))_{t\geq 0}$ on X. We assume without loss of generality that $0 \in \rho(A)$. Note that if the assumption $0 \in \rho(A)$ is not satisfied, one can substitute the operator A by the operator $(A - \sigma I)$ with $\sigma$ large enough such that $0 \in \rho(A - \sigma I)$. This allows us to define the fractional power $A^\alpha$ for $0 < \alpha < 1$, as a closed linear invertible operator with domain $D(A^\alpha)$ dense in X. The closeness of $A^\alpha$ implies that $D(A^\alpha)$, endowed with the graph norm of $A^\alpha$, $|x| = \|x\| + \|A^\alpha x\|$, is a Banach space. Since $A^\alpha$ is invertible, its graph norm $|.|$ is equivalent to the norm $|x|_\alpha = \|A^\alpha x\|$. Thus, $D(A^\alpha)$, equipped with the norm $|.|_\alpha$, is a Banach space, which we denote by $X_\alpha$. For $0 < \beta \leq \alpha < 1$, the imbedding $X_\alpha \hookrightarrow X_\beta$ is compact if the resolvent operator of A is compact. Also, the following properties are well known.

**Proposition 9.1 ([10])** *Let $0 < \alpha < 1$. Assume that the operator $-A$ is the infinitesimal generator of an analytic semigroup $(T(t))_{t\geq 0}$ on the Banach space X satisfying $0 \in \rho(A)$. Then we have:*

*(i)* $T(t) : X \to D(A^\alpha)$ *for every* $t > 0$.
*(ii)* $T(t)A^\alpha x = A^\alpha T(t)x$ *for every* $x \in D(A^\alpha)$ *and* $t \geq 0$.
*(iii)* *For every $t > 0$, $A^\alpha T(t)$ is bounded on X, and there exist $M_\alpha > 0$ and $\omega > 0$ such that*

$$\|A^\alpha T(t)\| \leq M_\alpha e^{-\omega t} t^{-\alpha} \text{ for } t > 0.$$

*(iv)* *If* $0 < \alpha \leq \beta < 1$, $D(A^\beta) \hookrightarrow D(A^\alpha)$.
*(v)* *There exists* $N_\alpha > 0$ *such that*

$$\|(T(t) - I)A^{-\alpha}\| \leq N_\alpha t^\alpha \text{ for } t > 0.$$

Recall that $A^{-\alpha}$ is given by the following formula:

$$A^{-\alpha} = \frac{1}{\Gamma(\delta)} \int_0^{+\infty} t^{\alpha-1} T(t)dt,$$

where the integral converges in the uniform operator topology for every $\alpha > 0$.

Consequently, if $T(t)$ is compact for each $t > 0$, then $A^{-\alpha}$ is compact.

## 9.3 Spectral Decomposition

To Eq. (9.1), we associate the following initial value problem:

$$\begin{cases} \frac{d}{dt} u(t) = -Au(t) + L(u_t) + f(t) \text{ for } t \geq 0 \\ \\ u_0 = \varphi \in C_\alpha, \end{cases} \tag{9.2}$$

where $f : \mathbb{R}^+ \to X$ is a continuous function.

For each $t \geq 0$, we define the linear operator $\mathcal{U}(t)$ on $C_\alpha$ by

$$\mathcal{U}(t)\varphi = v_t(., \varphi),$$

where $v(., \varphi)$ is the solution of the following homogeneous equation:

$$\begin{cases} \frac{d}{dt} v(t) = -Av(t) + L(v_t) \text{ for } t \geq 0 \\ \\ v_0 = \varphi \in C_\alpha. \end{cases}$$

**Proposition 9.2 ([3])** *Let $\mathcal{A}_\mathcal{U}$ defined on $C_\alpha$ by*

$$\begin{cases} D(\mathcal{A}_\mathcal{U}) = \Big\{ \varphi \in C_\alpha,\ \varphi' \in C_\alpha,\ \varphi(0) \in D(A), \\ \quad \varphi(0)' \in \overline{D(A)} \text{ and } \varphi(0)' = -A\varphi(0) + L(\varphi) \Big\} \\ \mathcal{A}_\mathcal{U}\varphi = \varphi' \text{ for } \varphi \in D(\mathcal{A}_\mathcal{U}). \end{cases}$$

*Then $\mathcal{A}_\mathcal{U}$ is the infinitesimal generator of the semigroup $(\mathcal{U}(t))_{t \geq 0}$ on $C_\alpha$.*

Let $\langle X_0 \rangle$ be the space defined by

$$\langle X_0 \rangle = \{X_0 c : \ c \in X\},$$

where the function $X_0c$ is defined by

$$(X_0c)(\theta) = \begin{cases} 0 & \text{if } \theta \in [-r, 0[, \\ c & \text{if } \theta = 0. \end{cases}$$

Consider the extension $\mathcal{A}_{\mathcal{U}}$ defined on $C_\alpha \oplus \langle X_0 \rangle$ by

$$\begin{cases} D(\widetilde{\mathcal{A}_{\mathcal{U}}}) = \left\{ \varphi \in C^1([-r, 0]; X_\alpha) : \varphi(0) \in D(A) \text{ and } \varphi(0)' \in \overline{D(A)} \right\} \\ \widetilde{\mathcal{A}_{\mathcal{U}}}\varphi = \varphi' + X_0(A\varphi(0) + L(\varphi) - \varphi(0)'). \end{cases}$$

We make the following assertion:

**(H$_0$)** The operator $-A$ is the infinitesimal generator of an analytic semigroup $(T(t))_{t \geq 0}$ on the Banach space X and satisfies $0 \in \rho(A)$.

**Lemma 9.1 ([4])** *Assume that* **(H$_0$)** *holds. Then, $\widetilde{\mathcal{A}_{\mathcal{U}}}$ satisfies the Hille–Yosida condition on $C_\alpha \oplus \langle X_0 \rangle$; there exist $\widetilde{M} \geq 0$, $\widetilde{\omega} \in \mathbb{R}$ such that $]\widetilde{\omega}, +\infty[ \subset \rho(\widetilde{\mathcal{A}_{\mathcal{U}}})$ and*

$$|(\lambda I - \widetilde{\mathcal{A}_{\mathcal{U}}})^{-n}| \leq \frac{\widetilde{M}}{(\lambda - \widetilde{\omega})^n} \quad \textit{for } n \in \mathbb{N} \textit{ and } \lambda > \widetilde{\omega}.$$

Now, we can state the variation of constants formula associated to Eq. (9.2).

**Theorem 9.1 ([3])** *Assume that* **(H$_0$)** *holds. Then for all $\varphi \in C_\alpha$, the solution $u$ of Eq.* (9.2) *is given by the following variation of constants formula*

$$u_t = \mathcal{U}(t)\varphi + \lim_{\lambda \to +\infty} \int_0^t \mathcal{U}(t-s)\widetilde{B}_\lambda(X_0 f(s))ds \quad \textit{for } t \geq 0,$$

*where* $\widetilde{B}_\lambda = \lambda(\lambda I - \widetilde{\mathcal{A}_{\mathcal{U}}})^{-1}$.

**Definition 9.1** We say that a semigroup $(\mathcal{U}(t))_{t \geq 0}$ is hyperbolic if

$$\sigma(\mathcal{A}_{\mathcal{U}}) \cap i\mathbb{R} = \emptyset.$$

For the sequel, we make the following assumption:

**(H$_1$)** $T(t)$ is compact on $\overline{D(A)}$ for every $t > 0$.

We get the following result on the spectral decomposition of the phase space $C_\alpha$.

**Proposition 9.3 ([3])** *Assume that* **(H$_0$)** *and* **(H$_1$)** *hold. If the semigroup $(\mathcal{U}(t))_{t \geq 0}$ is hyperbolic, then the space $C_\alpha$ is decomposed as a direct sum*

$$C_\alpha = S \oplus U$$

*of two $\mathcal{U}(t)$ invariant closed subspaces $S$ and $U$ such that the restriction of $(\mathcal{U}(t))_{t\geq 0}$ on $U$ is a group, and there exist positive constants $\overline{M}$ and $\omega$ such that*

$$|\mathcal{U}(t)\varphi|_{C_\alpha} \leq \overline{M}e^{-\omega t}|\varphi|_{C_\alpha} \ \textit{for } t \geq 0 \textit{ and } \varphi \in S$$

$$|\mathcal{U}(t)\varphi|_{C_\alpha} \leq \overline{M}e^{\omega t}|\varphi|_{C_\alpha} \quad \textit{for } t \leq 0 \textit{ and } \varphi \in U,$$

*where $S$ and $U$ are called, respectively, the stable and unstable spaces, and $\Pi^s$ and $\Pi^u$ denote, respectively, the projection operator on $S$ and $U$.*

## 9.4 $(\mu, \nu)$-Pseudo-almost Periodic Functions

In this section, we recall some properties about $\mu$-pseudo-almost periodic functions. The notion of $\mu$-pseudo-almost periodicity is a generalization of the pseudo-almost periodicity introduced by Zhang [13–15]; it is also a generalization of weighted pseudo- almost periodicity given by Diagana [7]. Let $BC(\mathbb{R}; X_\alpha)$ be the space of all bounded and continuous functions from $\mathbb{R}$ to $X_\alpha$ equipped with the uniform topology norm.

We denote by $\mathscr{B}$ the Lebesgue $\sigma$-field of $\mathbb{R}$ and by $\mathcal{M}$ the set of all positive measures $\mu$ on $\mathscr{B}$ satisfying $\mu(\mathbb{R}) = +\infty$ and $\mu([a, b]) < \infty$, for all $a, b \in \mathbb{R}$ $(a \leq b)$.

**Definition 9.2** A bounded continuous function $\phi : \mathbb{R} \to X$ is called almost periodic if for each $\varepsilon > 0$, there exists a relatively dense subset of $\mathbb{R}$ denoted by $\mathcal{K}(\varepsilon, \phi, X)$ such that $|\phi(t + \tau) - \phi(t)| < \varepsilon$ for all $(t, \tau) \in \mathbb{R} \times \mathcal{K}(\varepsilon, \phi, X)$.

We denote by $AP(\mathbb{R}; X)$ the space of all such functions.

**Definition 9.3** Let $X_1$ and $X_2$ be two Banach spaces. A bounded continuous function $\phi : \mathbb{R} \times X_1 \to X_2$ is called almost periodic in $t \in \mathbb{R}$ uniformly in $x \in X_1$ if for each $\varepsilon > 0$ and all compact $K \subset X_1$, there exists a relatively dense subset of $\mathbb{R}$ denoted by $\mathcal{K}(\varepsilon, \phi, K)$ such that $|\phi(t + \tau, x) - \phi(t, x)| < \varepsilon$ for all $t \in \mathbb{R},\ x \in K,\ \tau \in \mathcal{K}(\varepsilon, \phi, K)$.

We denote by $AP(\mathbb{R} \times X_1; X_2)$ the space of all such functions.

The next lemma is also a characterization of almost periodic functions.

**Lemma 9.2 ([12])** *A function $\phi \in C(\mathbb{R}, X)$ is almost periodic if and only if the space of functions $\{\phi_\tau : \tau \in \mathbb{R}\}$, where $(\phi_\tau)(t) = \phi(t + \tau)$ is relatively compact in $BC(\mathbb{R}; X)$.*

In the sequel, we recall some preliminary results concerning the $\alpha - (\mu, \nu)$-pseudo-almost periodic functions.

$\mathscr{E}(\mathbb{R}; X_\alpha, \mu, \nu)$ stands for the space of functions

$$\mathscr{E}(\mathbb{R}; X_\alpha, \mu, \nu) = \Big\{u \in BC(\mathbb{R}; X_\alpha) : \lim_{\tau\to+\infty} \frac{1}{\nu([-\tau,\tau])} \int_{-\tau}^{+\tau} |u(t)|_\alpha d\mu(t) = 0\Big\}.$$

To study delayed differential equations for which the history belongs to $C([-r,0]; X_\alpha)$, we need to introduce the space

$$\mathscr{E}(\mathbb{R}; X_\alpha, \mu, \nu, r) = \Big\{u \in BC(\mathbb{R}; X_\alpha) : \lim_{\tau\to+\infty} \frac{1}{\nu([-\tau,\tau])} \times \int_{-\tau}^{+\tau} \Big(\sup_{\theta\in[t-r,t]} |u(\theta)|_\alpha\Big) d\mu(t) = 0\Big\}.$$

In addition to the above-mentioned spaces, we consider the following spaces:

$$\mathscr{E}(\mathbb{R}\times X_\alpha, \mu, \nu) = \Big\{u \in BC(\mathbb{R}\times X_\alpha; X_\alpha) : \lim_{\tau\to+\infty} \frac{1}{\nu([-\tau,\tau])} \times \int_{-\tau}^{+\tau} |u(t,x)|_\alpha d\mu(t) = 0\Big\},$$

$$\mathscr{E}(\mathbb{R}\times X_\alpha, \mu, \nu, r) = \Big\{u \in BC(\mathbb{R}\times X_\alpha; X_\alpha) : \lim_{\tau\to+\infty} \frac{1}{\nu([-\tau,\tau])} \times \int_{-\tau}^{+\tau} \Big(\sup_{\theta\in[t-r,t]} |u(\theta,x)|_\alpha\Big) d\mu(t) = 0\Big\},$$

where in both cases the limit (as $\tau \to +\infty$) is uniform in compact subset of $X_\alpha$.

In view of previous definitions, it is clear that the space $\mathscr{E}(\mathbb{R}; X_\alpha, \mu, \nu, r)$ is continuously embedded in $\mathscr{E}(\mathbb{R}; X_\alpha, \mu, \nu)$.

On the other hand, one can observe that a $\rho$-weighted pseudo-almost periodic function is $\mu$-pseudo- almost periodic, where the measure $\mu$ is absolutely continuous with respect to the Lebesgue measure, and its Radon–Nikodym derivative is $\rho$:

$$d\mu(t) = \rho(t)dt,$$

and $\nu$ is the usual Lebesgue measure on $\mathbb{R}$, i.e., $\nu([-\tau,\tau]) = 2\tau$ for all $\tau \geq 0$.

*Example ([6])* Let $\rho$ be a nonnegative $\mathscr{B}$-measurable function. Denote by $\mu$ the positive measure defined by

$$\mu(A) = \int_A \rho(t)dt, \quad \text{for } A \in \mathscr{B}, \tag{9.3}$$

where $dt$ denotes the Lebesgue measure on $\mathbb{R}$. The function $\rho$, which occurs in Eq. (9.3), is called the Radon–Nikodym derivative of $\mu$ with respect to the Lebesgue measure on $\mathbb{R}$.

From $\mu, \nu \in \mathcal{M}$, we formulate the following hypothesis:

**(H₂)** Let $\mu, \nu \in \mathcal{M}$ be such that $\limsup_{\tau \to +\infty} \frac{\mu([-\tau, \tau])}{\nu([-\tau, \tau])} = \delta < \infty$.

We have the following result.

**Lemma 9.3** *Assume* $(\mathbf{H_2})$ *holds, and let* $f \in BC(\mathbb{R}; X_\alpha)$. *Then* $f \in \mathscr{E}(\mathbb{R}; X_\alpha, \mu, \nu)$ *if and only if for any* $\varepsilon > 0$,

$$\lim_{\tau \to +\infty} \frac{\mu(M_{\tau,\varepsilon}(f))}{\nu([-\tau, \tau]} = 0,$$

*where*

$$M_{\tau,\varepsilon}(f) = \{t \in [-\tau, \tau] : \ |f(t)|_\alpha \geq \varepsilon\}.$$

***Proof*** Suppose that $f \in \mathscr{E}(\mathbb{R}; X_\alpha, \mu, \nu)$. Then

$$\begin{aligned} \frac{1}{\nu([-\tau, \tau])} \int_{-\tau}^{+\tau} |f(t)|_\alpha d\mu(t) &= \frac{1}{\nu([-\tau, \tau])} \int_{M_{\tau,\varepsilon}(f)} |f(t)|_\alpha d\mu(t) \\ &\quad + \frac{1}{\nu([-\tau, \tau])} \int_{[-\tau,\tau] \setminus M_{\tau,\varepsilon}(f)} |f(t)|_\alpha d\mu(t) \\ &\geq \frac{1}{\nu([-\tau, \tau])} \int_{M_{\tau,\varepsilon}(f)} |f(t)|_\alpha d\mu(t) \\ &\geq \frac{\varepsilon \mu(M_{\tau,\varepsilon}(f))}{\nu([-\tau, \tau])}. \end{aligned}$$

Consequently,

$$\lim_{\tau \to +\infty} \frac{\mu(M_{\tau,\varepsilon}(f))}{\nu([-\tau, \tau]} = 0.$$

Suppose that $f \in BC(\mathbb{R}; X_\alpha)$ such that for any $\varepsilon > 0$,

$$\lim_{\tau \to +\infty} \frac{\mu(M_{\tau,\varepsilon}(f))}{\nu([-\tau, \tau]} = 0.$$

We can assume $|f(t)|_\alpha \leq N$ for all $t \in \mathbb{R}$. Using $(\mathbf{H_2})$, we have

$$\begin{aligned}\frac{1}{\nu([-\tau,\tau])}\int_{-\tau}^{+\tau}|f(t)|_\alpha d\mu(t) &= \frac{1}{\nu([-\tau,\tau])}\int_{M_{\tau,\varepsilon}(f)}|f(t)|_\alpha d\mu(t)\\ &\quad+\frac{1}{\nu([-\tau,\tau])}\int_{[-\tau,\tau]\setminus M_{\tau,\varepsilon}(f)}|f(t)|_\alpha d\mu(t)\\ &\le \frac{N}{\nu([-\tau,\tau])}\int_{M_{\tau,\varepsilon}(f)}d\mu(t)\\ &\quad+\frac{1}{\nu([-\tau,\tau])}\int_{[-\tau,\tau]\setminus M_{\tau,\varepsilon}(f)}|f(t)|_\alpha d\mu(t)\\ &\le \frac{N}{\nu([-\tau,\tau])}\int_{M_{\tau,\varepsilon}(f)}d\mu(t)\\ &\quad+\frac{\varepsilon}{\nu([-\tau,\tau])}\int_{[-\tau,\tau]}d\mu(t)\\ &\le \frac{N}{\nu([-\tau,\tau])}\mu(M_{\tau,\varepsilon}(f))+\frac{\varepsilon\mu([-\tau,\tau])}{\nu([-\tau,\tau])},\end{aligned}$$

which implies that

$$\lim_{\tau\to+\infty}\frac{1}{\nu([-\tau,\tau])}\int_{-\tau}^{+\tau}|f(t)|_\alpha d\mu(t)\le\delta\varepsilon \quad \text{for any } \varepsilon>0.$$

Therefore, $f\in\mathscr{E}(\mathbb{R};X_\alpha,\mu,\nu)$. ■

**Definition 9.4** Let $\mu,\nu\in\mathcal{M}$. A bounded continuous function $\phi:\mathbb{R}\to X_\alpha$ is called $(\mu,\nu)$-pseudo-almost periodic if $\phi=\phi_1+\phi_2$, where $\phi_1\in AP(\mathbb{R},X_\alpha)$ and $\phi_2\in\mathscr{E}(\mathbb{R};X_\alpha,\mu,\nu)$.

We denote by $PAP(\mathbb{R};X_\alpha,\mu,\nu)$ the space of all such functions.

**Definition 9.5** Let $\mu,\nu\in\mathcal{M}$. A bounded continuous function $\phi:\mathbb{R}\times X_\alpha\to X$ is called uniformly $(\mu,\nu)$-pseudo-almost periodic if $\phi=\phi_1+\phi_2$, where $\phi_1\in AP(\mathbb{R}\times X_\alpha;X_\alpha)$ and $\phi_2\in\mathscr{E}(\mathbb{R}\times X_\alpha,\mu,\nu)$.

We denote by $PAP(\mathbb{R}\times X_\alpha,\mu,\nu)$ the space of all such functions.

**Definition 9.6** Let $\mu,\nu\in\mathcal{M}$. A bounded continuous function $\phi:\mathbb{R}\to X_\alpha$ is called $(\mu,\nu)$-pseudo-almost periodic of class $r$ if $\phi=\phi_1+\phi_2$, where $\phi_1\in AP(\mathbb{R};X_\alpha)$ and $\phi_2\in\mathscr{E}(\mathbb{R};X_\alpha,\mu,\nu,r)$.

We denote by $PAP(\mathbb{R};X_\alpha,\mu,\nu,r)$ the space of all such functions.

**Definition 9.7** Let $\mu,\nu\in\mathcal{M}$. A bounded continuous function $\phi:\mathbb{R}\times X_\alpha\to X_\alpha$ is called uniformly $(\mu,\nu)$-pseudo-almost periodic of class $r$ if $\phi=\phi_1+\phi_2$, where $\phi_1\in AP(\mathbb{R}\times X_\alpha;X_\alpha)$ and $\phi_2\in\mathscr{E}(\mathbb{R}\times X_\alpha,\mu,\nu,r)$.

We denote by $PAP(\mathbb{R}\times X_\alpha,\mu,\nu,r)$ the space of all such functions.

## 9.5 Properties of $(\mu, \nu)$-Pseudo-almost Periodic Functions of Class $r$

**Lemma 9.4** *Assume that* ($\mathbf{H_2}$) *holds. The space* $\mathscr{E}(\mathbb{R}; X_\alpha, \mu, \nu, r)$ *endowed with the uniform topology norm is a Banach space.*

***Proof*** We can see that $\mathscr{E}(\mathbb{R}; X_\alpha, \mu, \nu, r)$ is a vector subspace of $BC(\mathbb{R}; X_\alpha)$. To complete the proof, it is enough to prove that $\mathscr{E}(\mathbb{R}; X_\alpha, \mu, \nu, r)$ is closed in $BC(\mathbb{R}; X_\alpha)$. Let $(z_n)_n$ be a sequence in $\mathscr{E}(\mathbb{R}; X_\alpha, \mu, \nu, r)$ such that $\lim_{n\to+\infty} z_n = z$ uniformly in $\mathbb{R}$. From $\nu(\mathbb{R}) = +\infty$, it follows $\nu([-\tau, \tau]) > 0$ for $\tau$ sufficiently large. Let $\|z\|_{\infty,\alpha} = \sup_{t\in\mathbb{R}} |z(t)|_\alpha$ and $n_0 \in \mathbb{N}$ such that for all $n \geq n_0$, $\|z_n - z\|_{\infty,\alpha} < \varepsilon$. Let $n \geq n_0$; then we have

$$
\begin{aligned}
&\frac{1}{\nu([-\tau,\tau])}\int_{-\tau}^{+\tau}\Big(\sup_{\theta\in[t-r,t]}|z(\theta)|_\alpha\Big)d\mu(t)\\
&\leq \frac{1}{\nu([-\tau,\tau])}\int_{-\tau}^{+\tau}\Big(\sup_{\theta\in[t-r,t]}|z_n(\theta)-z(\theta)|_\alpha\Big)d\mu(t)\\
&\quad+\frac{1}{\nu([-\tau,\tau])}\int_{-\tau}^{+\tau}\Big(\sup_{\theta\in[t-r,t]}|z_n(\theta)|_\alpha\Big)d\mu(t)\\
&\leq \frac{1}{\nu([-\tau,\tau])}\int_{-\tau}^{+\tau}\Big(\sup_{t\in\mathbb{R}}|z_n(t)-z(t)|_\alpha\Big)d\mu(t)\\
&\quad+\frac{1}{\nu([-\tau,\tau])}\int_{-\tau}^{+\tau}\Big(\sup_{\theta\in[t-r,t]}|z_n(\theta)|_\alpha\Big)d\mu(t)\\
&\leq \|z_n - z\|_{\infty,\alpha}\times\frac{\mu([-\tau,\tau])}{\nu([-\tau,\tau])}\\
&\quad+\frac{1}{\nu([-\tau,\tau])}\int_{-\tau}^{+\tau}\Big(\sup_{\theta\in[t-r,t]}|z_n(\theta)|_\alpha\Big)d\mu(t).
\end{aligned}
$$

We deduce that

$$
\limsup_{\tau\to+\infty}\frac{1}{\nu([-\tau,\tau])}\int_{-\tau}^{+\tau}\Big(\sup_{\theta\in[t-r,t]}|z(\theta)|_\alpha\Big)d\mu(t) \leq \delta\varepsilon \quad \text{for any } \varepsilon > 0.
$$

■

From the definition of $PAP(\mathbb{R}; X_\alpha, \mu, \nu, r)$, we deduce the following result.

**Proposition 9.4** *Assume that* ($\mathbf{H_2}$) *holds, and let* $\mu, \nu \in \mathcal{M}$. *The space* $PAP(\mathbb{R}; X_\alpha, \mu, \nu, r)$ *endowed with the uniform topology norm is a Banach space.*

Next result is a characterization of $\alpha - (\mu, \nu)$-ergodic functions of class $r$.

**Theorem 9.2** *Assume that* ($\mathbf{H_2}$) *holds, and let* $\mu, \nu \in \mathcal{M}$ *and* $I$ *be a bounded interval (eventually* $I = \emptyset$*). Assume that* $f \in BC(\mathbb{R}, X_\alpha)$*. Then the following assertions are equivalent:*

*(i)* $f \in \mathscr{E}(\mathbb{R}, X_\alpha, \mu, \nu, r)$.

*(ii)* $\lim_{\tau\to+\infty} \frac{1}{\nu([-\tau,\tau]\setminus I)} \int_{[-\tau,\tau]\setminus I} \Big(\sup_{\theta\in[t-r,t]} |f(\theta)|_\alpha\Big) d\mu(t) = 0.$

*(iii) For any* $\varepsilon > 0$, $\lim_{\tau\to+\infty} \frac{\mu\Big(\Big\{t \in [-\tau,\tau]\setminus I : \sup_{\theta\in[t-r,t]} |f(\theta)|_\alpha > \varepsilon\Big\}\Big)}{\nu([-\tau,\tau]\setminus I)} = 0.$

***Proof*** $(i) \Leftrightarrow (ii)$ Denote by $A = \nu(I)$, $B = \int_I \Big(\sup_{\theta\in[t-r,t]} |f(\theta)|_\alpha\Big) d\mu(t)$. We have $A$ and $B \in \mathbb{R}$, since the interval $I$ is bounded and the function $f$ is bounded and continuous. For $\tau > 0$ such that $I \subset [-\tau,\tau]$ and $\nu([-\tau,\tau]\setminus I) > 0$, we have

$$\begin{aligned}
&\frac{1}{\nu([-\tau,\tau]\setminus I)} \int_{[-\tau,\tau]\setminus I} \Big(\sup_{\theta\in[t-r,t]} |f(\theta)|_\alpha\Big) d\mu(t)\\
&= \frac{1}{\nu([-\tau,\tau]) - A}\Big[\int_{[-\tau,\tau]} \Big(\sup_{\theta\in[t-r,t]} |f(\theta)|_\alpha\Big) d\mu(t) - B\Big]\\
&= \frac{\nu([-\tau,\tau])}{\nu([-\tau,\tau]) - A}\Big[\frac{1}{\nu([-r,r])} \int_{[-\tau,\tau]} \Big(\sup_{\theta\in[t-r,t]} |f(\theta)|_\alpha\Big) d\mu(t) - \frac{B}{\nu([-\tau,\tau])}\Big].
\end{aligned}$$

From the above equalities and the fact that $\nu(\mathbb{R}) = +\infty$, we deduce that $(ii)$ is equivalent to

$$\lim_{\tau\to+\infty} \frac{1}{\nu([-\tau,\tau])} \int_{-\tau}^{+\tau} \Big(\sup_{\theta\in[t-r,t]} |f(\theta)|_\alpha\Big) d\mu(t) = 0,$$

that is $(i)$.

$(iii) \Rightarrow (ii)$ Denote by $A_\tau^\varepsilon$ and $B_\tau^\varepsilon$ the following sets

$$A_\tau^\varepsilon = \Big\{t \in [-\tau,\tau]\setminus I : \sup_{\theta\in[t-r,t]} |f(\theta)|_\alpha > \varepsilon\Big\} \text{ and}$$

$$B_\tau^\varepsilon = \Big\{t \in [-\tau,\tau]\setminus I) : \sup_{\theta\in[t-r,t]} |f(\theta)|_\alpha \le \varepsilon\Big\}.$$

Assume that $(iii)$ holds, that is

$$\lim_{\tau\to+\infty} \frac{\mu(A_\tau^\varepsilon)}{\nu([-\tau,\tau]\setminus I)} = 0. \tag{9.4}$$

From the equality

$$\int_{[-\tau,\tau]\setminus I}\left(\sup_{\theta\in[t-r,t]}|f(\theta)|_\alpha\right)d\mu(t) = \int_{A_\tau^\varepsilon}\left(\sup_{\theta\in[t-r,t]}|f(\theta)|_\alpha\right)d\mu(t)$$
$$+\int_{B_\tau^\varepsilon}\left(\sup_{\theta\in[t-r,t]}|f(\theta)|\right)d\mu(t),$$

we deduce that for $\tau$ sufficiently large

$$\frac{1}{\nu([-\tau,\tau]\setminus I)}\int_{[-\tau,\tau]\setminus I}\left(\sup_{\theta\in[t-r,t]}|f(\theta)|_\alpha\right)d\mu(t) \le \|f\|_{\infty,\alpha}\frac{\mu(A_\tau^\varepsilon)}{\nu([-\tau,\tau]\setminus I)}$$
$$+\varepsilon\frac{\mu(B_\tau^\varepsilon)}{\nu([-\tau,\tau]\setminus I)}.$$

By using $(\mathbf{H_2})$, it follows that

$$\lim_{\tau\to+\infty}\frac{1}{\nu([-\tau,\tau])}\int_{-\tau}^{+\tau}\left(\sup_{\theta\in[t-r,t]}|f(\theta)|_\alpha\right)d\mu(t) \le \delta\varepsilon, \quad \text{for any } \varepsilon>0,$$

and consequently (ii) holds.
$(ii) \Rightarrow (iii)$ Assume that $(ii)$ holds. From the following inequality

$$\int_{[-\tau,\tau]\setminus I}\left(\sup_{\theta\in[t-r,t]}|f(\theta)|_\alpha\right)d\mu(t) \ge \int_{A_\tau^\varepsilon}\left(\sup_{\theta\in[t-r,t]}|f(\theta)|_\alpha\right)d\mu(t)$$
$$\frac{1}{\nu([-\tau,\tau]\setminus I)}\int_{[-\tau,\tau]\setminus I}\left(\sup_{\theta\in[t-r,t]}|f(\theta)|_\alpha\right)d\mu(t) \ge \varepsilon\frac{\mu(A_\tau^\varepsilon)}{\nu([-\tau,\tau]\setminus I)}$$
$$\frac{1}{\varepsilon\nu([-\tau,\tau]\setminus I)}\int_{[-\tau,\tau]\setminus I}\left(\sup_{\theta\in[t-r,t]}|f(\theta)|_\alpha\right)d\mu(t) \ge \frac{\mu(A_\tau^\varepsilon)}{\nu([-\tau,\tau]\setminus I)},$$

for $\tau$ sufficiently large, we obtain Eq. (9.4), that is, $iii$). ■

From $\mu \in \mathcal{M}$, we formulate the following hypotheses:

$(\mathbf{H_3})$ For all $a$, $b$, and $c \in \mathbb{R}$, such that $0 \le a < b \le c$, there exist $\delta_0$ and $\alpha_0 > 0$ such that

$$|\delta| \ge \delta_0 \Rightarrow \mu(a+\delta, b+\delta) \ge \alpha_0\mu(\delta, c+\delta).$$

$(\mathbf{H_4})$ For all $\tau \in \mathbb{R}$, there exist $\beta > 0$ and a bounded interval $I$ such that

$$\mu(\{a+\tau : a \in A\} \le \beta\mu(A) \text{ when } A \in \mathcal{B} \text{ satisfies } A\cap I = \emptyset.$$

We have the following results due to [6].

**Lemma 9.5 ([6])** *Hypothesis* $(\mathbf{H_4})$ *implies* $(\mathbf{H_3})$.

**Proposition 9.5 ([5, 8])** *$\mu, \nu \in \mathcal{M}$ satisfy* **(H3)** *and $f \in PAP(\mathbb{R}; X, \mu, \nu)$ be such that*

$$f = g + h,$$

*where $g \in AP(\mathbb{R}, X)$ and $h \in \mathcal{E}(\mathbb{R}, X, \mu, \nu)$. Then*

$$\{g(t),\ t \in \mathbb{R}\} \subset \overline{\{f(t),\ t \in \mathbb{R}\}} \text{ (the closure of the range of} f).$$

**Corollary 9.1 ([8])** *Assume that* **(H3)** *holds. Then the decomposition of a $(\mu, \nu)$-pseudo-almost periodic function in the form $f = g + \phi$, where $g \in AP(\mathbb{R}; X)$ and $\phi \in \mathcal{E}(\mathbb{R}; X, \mu, \nu)$ is unique.*

The following corollary is a consequence of Theorem 9.2.

**Proposition 9.6** *Let $\mu, \nu \in \mathcal{M}$. Assume* **(H3)** *holds. Then the decomposition of a $\alpha - (\mu, \nu)$-pseudo-almost periodic function $\phi = \phi_1 + \phi_2$, where $\phi_1 \in AP(\mathbb{R}; X_\alpha)$ and $\phi_2 \in \mathcal{E}(\mathbb{R}; X_\alpha, \mu, \nu, r)$, is unique.*

***Proof*** In fact, since as a consequence of Corollary 9.1, the decomposition of a $(\mu, \nu)$-pseudo-almost periodic function $\phi = \phi_1 + \phi_2$, where $\phi_1 \in AP(\mathbb{R}; X_\alpha)$ and $\phi_2 \in \mathcal{E}(\mathbb{R}; X_\alpha, \mu, \nu)$, is unique. Since $PAP(\mathbb{R}; X_\alpha, \mu, \nu) \subset PAP(\mathbb{R}; X, \mu, \nu)$ and $PAP(\mathbb{R}; X_\alpha, \mu, \nu, r) \subset PAP(\mathbb{R}; X_\alpha, \mu, \nu)$, we get the desired result. ■

**Definition 9.8** Let $\mu_1, \mu_2 \in \mathcal{M}$. We say that $\mu_1$ is equivalent to $\mu_2$, denoting this as $\mu_1 \sim \mu_2$ if there exist constants $\alpha$ and $\beta > 0$ and a bounded interval I (eventually $I = \emptyset$) such that

$$\alpha\mu_1(A) \le \mu_2(A) \le \beta\mu_1(A), \quad \text{when } A \in \mathcal{B} \text{ satisfies } A \cap I = \emptyset.$$

From [6] $\sim$ is a binary equivalence relation on $\mathcal{M}$. The equivalence class of a given measure $\mu \in \mathcal{M}$ will then be denoted by

$$cl(\mu) = \{\varpi \in \mathcal{M} : \ \mu \sim \varpi\}.$$

**Theorem 9.3** *Let $\mu_1, \mu_2, \nu_1, \nu_2 \in \mathcal{M}$. If $\mu_1 \sim \mu_2$ and $\nu_1 \sim \nu_2$, then $PAP(\mathbb{R}; X_\alpha, \mu_1, \nu_1, r) = PAP(\mathbb{R}; X_\alpha, \mu_2, \nu_2, r)$.*

***Proof*** Since $\mu_1 \sim \mu_2$ and $\nu_1 \sim \nu_2$, there exist some constants $\alpha_1, \alpha_2, \beta_1, \beta_2 > 0$, and a bounded interval I (eventually $I = \emptyset$) such that $\alpha_1\mu_1(A) \le \mu_2(A) \le \beta_1\mu_1(A)$ and $\alpha_2\nu_1(A) \le \nu_2(A) \le \beta_2\nu_1(A)$ for each $A \in \mathcal{B}$ satisfies $A \cap I = \emptyset$ i.e.,

$$\frac{1}{\beta_2\nu_1(A)} \le \frac{1}{\nu_2(A)} \le \frac{1}{\alpha_2\nu_1(A)}.$$

Since $\mu_1 \sim \mu_2$ and $\mathscr{B}$ is the Lebesgue $\sigma$-field, we obtain for $\tau$ sufficiently large that

$$\frac{\alpha_1\mu_1\Big(\Big\{t \in [-\tau,\tau]\setminus I: \sup_{\theta\in[t-r,t]} |f(\theta)|_\alpha > \varepsilon\Big\}\Big)}{\beta_2\nu_1([-\tau,\tau]\setminus I)}$$
$$\leq \frac{\mu_2\Big(\Big\{t \in [-\tau,\tau]\setminus I: \sup_{\theta\in[t-r,t]} |f(\theta)|_\alpha > \varepsilon\Big\}\Big)}{\nu_2([-\tau,\tau]\setminus I)}$$
$$\leq \frac{\beta_1\mu_1\Big(\Big\{t \in [-\tau,\tau]\setminus I: \sup_{\theta\in[t-r,t]} |f(\theta)|_\alpha > \varepsilon\Big\}\Big)}{\alpha_2\nu_1([-\tau,\tau]\setminus I)}.$$

By using Theorem 9.2, we deduce that $\mathscr{E}(\mathbb{R}, X_\alpha, \mu_1, \nu_1, r) = \mathscr{E}(\mathbb{R}, X_\alpha, \mu_2, \nu_2, r)$. From the definition of a $(\mu,\nu)$-pseudo-almost periodic function, we deduce that $PAP(\mathbb{R}; X_\alpha, \mu_1, \nu_1, r) = PAP(\mathbb{R}; X_\alpha, \mu_2, \nu_2, r)$. ■

Let $\mu, \nu \in \mathcal{M}$. We denote by

$$cl(\mu,\nu) = \{\varpi_1, \varpi_2 \in \mathcal{M}: \ \mu \sim \varpi_2 \text{ and } \nu \sim \varpi_2\}.$$

In what follows, we prove some preliminary results concerning the composition of $(\mu,\nu)$-pseudo-almost periodic functions of class $r$.

**Theorem 9.4** *Let $\mu, \nu \in \mathcal{M}$, $\phi \in PAP(\mathbb{R}\times X_\alpha, \mu, \nu, r)$, and $h \in PAP(\mathbb{R}; X_\alpha, \mu, \nu, r)$. Assume that there exists a function $L_\phi : \mathbb{R} \to [0, +\infty[$ such that*

$$|\phi(t,x_1) - \phi(t,x_2)| \leq L_\phi(t)|x_1 - x_2|_\alpha \quad \textit{for} \quad t \in \mathbb{R} \quad \textit{and for} \quad x_1, x_2 \in X_\alpha. \tag{9.5}$$

*If*

$$\limsup_{\tau\to+\infty} \frac{1}{\nu([-\tau,\tau])}\int_{-\tau}^{\tau}\Big(\sup_{\theta\in[t-r,t]} L_\phi(\theta)\Big)d\mu(t) < \infty \quad \textit{and}$$
$$\lim_{\tau\to+\infty} \frac{1}{\nu([-\tau,\tau])}\int_{-\tau}^{+\tau}\Big(\sup_{\theta\in[t-r,t]} L_\phi(\theta)\Big)\xi(t)d\mu(t) = 0 \tag{9.6}$$

*for each $\xi \in \mathscr{E}(\mathbb{R}, \mathbb{R}, \mu, \nu)$, then the function $t \to \phi(t, h(t))$ belongs to $PAP(\mathbb{R}; X_\alpha, \mu, \nu, r)$.*

***Proof*** Assume that $\phi = \phi_1 + \phi_2$, $h = h_1 + h_2$, where $\phi_1 \in AP(\mathbb{R}\times X_\alpha; X_\alpha)$, $\phi_2 \in \mathscr{E}(\mathbb{R}\times X_\alpha, \mu, \nu, r)$ and $h_1 \in AP(\mathbb{R}; X_\alpha)$, $h_2 \in \mathscr{E}(\mathbb{R}; X_\alpha, \mu, \nu, r)$. Consider the following decomposition:

$$\phi(t,h(t)) = \phi_1(t,h_1(t)) + [\phi(t,h(t)) - \phi(t,h_1(t))] + \phi_2(t,h_1(t)).$$

Since $h_1 \in AP(\mathbb{R}; X_\alpha)$, Lemma 9.2 implies that the set $K = \{h_{1_\tau} : \tau \in \mathbb{R}\}$ is relatively compact in $BC(\mathbb{R}, X_\alpha)$. Consequently, since $\phi_1 \in AP(\mathbb{R} \times X_\alpha; X_\alpha)$, then for all $\varepsilon > 0$, there exists a relatively dense subset of $\mathbb{R}$ denoted by $\mathcal{K}(\varepsilon, \phi_1, K)$ such that $|\phi(t+\tau, x) - \phi(t, x)|_\alpha < \varepsilon$ for all $t \in \mathbb{R},\ x \in K,\ \tau \in \mathcal{K}(\varepsilon, \phi_1, K)$. It follows that $\phi_1(., h_1(.)) \in AP(\mathbb{R}; X_\alpha)$. It remains to prove that both $\phi(., h(.)) - \phi(., h_1(.))$ and $\phi_2(., h_1(.))$ belong to $\mathcal{E}(\mathbb{R}; X_\alpha, \mu, \nu, r)$. Consequently, using inequality (9.5), it follows that

$$\frac{\mu\left(\left\{t \in [-\tau, \tau] : \sup\limits_{\theta \in [t-r, t]} |\phi(\theta, h(\theta)) - \phi(\theta, h_1(\theta))|_\alpha > \varepsilon\right\}\right)}{\nu([-\tau, \tau])}$$

$$\leq \frac{\mu\left(\left\{t \in [-\tau, \tau] : \sup\limits_{\theta \in [t-r, t]} (L_\phi(\theta)|h_2(\theta)|_\alpha) > \varepsilon\right\}\right)}{\nu([-\tau, \tau])}$$

$$\leq \frac{\mu\left(\left\{t \in [-\tau, \tau] : \left(\sup\limits_{\theta \in [t-r, t]} L_\phi(\theta)\right)\left(\sup\limits_{\theta \in [t-r, t]} |h_2(\theta)|_\alpha\right) > \varepsilon\right\}\right)}{\nu([-\tau, \tau])}.$$

Since $h_2$ is $(\mu, \nu)$-ergodic of class $r$, Theorem 9.2 and Eq. (9.6) yield that for the above-mentioned $\varepsilon$, we have

$$\lim_{\tau \to +\infty} \frac{\mu\left(\left\{t \in [-\tau, \tau] : \left(\sup\limits_{\theta \in [t-r, t]} L_\phi(\theta)\right)\left(\sup\limits_{\theta \in [t-r, t]} |h_2(\theta)|_\alpha\right) > \varepsilon\right\}\right)}{\nu([-\tau, \tau])} = 0,$$

and then we obtain

$$\lim_{\tau \to +\infty} \frac{\mu\left(\left\{t \in [-\tau, \tau] : \sup\limits_{\theta \in [t-r, t]} |\phi(\theta, h(\theta)) - \phi(\theta, h_1(\theta))|_\alpha > \varepsilon\right\}\right)}{\nu([-\tau, \tau])} = 0. \tag{9.7}$$

By Theorem 9.2, Eq. (9.7) shows that $t \mapsto \phi(t, h(t)) - \phi(t, h_1(t))$ is $(\mu, \nu)$-ergodic of class $r$.

Now to complete the proof, it is enough to prove that $t \mapsto \phi_2(t, h(t))$ is $(\mu, \nu)$-ergodic of class $r$. Since $\phi_2$ is uniformly continuous on the compact set $K = \overline{\{h_1(t) : t \in \mathbb{R}\}}$ with respect to the second variable $x$, we deduce that for given $\varepsilon > 0$, there exists $\delta > 0$ such that, for all $t \in \mathbb{R}$, $\xi_1$ and $\xi_2 \in K$, one has

$$|\xi_1 - \xi_2| \leq \delta \Rightarrow |\phi_2(t, \xi_1) - \phi_2(t, \xi_2)|_\alpha \leq \varepsilon.$$

Therefore, there exist $n(\varepsilon)$ and $\{z_i\}_{i=1}^{n(\varepsilon)} \subset K$, such that

$$K \subset \bigcup_{i=1}^{n(\varepsilon)} B_\delta(z_i, \delta),$$

and then

$$\|\phi_2(t, h_1(t))\| \leq \varepsilon + \sum_{1}^{n(\varepsilon)} \|\phi_2(t, z_i)\|.$$

Since

$$\forall i \in \{1, \ldots, n(\varepsilon)\}, \quad \lim_{\tau \to +\infty} \frac{1}{\nu([-\tau, \tau])} \int_{-\tau}^{\tau} \Big( \sup_{\theta \in [t-r,t]} |\phi_2(\theta, z_i)|_\alpha \Big) d\mu(t) = 0,$$

we deduce that

$$\forall \varepsilon > 0, \quad \limsup_{\tau \to +\infty} \frac{1}{\nu([-\tau, \tau])} \int_{-\tau}^{\tau} \Big( \sup_{\theta \in [t-r,t]} |\phi_2(\theta, h_1(t))|_\alpha \Big) d\mu(t) \leq \varepsilon\delta,$$

which implies

$$\lim_{\tau \to +\infty} \frac{1}{\nu([-\tau, \tau])} \int_{-\tau}^{\tau} \Big( \sup_{\theta \in [t-r,t]} |\phi_2(\theta, h_1(\theta))|_\alpha \Big) d\mu(t) = 0.$$

Consequently, $t \mapsto \phi_2(t, h(t))$ is $(\mu, \nu)$-ergodic of class $r$. ■

For $\mu \in \mathcal{M}$ and $\delta \in \mathbb{R}$, we denote $\mu_\delta$ the positive measure on $(\mathbb{R}, \mathcal{B})$ defined by

$$\mu_\delta(A) = \mu([a + \delta : a \in A]). \tag{9.8}$$

**Lemma 9.6 ([6])** *Let $\mu \in \mathcal{M}$ satisfy* **(H$_4$)**. *Then the measures $\mu$ and $\mu_\delta$ are equivalent for all $\delta \in \mathbb{R}$.*

**Lemma 9.7 ([6])** **(H$_4$)** *implies*

$$\textit{for all } \sigma > 0 \ \limsup_{\tau \to +\infty} \frac{\mu([-\tau - \sigma, \tau + \sigma])}{\mu([-\tau, \tau])} < +\infty.$$

We have the following result.

**Theorem 9.5** *Assume that* **(H$_4$)** *holds. Let $\mu, \nu \in \mathcal{M}$ and $\phi \in PAP(\mathbb{R}; X_\alpha, \mu, \nu, r)$; then the function $t \to \phi_t$ belongs to $PAP(C_\alpha, \mu, \nu, r)$.*

***Proof*** Assume that $\phi = g + h$, where $g \in AP(\mathbb{R}; X_\alpha)$ and $h \in \mathcal{E}(\mathbb{R}; X_\alpha, \mu, \nu, r)$. Then we can see that $\phi_t = g_t + h_t$, and $g_t$ is almost periodic. Let us denote by

$$M_\delta(\tau) = \frac{1}{\nu_\delta([-\tau, \tau])} \int_{-\tau}^{+\tau} \Big( \sup_{\theta \in [t-r,t]} |h(\theta)|_\alpha \Big) d\mu_\delta(t),$$

where $\mu_\delta$ and $\nu_\delta$ are the positive measures defined by Eq. (9.8). By using Lemma 9.6, it follows that $\mu_\delta$ and $\mu$ are equivalent and $\nu_\delta$ and $\nu$ are also equivalent.

Then by using Theorem 9.3, we have $\mathscr{E}(\mathbb{R}; X_\alpha, \mu_\delta, \nu_\delta, r) = \mathscr{E}(\mathbb{R}; X_\alpha, \mu, \nu, r)$; therefore, $h \in \mathscr{E}(\mathbb{R}; X, \mu_\delta, \nu_\delta, r)$, that is,

$$\lim_{\tau \to +\infty} M_\delta(\tau) = 0, \quad \text{for all } \delta \in \mathbb{R}.$$

On the other hand, for $r > 0$, we have

$$\begin{aligned}
&\frac{1}{\nu([-\tau, \tau])} \int_{-\tau}^{+\tau} \Big( \sup_{\theta \in [t-r,t]} \Big[ \sup_{\xi \in [-r,0]} |h(\theta + \xi)|_\alpha \Big] \Big) d\mu(t) \\
&\le \frac{1}{\nu([-\tau, \tau])} \int_{-\tau}^{+\tau} \Big( \sup_{\theta \in [t-2r,t]} |h(\theta)|_\alpha \Big) d\mu(t) \\
&\le \frac{1}{\nu([-\tau, \tau])} \int_{-\tau}^{+\tau} \Big( \sup_{\theta \in [t-2r,t-r]} |h(\theta)|_\alpha + \sup_{\theta \in [t-r,t]} |h(\theta)|_\alpha \Big) d\mu(t) \\
&\le \frac{1}{\nu([-\tau, \tau])} \int_{-\tau-r}^{+\tau-r} \Big( \sup_{\theta \in [t-r,t]} |h(\theta)|_\alpha \Big) d\mu(t+r) \\
&\quad + \frac{1}{\nu([-\tau, \tau])} \int_{-\tau}^{+\tau} \Big( \sup_{\theta \in [t-r,t]} |h(\theta)|_\alpha \Big) d\mu(t) \\
&\le \frac{1}{\nu([-\tau, \tau])} \int_{-\tau-r}^{+\tau+r} \Big( \sup_{\theta \in [t-r,t]} |h(\theta)|_\alpha \Big) d\mu(t+r) \\
&\quad + \frac{1}{\nu([-\tau, \tau])} \int_{-\tau}^{+\tau} \Big( \sup_{\theta \in [t-r,t]} |h(\theta)|_\alpha \Big) d\mu(t) \\
&\le \Big[ \frac{\nu([-\tau - r, \tau + r])}{\nu([-\tau, \tau])} \Big] \\
&\quad \times \frac{1}{\nu([-\tau - r, \tau + r]} \int_{-\tau-r}^{+\tau+r} \Big( \sup_{\theta \in [t-r,t]} |h(\theta)|_\alpha \Big) d\mu(t+r) \\
&\quad + \frac{1}{\nu([-\tau, \tau])} \int_{-\tau}^{+\tau} \Big( \sup_{\theta \in [t-r,t]} |h(\theta)|_\alpha \Big) d\mu(t).
\end{aligned}$$

Consequently,

$$\begin{aligned}
&\frac{1}{\nu([-\tau, \tau])} \int_{-\tau}^{+\tau} \Big( \sup_{\theta \in [t-r,t]} \Big[ \sup_{\xi \in [-r,0]} |h(\theta + \xi)|_\alpha \Big] \Big) d\mu(t) \\
&\le \Big[ \frac{\nu([-\tau - r, \tau + r])}{\nu([-\tau, \tau])} \Big] \times M_r(\tau + r) \\
&\quad + \frac{1}{\nu([-\tau, \tau])} \int_{-\tau}^{+\tau} \Big( \sup_{\theta \in [t-r,t]} |h(\theta)|_\alpha \Big) d\mu(t),
\end{aligned}$$

which shows using Lemmas 9.6 and 9.7 that $\phi_t$ belongs to $PAP(C_\alpha, \mu, \nu, r)$. Thus, we obtain the desired result. ∎

**Lemma 9.8 ([8])** *Let $\mu, \nu \in \mathcal{M}$ satisfy* (**H**$_4$). *Then $PAP(\mathbb{R}, X, \mu, \nu)$ is invariant by translation, that is, $f \in PAP(\mathbb{R}, X, \mu, \nu)$ implies $f_\gamma \in PAP(\mathbb{R}, X, \mu, \nu)$ for all $\gamma \in \mathbb{R}$.*

**Corollary 9.2** *Let $\mu, \nu \in \mathcal{M}$ satisfy* (**H**$_4$). *Then $PAP(\mathbb{R}, X, \mu, \nu, r)$ is invariant by translation, that is, $f \in PAP(\mathbb{R}, X, \mu, \nu, r)$ implies $f_\gamma \in PAP(\mathbb{R}, X, \mu, \nu, r)$ for all $\gamma \in \mathbb{R}$.*

***Proof*** It suffices to prove that $\mathcal{E}(\mathbb{R}, X, \mu, \nu, r)$ is invariant by translation. Let $f \in \mathcal{E}(\mathbb{R}, X, \mu, \nu, \infty)$ and $F^t(\theta) = \sup_{\theta \in [t-r,t]} |f(\theta)|$. Then $F^t \in \mathcal{E}(\mathbb{R}, \mathbb{R}, \mu, \nu)$, but since $\mathcal{E}(\mathbb{R}, \mathbb{R}, \mu, \nu)$ is invariant by translation, it follows that

$$\lim_{\tau \to +\infty} \frac{1}{\nu([-\tau, \tau])} \int_{-\tau}^{\tau} F^t(\theta + \gamma) d\mu(t) = \lim_{\tau \to +\infty} \frac{1}{\nu([-\tau, \tau])} \int_{-\tau}^{\tau} \sup_{\theta \in [t-r,t]} |f(\theta + \gamma)| d\mu(t) = 0,$$

which implies that $f(. + \gamma) \in PAP(\mathbb{R}, X, \mu, \nu, r)$. ∎

## 9.6 $(\mu, \nu)$-Pseudo-almost Periodic Solutions of Class $r$

In what follows, we will be looking at the existence of bounded integral solutions of class $r$ of Eq. (9.1).

**Proposition 9.7** *[3] Assume that* (**H**$_0$) *and* (**H**$_1$) *hold and the semigroup $(\mathcal{U}(t))_{t \geq 0}$ is hyperbolic. If $f$ is bounded on $\mathbb{R}$, then there exists a unique bounded solution $u$ of Eq.* (9.1) *on $\mathbb{R}$, given by*

$$u_t = \lim_{\lambda \to +\infty} \int_{-\infty}^{t} \mathcal{U}^s(t-s)\Pi^s(\widetilde{B}_\lambda X_0 f(s))ds + \lim_{\lambda \to +\infty} \int_{+\infty}^{t} \mathcal{U}^u(t-s)\Pi^u(\widetilde{B}_\lambda X_0 f(s))ds \quad \text{for } t \in \mathbb{R},$$

*where $\widetilde{B}_\lambda = \lambda(\lambda I - \mathcal{A}_\mathcal{U})^{-1}$ for $\lambda > \widetilde{\omega}$, and $\Pi^s$ and $\Pi^u$ are the projections of $C_\alpha$ onto the stable and unstable subspaces, respectively.*

**Proposition 9.8** *Let $h \in AP(\mathbb{R}; X)$ and $\Gamma$ be the mapping defined for $t \in \mathbb{R}$ by*

$$\Gamma h(t) = \Big[\lim_{\lambda \to +\infty} \int_{-\infty}^{t} \mathcal{U}^s(t-s)\Pi^s(\widetilde{B}_\lambda X_0 h(s))ds + \lim_{\lambda \to +\infty} \int_{+\infty}^{t} \mathcal{U}^u(t-s)\Pi^u(\widetilde{B}_\lambda X_0 h(s))ds\Big](0).$$

*Then $\Gamma h \in AP(\mathbb{R}, X_\alpha)$.*

***Proof*** We can see that $\Gamma h \in BC(\mathbb{R}; X_\alpha)$. In fact,

$$
\begin{aligned}
|\Gamma h(t)|_\alpha &\leq \lim_{\lambda\to+\infty} \int_{-\infty}^{t} |\mathcal{U}^s(t-s)\Pi^s(\widetilde{B}_\lambda X_0 h(s))|_\alpha ds \\
&\quad + \lim_{\lambda\to+\infty} \int_{t}^{+\infty} |\mathcal{U}^u(t-s)\Pi^u(\widetilde{B}_\lambda X_0 h(s))|_\alpha ds\Big](0) \\
&\leq \lim_{\lambda\to+\infty} \int_{-\infty}^{t} \|\mathcal{A}_{\mathcal{U}}^{\alpha}\mathcal{U}^s(t-s)\Pi^s(\widetilde{B}_\lambda X_0 h(s))\| ds \\
&\quad + \lim_{\lambda\to+\infty} \int_{t}^{+\infty} \|\mathcal{A}_{\mathcal{U}}^{\alpha}\mathcal{U}^u(t-s)\Pi^u(\widetilde{B}_\lambda X_0 h(s))\| ds\Big](0) \\
&\leq \overline{M}\widetilde{M}\int_{-\infty}^{t} \frac{e^{-\omega(t-s)}}{(t-s)^\alpha}|\Pi^s|\,\|h(s)\|ds \\
&\quad +\overline{M}\widetilde{M}\int_{t}^{+\infty} \frac{e^{\omega(t-s)}}{(s-t)^\alpha}|\Pi^u|\,\|h(s)\|ds \\
&\leq \overline{M}\widetilde{M}\int_{-\infty}^{t} \frac{e^{-\omega(t-s)}}{(t-s)^\alpha}|\Pi^s|\,\|h(s)\|ds \\
&\quad +\overline{M}\widetilde{M}\int_{t}^{+\infty} \frac{e^{\omega(t-s)}}{(s-t)^\alpha}|\Pi^u|\,\|h(s)\|ds \\
&\leq \overline{M}\widetilde{M}\int_{-\infty}^{t} \frac{e^{-\omega(t-s)}}{(t-s)^\alpha}|\Pi^s|\,\|h(s)\|ds \\
&\quad +\overline{M}\widetilde{M}\int_{t}^{+\infty} \frac{e^{-\omega(s-t)}}{(s-t)^\alpha}|\Pi^u|\,\|h(s)\|ds \\
&\leq \frac{2K\|h\|_\infty}{\omega^{1-\alpha}}\int_{0}^{+\infty} e^{-s}s^{-\alpha}ds = \frac{2K\|h\|_\infty\Gamma(1-\alpha)}{\omega^{1-\alpha}} < \infty,
\end{aligned}
$$

where $K = \max(\overline{M}\widetilde{M}|\Pi^s|, \overline{M}\widetilde{M}|\Pi^u|)$. Since $h$ is an almost periodic function, then the set of functions $\{h_\tau : \delta \in \mathbb{R}\}$, $h_\tau(t) = h(t+\tau)$, is precompact in $BC(\mathbb{R}; X)$. On the other hand, we have

$$
\begin{aligned}
(\Gamma h)_\tau(t) &= (\Gamma h)(t+\tau) \\
&= \Big[\lim_{\lambda\to+\infty} \int_{-\infty}^{t+\tau} \mathcal{U}^s(t+\tau-s)\Pi^s(\widetilde{B}_\lambda X_0 h(s))ds \\
&\quad + \lim_{\lambda\to+\infty} \int_{+\infty}^{t+\tau} \mathcal{U}^u(t+\tau-s)\Pi^u(\widetilde{B}_\lambda X_0 h(s))ds\Big] \\
&= \Big[\lim_{\lambda\to+\infty} \int_{-\infty}^{t} \mathcal{U}^s(t-s)\Pi^s(\widetilde{B}_\lambda X_0 h(s+\tau))ds
\end{aligned}
$$

$$
\begin{aligned}
&+ \lim_{\lambda\to+\infty}\int_{+\infty}^{t} \mathcal{U}^u(t-s)\Pi^u(\widetilde{B}_\lambda X_0 h(s+\tau))ds\Big]\\
&= \Big[\lim_{\lambda\to+\infty}\int_{-\infty}^{t} \mathcal{U}^s(t-s)\Pi^s(\widetilde{B}_\lambda X_0 h_\tau(s))ds\\
&\quad + \lim_{\lambda\to+\infty}\int_{+\infty}^{t} \mathcal{U}^u(t-s)\Pi^u(\widetilde{B}_\lambda X_0 h_\tau(s))ds\Big]\\
&= (\Gamma h_\tau)(t) \text{ for all } t\in\mathbb{R}.
\end{aligned}
$$

Thus $(\Gamma h)_\tau = (\Gamma h_\tau)$, which implies that $\{(\Gamma h)_\delta : \delta \in \mathbb{R}\}$ is relatively compact in $BC(\mathbb{R}; X_\alpha)$ since $\Gamma$ is continuous from $BC(\mathbb{R}; X_\alpha)$ into $BC(\mathbb{R}; X_\alpha)$. Thus, $\Gamma h \in AP(\mathbb{R}, X_\alpha)$. ■

**Theorem 9.6** *Let $\mu, \nu \in \mathcal{M}$ satisfy* **(H$_4$)** *and $g \in \mathcal{E}(\mathbb{R}; X, \mu, \nu, r)$. Then $\Gamma g \in \mathcal{E}(\mathbb{R}; X_\alpha, \mu, \nu, r)$.*

***Proof*** In fact, for $\tau > 0$, we get

$$
\begin{aligned}
&\int_{-\tau}^{\tau}\Big(\sup_{\theta\in[t-r,t]} |\Gamma g(\theta)|_\alpha\Big)d\mu(t)\\
&\le \int_{-\tau}^{\tau}\Big(\sup_{\theta\in[t-r,t]}\Big[\lim_{\lambda\to+\infty}\int_{-\infty}^{\theta} |\mathcal{U}^s(\theta-s)\Pi^s(\widetilde{B}_\lambda X_0 g(s))|_\alpha ds\\
&\quad + \lim_{\lambda\to+\infty}\int_{\theta}^{+\infty} |\mathcal{U}^u(\theta-s)\Pi^u(\widetilde{B}_\lambda X_0 g(s))|_\alpha ds\Big](0)\Big)d\mu(t)\\
&\le \int_{-\tau}^{\tau}\Big(\sup_{\theta\in[t-r,t]}\Big[\lim_{\lambda\to+\infty}\int_{-\infty}^{\theta} \|\mathcal{A}_{\mathcal{U}}^{\alpha}\mathcal{U}^s(\theta-s)\Pi^s(\widetilde{B}_\lambda X_0 g(s))\| ds\\
&\quad + \lim_{\lambda\to+\infty}\int_{\theta}^{+\infty} \|\mathcal{A}_{\mathcal{U}}^{\alpha}\mathcal{U}^u(\theta-s)\Pi^u(\widetilde{B}_\lambda X_0 g(s))\| ds\Big](0)\Big)d\mu(t)\\
&\le \overline{M}\widetilde{M}\int_{-\tau}^{\tau}\Big(\sup_{\theta\in[t-r,t]}\int_{-\infty}^{\theta}\frac{e^{-\omega(\theta-s)}}{(\theta-s)^\alpha}|\Pi^s|\,|g(s)|ds\Big)d\mu(t)\\
&\quad + \overline{M}\widetilde{M}\int_{-\tau}^{\tau}\Big(\sup_{\theta\in[t-r,t]}\int_{\theta}^{+\infty}\frac{e^{\omega(\theta-s)}}{(s-\theta)^\alpha}|\Pi^u|\,|g(s)|ds\Big)d\mu(t)\\
&\le K\Big[\int_{-\tau}^{\tau}\Big(\sup_{\theta\in[t-r,t]}\int_{-\infty}^{\theta}\frac{e^{-\omega(\theta-s)}}{(\theta-s)^\alpha}|g(s)|ds\Big)d\mu(t)\\
&\quad + \int_{-\tau}^{\tau}\Big(\sup_{\theta\in[t-r,t]}\int_{\theta}^{+\infty}\frac{e^{\omega(\theta-s)}}{(s-\theta)^\alpha}|g(s)|ds\Big)d\mu(t)\Big],
\end{aligned}
$$

where $K = \max(\overline{M}\widetilde{M}|\Pi^s|, \overline{M}\widetilde{M}|\Pi^u|)$. On the one hand, using Fubini's theorem, we have

$$\begin{aligned}
&\int_{-\tau}^{\tau}\Big(\sup_{\theta\in[t-r,t]}\int_{-\infty}^{\theta}\frac{e^{-\omega(\theta-s)}}{(\theta-s)^{\alpha}}|g(s)|ds\Big)d\mu(t)\\
&\le \int_{-\tau}^{\tau}\Big(\sup_{\theta\in[t-r,t]}\int_{0}^{+\infty}\frac{e^{-\omega s}}{s^{\alpha}}|g(\theta-s)|ds\Big)d\mu(t)\\
&\le \int_{0}^{+\infty}\frac{e^{-\omega s}}{s^{\alpha}}\Big(\sup_{\theta\in[t-r,t]}\int_{-\tau}^{\tau}|g(\theta-s)|d\mu(t)\Big)ds.
\end{aligned}$$

By the Lebesgue dominated convergence theorem and by using Corollary 9.2, it follows that

$$\lim_{\tau\to+\infty}\int_{0}^{+\infty}\frac{e^{-\omega s}}{s^{\alpha}}\frac{1}{\nu([-\tau,\tau])}\Big(\sup_{\theta\in[t-r,t]}\int_{-\tau}^{\tau}|g(\theta-s)|d\mu(t)\Big)ds = 0.$$

On the other hand by Fubini's theorem, we also have

$$\begin{aligned}
&\int_{-\tau}^{\tau}\Big(\sup_{\theta\in[t-r,t]}\int_{\theta}^{+\infty}\frac{e^{\omega(\theta-s)}}{(s-\theta)^{\alpha}}|g(s)|ds\Big)d\mu(t)\\
&\le \int_{-\tau}^{\tau}\Big(\sup_{\theta\in[t-r,t]}\int_{0}^{+\infty}\frac{e^{-\omega s}}{s^{\alpha}}|g(s+\theta)|ds\Big)d\mu(t)\\
&\le \int_{0}^{+\infty}\frac{e^{-\omega s}}{s^{\alpha}}\Big(\sup_{\theta\in[t-r,t]}\int_{-\tau}^{\tau}|g(s+\theta)|d\mu(t)\Big)ds.
\end{aligned}$$

Reasoning like above, it follows that

$$\lim_{\tau\to+\infty}\int_{0}^{+\infty}\frac{e^{-\omega s}}{s^{\alpha}}\Big(\sup_{\theta\in[t-r,t]}\int_{-\tau}^{\tau}|g(s+\theta)|d\mu(t)\Big)ds = 0.$$

Consequently,

$$\lim_{\tau\to+\infty}\frac{1}{\nu([-\tau,\tau])}\int_{-\tau}^{\tau}\Big(\sup_{\theta\in[t-r,t]}|(\Gamma g)(\theta)|_{\alpha}\Big)d\mu(t) = 0.$$

Thus, we obtain the desired result. ∎

For the existence of $(\mu,\nu)$-pseudo-almost periodic solution of class $r$, we make the following assumption.

**(H$_5$)** $f:\mathbb{R}\to X$ is in $cl(\mu,\nu)$-pseudo-almost periodic of class $r$.

**Proposition 9.9** *Assume* $(\mathbf{H_0})$, $(\mathbf{H_1})$, $(\mathbf{H_3})$, *and* $(\mathbf{H_5})$ *hold. Then Eq.* (9.1) *has a unique* $\alpha - cl(\mu, \nu)$*-pseudo-almost periodic solution of class* $r$.

***Proof*** Since $f$ is a $(\mu, \nu)$-pseudo-almost periodic function, $f$ has a decomposition $f = f_1 + f_2$, where $f_1 \in AP(\mathbb{R}; X)$ and $f_2 \in \mathcal{E}(\mathbb{R}; X, \mu, \nu, r)$. Using Propositions 9.7, 9.8 and Theorem 9.6, we get the desired result. ■

Our next objective is to show the existence of $(\mu, \nu)$-pseudo-almost periodic solutions of class $r$ for the following problem:

$$u'(t) = -Au(t) + L(u_t) + f(t, u_t) \text{ for } t \in \mathbb{R}, \tag{9.9}$$

where $f : \mathbb{R} \times C_\alpha \to X$ is continuous.

For the sequel, we make the following assumption.

$(\mathbf{H_6})$ Let $\mu, \nu \in \mathcal{M}$ and $f : \mathbb{R} \times C_\alpha \to X$ $cl(\mu, \nu)$-pseudo-almost periodic of class $r$ such that there exists a positive constant $L_f$ such that

$$\|f(t, \varphi_1) - f(t, \varphi_2)\| \le L_f \|\varphi_1 - \varphi_2\|_{C_\alpha} \quad \text{for all } t \in \mathbb{R} \quad \text{and } \varphi_1, \varphi_2 \in C_\alpha,$$

and $L_f$ satisfies (9.6).

**Theorem 9.7** *Assume* $(\mathbf{H_0})$, $(\mathbf{H_1})$, $(\mathbf{H_2})$, $(\mathbf{H_3})$, *and* $(\mathbf{H_6})$ *hold. If*

$$\frac{2KL_f\Gamma(1-\alpha)}{\omega^{1-\alpha}} < 1,$$

*then Eq.* (9.9) *has a unique* $\alpha - cl(\mu, \nu)$*-pseudo-almost periodic solution of class* $r$.

***Proof*** Let $x$ be a function in $PAP(\mathbb{R}; X, \mu, \nu, r)$, and from Theorem 9.5, the function $t \to x_t$ belongs to $PAP(C_\alpha, \mu, r)$. Hence, Theorem 9.4 implies that the function $g(.) := f(., x_.)$ is in $PAP(\mathbb{R}; X, \mu, r)$. Consider the mapping

$$\mathcal{H} : PAP(\mathbb{R}; X_\alpha, \mu, \nu, r) \to PAP(\mathbb{R}; X_\alpha, \mu, \nu, r)$$

defined for $t \in \mathbb{R}$ by

$$\begin{aligned}(\mathcal{H}x)(t) = \Big[ & \lim_{\lambda \to +\infty} \int_{-\infty}^{t} \mathcal{U}^s(t-s)\Pi^s(\widetilde{B}_\lambda X_0 f(s, x_s))ds \\ & + \lim_{\lambda \to +\infty} \int_{+\infty}^{t} \mathcal{U}^u(t-s)\Pi^u(\widetilde{B}_\lambda X_0 f(s, x_s))ds \Big](0).\end{aligned}$$

From Propositions 9.7, 9.8 and taking into account Theorem 9.6, it suffices now to show that the operator $\mathcal{H}$ has a unique fixed point in $PAP(\mathbb{R}; X_\alpha, \mu, r)$. Let $x_1, x_2 \in PAP(\mathbb{R}; X_\alpha, \mu, \nu, r)$. Then we have

$$
\begin{aligned}
|\mathcal{H}x_1(t) - \mathcal{H}x_2(t)|_\alpha &\leq \Big| \lim_{\lambda \to +\infty} \int_{-\infty}^{t} \mathcal{U}^s(t-s)\Pi^s(\widetilde{B}_\lambda X_0[f((s, x_{1s}))\\
&\quad - f((s, x_{1s}))]ds \Big|_\alpha\\
&\quad + \Big| \lim_{\lambda \to +\infty} \int_{+\infty}^{t} \mathcal{U}^s(t-s)\Pi^u(\widetilde{B}_\lambda X_0[f((s, x_{2s}))\\
&\quad - f((s, x_{2s}))]ds \Big|_\alpha\\
&\leq KL_f\Big( \int_{-\infty}^{t} \frac{e^{-\omega(t-s)}}{(t-s)^\alpha} |x_{1s} - x_{2s}| ds\\
&\quad + \int_{t}^{+\infty} \frac{e^{-\omega(s-t)}}{(s-t)^\alpha} |x_{1s} - x_{2s}| ds \Big)\\
&\leq KL_f\Big( \int_{-\infty}^{t} \frac{e^{-\omega(t-s)}}{(t-s)^\alpha} ds\\
&\quad + \int_{t}^{+\infty} \frac{e^{-\omega(s-t))}}{(s-t)^\alpha} ds \Big) |x_1 - x_2|\\
&\leq KL_f\Big( \frac{1}{\omega^{1-\alpha}} \int_{0}^{+\infty} \frac{e^{-s}}{s^\alpha} ds\\
&\quad + \frac{1}{\omega^{1-\alpha}} \int_{0}^{+\infty} \frac{e^{-s}}{s^\alpha} ds \Big) |x_1 - x_2|\\
&\leq \frac{2KL_f}{\omega^{1-\alpha}} \Big( \int_{0}^{+\infty} e^{-s} s^{-\alpha} ds \Big) |x_1 - x_2|\\
&\leq \frac{2KL_f \Gamma(1-\alpha)}{\omega^{1-\alpha}} |x_1 - x_2|.
\end{aligned}
$$

This means that $\mathcal{H}$ is a strict contraction. Thus by Banach's fixed point theorem, $\mathcal{H}$ has a unique fixed point $u$ in $PAP(\mathbb{R}; X, \mu, \nu, r)$. We conclude that Eq. (9.9) has one and only one $cl(\mu, \nu)$-pseudo-almost periodic solution of class $r$. ■

## 9.7 Application

For illustration, we propose to study the existence of solutions for the following model:

$$\begin{cases} \frac{\partial}{\partial t}z(t,x) = \frac{\partial^2}{\partial x^2}z(t,x) + \int_{-r}^{0} G(\theta)z(t+\theta,x)d\theta - \cos t - \frac{1}{\sqrt{2}}\cos(\sqrt{2}t) \\ \quad + \arctan(t) + h\Big(t, \frac{\partial}{\partial x}z(t+\theta,x)\Big) \text{ for } t \in \mathbb{R} \text{ and } x \in [0,\pi] \\ z(t,0) = z(t,\pi) = 0 \text{ for } t \in \mathbb{R}, \end{cases} \tag{9.10}$$

where $G : [-r, 0] \to \mathbb{R}$ is a continuous function and $h : \mathbb{R} \times \mathbb{R} \to \mathbb{R}$ is Lipschitz continuous with respect to the second argument. To rewrite equation (9.10) in the abstract form, we introduce the space $X = L^2([0, \pi]; \mathbb{R})$ vanishing at 0 and $\pi$, equipped with the $L^2$ norm that is to say for all $x \in X$,

$$\|x\|_{L^2} = \Big(\int_0^{\pi} |x(s)|^2 ds\Big)^{\frac{1}{2}}.$$

Let $A : X \to X$ be defined by

$$\begin{cases} D(A) = H^2(0,\pi) \cap H_0^1(0,\pi) \\ Ay = y''. \end{cases}$$

Then the spectrum $\sigma(A)$ of A equals to the point spectrum $\sigma_p(A)$ and is given by

$$\sigma(A) = \sigma_p(A) = \{-n^2 : n \geq 1\},$$

and the associated eigenfunctions $(e_n)_{n\geq 1}$ are given by

$$e_n(s) = \sqrt{\frac{2}{\pi}} \sin(ns), \ s \in [0, \pi].$$

Then the operator is computed by

$$Ay = \sum_{n=1}^{+\infty} n^2(y, e_n)e_n, \ y \in D(A).$$

For each $y \in D(A^{\frac{1}{2}}) = \{y \in X : \sum_{n=1}^{+\infty} n(y, e_n)e_n \in X\}$, the operator $A^{\frac{1}{2}}$ is given by

$$A^{\frac{1}{2}}y = \sum_{n=1}^{+\infty} n(y, e_n)e_n, \ y \in D(A).$$

**Lemma 9.9 ([11])** *If* $y \in D(A^{\frac{1}{2}})$, *then* $y$ *is absolutely continuous,* $y' \in X$, *and* $|y'| = |A^{\frac{1}{2}} y|$.

It is well known that $-A$ is the generator of a compact analytic semigroup $(T(t))_{t\geq 0}$ on $X$ that is given by

$$T(t)x = \sum_{n=1}^{+\infty} e^{-n^2 t}(x, e_n)e_n, \ x \in X.$$

Then $(\mathbf{H_0})$ and $(\mathbf{H_1})$ are satisfied. Here we choose $\alpha = \frac{1}{2}$.

We define $f : \mathbb{R} \times C_{\frac{1}{2}} \to X$ and $L : C_{\frac{1}{2}} \to X$ as follows:

$$\begin{aligned} f(t,\varphi)(x) &= -\cos t - \frac{1}{\sqrt{2}}\cos(\sqrt{2}t) + \arctan(t) \\ &\quad + h\Big(t, \frac{\partial}{\partial x}\varphi(\theta, x)\Big) \ \text{ for } x \in [0,\pi] \ \text{ and } \ t \in \mathbb{R}, \end{aligned}$$

$$L(\varphi)(x) = \int_{-r}^{0} G(\theta)\varphi(\theta, x))d\theta \ \text{ for } \ -r \leq \theta \leq 0 \ \text{ and } \ x \in [0,\pi].$$

Let us pose $v(t) = z(t, x)$. Then Eq. (9.10) takes the following abstract form:

$$v'(t) = Av(t) + L(v_t) + f(t, v_t) \text{ for } t \in \mathbb{R}. \tag{9.11}$$

Consider the measures $\mu$ and $\nu$ where its Radon–Nikodym derivatives are, respectively, $\rho_1, \rho_2 : \mathbb{R} \to \mathbb{R}$ defined by

$$\rho_1(t) = \begin{cases} 1 \ \text{ for } \ t > 0 \\ e^t \ \text{ for } \ t \leq 0 \end{cases}$$

and

$$\rho_2(t) = |t| \ \text{ for } \ t \in \mathbb{R},$$

i.e., $d\mu(t) = \rho_1(t)dt$ and $d\nu(t) = \rho_2(t)dt$ where $dt$ denotes the Lebesgue measure on $\mathbb{R}$ and

$$\mu(A) = \int_A \rho_1(t)dt \ \text{ for } \ \nu(A) = \int_A \rho_2(t)dt \ \text{ for } \ A \in \mathscr{B}.$$

From [6], $\mu, \nu \in \mathcal{M}$, and $\mu, \nu$ satisfy Hypothesis $(\mathbf{H_4})$.

We have

$$\limsup_{\tau\to+\infty}\frac{\mu([-\tau,\tau])}{\nu([-\tau,\tau])}=\limsup_{\tau\to+\infty}\frac{\displaystyle\int_{-\tau}^{0}e^{t}dt+\int_{0}^{\tau}dt}{2\displaystyle\int_{0}^{\tau}tdt}$$

$$=\limsup_{\tau\to+\infty}\frac{1-e^{-\tau}+\tau}{\tau^{2}}=0<\infty,$$

which implies that $(\mathbf{H_2})$ is satisfied.

Since $A^{\frac{1}{2}}\Big(-\cos t-\dfrac{1}{\sqrt{2}}\cos(\sqrt{2}t)\Big)=\sin t+\sin(\sqrt{2}t)$ and $t\to\sin t+\sin(\sqrt{2}t)$ belongs to $AP(\mathbb{R},X)$, it follows that $t\to\Big(-\cos t-\dfrac{1}{\sqrt{2}}\cos(\sqrt{2}t)\Big)$ belongs to $AP(\mathbb{R},X_{\frac{1}{2}})$. On the other hand, we have the following:

$$\frac{1}{\nu([-\tau,\tau])}\int_{-\tau}^{+\tau}\sup_{\theta\in[t-r,t]}|\arctan(\theta)|_{\frac{1}{2}}dt=\frac{1}{\nu([-\tau,\tau])}$$

$$\times\int_{-\tau}^{+\tau}\sup_{\theta\in[t-r,t]}|A^{\frac{1}{2}}\arctan(\theta)|dt$$

$$=\frac{1}{\nu([-\tau,\tau])}\int_{-\tau}^{+\tau}\sup_{\theta\in[t-r,t]}\Big(\frac{1}{1+\theta^{2}}\Big)dt$$

$$\leq\frac{\mu([-\tau,\tau])}{\nu([-\tau,\tau])}\to 0 \text{ as } \tau\to+\infty.$$

It follows that $t\mapsto\arctan t$ is $(\mu,\nu)$-ergodic of class $r$; consequently, $f$ is uniformly $(\mu,\nu)$-pseudo-almost periodic of class $r$. Moreover, $L$ is a bounded linear operator from $C_{\frac{1}{2}}$ to $X$.

Let $k$ be the Lipschitz constant of $h$; then for every $\varphi_1,\varphi_2\in C_{\frac{1}{2}}$ and $t\geq 0$, we have

$$\|f(t,\varphi_1)(x)-f(t,\varphi_2)(x)\|=\Big(\int_0^{\pi}\Big[h\Big(\theta,\frac{\partial}{\partial x}\varphi_1(\theta,x)\Big)$$

$$-h\Big(t,\frac{\partial}{\partial x}\varphi_1(t,x)\Big)\Big]^{2}dx\Big)^{\frac{1}{2}}$$

$$\leq L_h\Big[\int_0^{\pi}\Big(\frac{\partial}{\partial x}\varphi_1(\theta,x)-\frac{\partial}{\partial x}\varphi_2(\theta,x)\Big)^{2}dx\Big]^{\frac{1}{2}}$$

$$\leq L_h \sup_{-r\leq\theta\leq 0}\left[\int_0^{\pi}\left(\frac{\partial}{\partial x}\varphi_1(\theta,x)\right.\right.$$
$$\left.\left.-\frac{\partial}{\partial x}\varphi_2(\theta,x)\right)^2 dx\right]^{\frac{1}{2}}$$

$$\leq L_h\|\varphi_1-\varphi_2\|_{C_\alpha}.$$

Consequently, we conclude that $f$ is Lipschitz continuous and $cl(\mu,\nu)$-pseudo-almost periodic of class $r$.

**Lemma 9.10 ([9])** *If* $\int_{-r}^{0}|G(\theta)|d\theta < 1$, *then the semigroup* $(\mathcal{U}(t))_{t\geq 0}$ *is hyperbolic.*

For example, let us pose $G(\theta)=\dfrac{\theta^2-1}{(\theta^2+1)^2}$ for $\theta\in[-r,0]$. Then we can see that

$$\int_{-r}^{0}|G(\theta)|d\theta=\int_{-r}^{0}\left|\frac{\theta^2-1}{(\theta^2+1)^2}\right|d\theta=\left[\frac{\theta}{\theta^2+1}\right]_{-r}^{0}=\frac{r}{r^2+1}<1 \text{ if } r<1$$

and

$$\int_{-r}^{0}|G(\theta)|d\theta=\int_{-r}^{0}\left|\frac{\theta^2-1}{(\theta^2+1)^2}\right|d\theta=\int_{-r}^{-1}\frac{\theta^2-1}{(\theta^2+1)^2}d\theta+\int_{-1}^{0}\frac{-\theta^2+1}{(\theta^2+1)^2}d\theta$$
$$=1-\frac{r}{r^2+1}<1 \text{ if } r\geq 1.$$

By Proposition 9.7, we deduce the following result.

**Theorem 9.8** *Under the above assumptions, if $Lip(h)$ is small enough, then Eq.* (9.11) *has a unique $cl(\mu,\nu)$-pseudo-almost periodic solution $v$ of class $r$.*

**Acknowledgments** The authors would like to thank the referees for a careful reading and several constructive comments and making some useful corrections that have improved the presentation of the paper.

## References

1. M. Adimy, K. Ezzinbi, Existence and linearized stability for partial neutral functional differential equations. Differ. Equ. Dyn. Syst. **7**, 371–417 (1999)
2. M. Adimy, K. Ezzinbi, Existence and stability in the $\alpha$-norm for partial functional differential equations of neutral type. Annali di Matematica Pura ed Applicata **185**(3), 437–460 (2006)

3. M. Adimy, A. Elazzouzi, K. Ezzinbi, Reduction principle and dynamic behaviors for a class of partial functional differential equations. Nonlinear Analy. Theory Methods Appl. **71**, 1709–1727 (2009)
4. M. Adimy, K. Ezzinbi, M. Laklach, Spectral decomposition for partial neutral functional differential equations. Can. Appl. Math. Quart. **9**(1), 1–34 (2001)
5. J. Blot, P. Cieutat, K. Ezzinbi, Measure theory and pseudo almost automorphic functions: new developments and applications. Nonlinear Analy. **75**, 2426–2447 (2012)
6. J. Blot, P. Cieutat, K. Ezzinbi, New approach for weighted pseudo almost periodic functions under the light of measure theory, basic results and applications. Appl. Analy. **92**, 493–526 (2013)
7. T. Diagana, Weighted pseudo-almost periodic solutions to some differential equations. Nonlinear Anal. Theory Methods Appl. **68**(8), 2250–2260 (2008)
8. T. Diagana, K. Ezzinbi, M. Miraoui, Pseudo-almost periodic and pseudo-almost automorphic solutions to some evolution equations involving theoretical measure theory. CUBO A Math. J. **16**(02), 01–31 (2014)
9. K. Ezzinbi, S. Fatajou, G.M. N'Guérékata, $C^n$-almost automorphic solutions for partial neutral functional differential equations. Appl. Analy. **86**(9), 1127–1146 (2007)
10. A. Pazy, *Semigroups of Linear Operators and Application to Partial Differential Equation, Applied Mathematical Sciences*, vol. 44 (Springer, New York, 1983)
11. C.C. Travis, G.F. Webb, Existence, stability, and compactness in the $\alpha$-norm for partial functional differential equations. Trans. Amer. Math. Soc. **240**, 129–143 (1978)
12. S. Zaidman, *Almost-Periodic Functions in Abstract Spaces*. Research Notes in Mathematics, vol. 126 (Pitman (Advanced Publishing Program), Boston, 1985)
13. C. Zhang, Integration of vector-valued pseudo-almost periodic functions. Proc. Am. Math. Soc. **121**(1), 167–174 (1994)
14. C. Zhang, Pseudo almost periodic solutions of some differential equations. J. Math. Anal. Appl. **181**(1), 62–76 (1994)
15. C. Zhang, Pseudo almost periodic solutions of some differential equations II. J. Math. Anal. Appl. **192**(2), 543–561 (1995)

# Chapter 10
# Global Stability for a Delay SIR Epidemic Model with General Incidence Function, Observers Design

**Aboudramane Guiro, Dramane Ouedraogo, and Harouna Ouedraogo**

**Abstract** In (Connell McCuskey, Nonlinear Anal RWA 11:3106–3109, 2010), the authors presented an SIR model of disease transmission with delay in a particular nonlinear incidence. In their work, they showed the global stability of the endemic equilibrium for the reproduction number $R_0$ is greater than one. In this chapter, we reviewed on the same model with delay in general incidence function. The global stability of the endemic equilibrium is studied for $R_0 > 1$ by using a Lyapunov functional. With supposed well-known parameters, we built simple observer and a high-gain observer using a canonical controller form. Then, we proposed nonlinear auxilary dynamical systems which are used for the implementation. Numerical simulations are included in order to test the behaviour and the performance of the given observers.

**Keywords** Epidemic model · SIR · Delays · Global stability · Lyapunov function · Reproduction number · General incidence function · Observability · Observer · High gain

## 10.1 Introduction

In the modelling of the dynamics of infectious diseases such as dengue [7] and chikungunya [14], a common model structure involves dividing the population into three classes: susceptible, infectious, and recovered individuals. In this chapter, we consider an SIR model of disease transmission in [4]. In this chapter, it is a

A. Guiro (✉)
Département de Mathématiques, Université Nazi Boni, Bobo-Dioulasso, Burkina Faso

D. Ouedraogo
Centre Universitaire de Banfora, Université Nazi Boni, Bobo-Dioulasso, Burkina Faso

H. Ouedraogo
Département de Mathématiques, Université Joseph Ki-Zerbo, Ouagadougou, Burkina Faso



T. Diagana et al. (eds.), *Partial Differential Equations and Applications*,
Springer Proceedings in Mathematics & Statistics 420,
https://doi.org/10.1007/978-3-031-27661-3_10

refinement of hypothesis and a genraralization of earlier models based on incidence function $\dfrac{\beta S I_\tau}{1+\alpha I_\tau}$. The model in [4] keeps pointing out the saturation in the force of infection by using the general incidence function $f(S, I_\tau)$.

In [4], a detailed analysis of the current model is presented. It is shown that if the basic reproduction number $R_0$ is less or equal to one, then the disease-free equilibrium is globally asymptotically stable. If $R_0 > 1$, then the endemic equilibrium is globally asymptotically stable, without any further conditions on the parameters.

Our approach is to use here a Lyapunov functional similar to those used in [5, 6, 16] for various mass action types.

In this chapter, we show that when $R_0 > 1$, the endemic equilibrium is globally asymptotically stable by using a suitable Lyapunov functional.

Now by considering the system

$$\dot{x} = g_1(x) \tag{10.1}$$

and assuming that (1) is a good model of the system under consideration, when it is possible to get the value of the state at some time $t_0$, then it is possible to compute $x(t)$ for all $t \geq t_0$ by integrating the differential equation with the initial condition $x(t_0)$. Unfortunately, it is often not possible to measure the whole state at a given time, and therefore, it is not possible to integrate the differential equation because one does not know an initial condition. One can only have a partial information of the state, and this partial information is precisely given by $y(t)$ the output of the system. Therefore, we shall show how to use this partial information $y(t)$ with the given model in order to have a dynamical estimate $\hat{x}(t)$ of the real unknown state variable $x(t)$. This estimate will be produced by an auxiliary dynamical system that uses the information $y(t)$ given by the system (10.4). This auxiliary dynamical system form is

$$\dot{\hat{x}} = g_2(\hat{x}, y). \tag{10.2}$$

The estimated error is $e(t) = \hat{x}(t) - x(t)$ and satisfies the following equation given by

$$\dot{e} = g_2(\hat{x}, y) - g_1(x). \tag{10.3}$$

$e(t) = x(t) - \hat{x}(t) \to 0$ when $t \to +\infty$, regardless of initial conditions of system (10.1) and system (10.2). A dynamical system (10.2) satisfying these conditions is called an observer for system (10.1). When the convergence of $\hat{x}(t)$ towards $x(t)$ is exponential, the system (10.1) is an exponential observer. More precisely, system (10.2) is an exponential observer for system (10.1) if there exists $\lambda > 0$ such that, for all $t \geq 0$ and for initial conditions $(x(0), \hat{x}(0))$, the corresponding solutions of (10.1) and (10.2) satisfy $\|\hat{x}(t) - x(t)\| \leq \exp(-\lambda t)\|\hat{x}(0) - x(0)\|$. In this condition, a good estimate of the real unmeasured state is rapidly obtained. One must notice that the observer is not linked to the choice of initial condition of the observer.

The use of observer theory in biological system is scarce; however, some authors have reviewed on that (see Ngom et al. [15]; Guiro et al. [10]). In [15], an observer has been constructed for a stage-structured discrete-time fishery model that exhibits an unknown recruitment function. In [10], a stage-structured continuous model is considered, and it is assumed that only the last class (mature individuals) is harvested. So, in this present work, we use control theory (observability theory) to construct an estimator (observer) for the system (10.4) when the measured (output) variable is the infectious population, i.e., $y(t) = I(t)$.

In this chapter, we construct a so-called simple observer as opposed to Gauthier–Kupka-type observers [1, 8, 9] who require an extension by continuity of compact space at the risk of the explosion of the system. The observer constructed here is quite simple and is a copy of the original system augmented by a corrective term that gives satisfactory results for the estimation of the states.

This chapter is organized as follows: The model description and basic results are given in Sect. 10.2. In Sect. 10.3, the basic reproduction number and equilibria are presented. Section 10.4 contains the local and global stabilities of the free-equilibrium point; in this part, we also study the global stability of the endemic equilibrium. Section 10.5 introduces methods for designing an observer. Section 10.6 is devoted to numerical simulation. Finally, in Sect. 10.7, we give conclusions.

## 10.2 Model Presentation

In this model, the population is divided into susceptible, infectious, and recovered classes with sizes $S$, $I$, and $R$, respectively. Recruitment of new individuals is into the susceptible class, at a constant rate $B$. The death rates for the classes are $\mu_1$, $\mu_2$, and $\mu_3$, respectively. The average time spent in class $I$ before recovery is $1/\gamma$. For biological reasons, we assume that $\mu_1 \leq \mu_2 + \gamma$; that is, removal of infectives is at least as fast as removal of susceptibles. Transmission of the disease is done through vectors that undergo fast dynamics and a fixed latent period $\tau$. In order to avoid excessive use of parameters $(t, t - \tau)$, we use the following convention: $S = S(t)$, $I = I(t)$, and $I_\tau = I(t - \tau)$; then, we generalize the problem from [4] as follows:

$$\begin{cases} \dot{S} = B - \mu_1 S - \dfrac{\beta S I_\tau}{1 + \alpha I_\tau}, \\ \dot{I} = \dfrac{\beta S I_\tau}{1 + \alpha I_\tau} - (\mu_2 + \gamma) I, \end{cases} \tag{10.4}$$

and

$$\dot{R} = \gamma I - \mu_3 R.$$

Our aim is to study the same system with general incidence function $f(S, I_\tau)$.

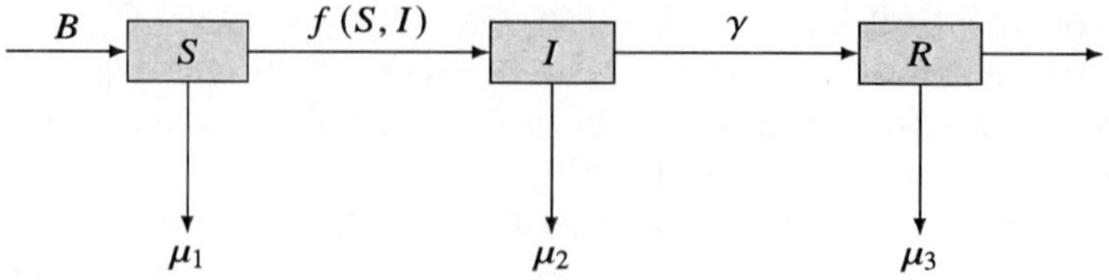


**Fig. 10.1** The compartmental diagram for the SIR model

As general as possible, the incidence function $f$ must satisfy technical conditions. Thus, we assume that:

**H1** $f$ is non-negative $C^1$ functions on the non-negative quadrant.
**H2** For all $(S, I) \in \mathbb{R}_+^2$, $f(S, 0) = f(0, I) = 0$.

*Remark 10.2.1* $f$ is an incidence function that explains the contact between two species. Therefore, **H2** is a natural assumption that means that if there is not a new infection when there is not an infectious human or a susceptible human.

Let us denote by $f_1$ and $f_2$ the partial derivatives of $f$ with respect to the first $(S)$ and second $(I)$ variables. Using the same notations, the model is given by the following system (Fig. 10.1):

$$\begin{cases} \dot{S} = B - \mu_1 S - f(S, I_\tau), \\ \dot{I} = f(S, I_\tau) - (\mu_2 + \gamma)I, \end{cases} \tag{10.5}$$

and

$$\dot{R} = \gamma I - \mu_3 R. \tag{10.6}$$

Since $R$ does not appear in the equations for $\dot{S}$ and $\dot{I}$, it is sufficient to analyse the behaviour of solutions to (10.5).

We assume that system (10.5) holds with given initial conditions

$$S(0) \in \mathbb{R}_+ \text{ and } I(\theta) = \phi(\theta) \text{ for } \theta \in [-\tau, 0],$$

where $\phi \in C([-\tau, 0], \mathbb{R}_+)$, the space of continuous functions from $[-\tau, 0]$ to $\mathbb{R}_+$. □

**Theorem 10.2.1** *The positive orthant*

$$\{(S, I, R) \in \mathbb{R}^3 : S \geq 0, I \geq 0, R \geq 0\}$$

*is positively invariant for system* (10.5). □

To prove Theorem 10.2.1, we need the following lemma.

**Lemma 10.2.1 ([3])** *Let $L : \mathbb{R}^n \to \mathbb{R}$ be a differentiable function, and let $a \in \mathbb{R}$. Let $X(x)$ be the vector field, and let G be the closed set $G = \{x \in \mathbb{R}^n : L(x) \leq a\}$ such that $\nabla L(x) \neq 0$ for all $x \in L^{-1}(a) = \{x \in \mathbb{R}^n : L(x) = a\}$. If $< X(x), \nabla L(x) > \leq 0$ for all $x \in L^{-1}(a)$, then the set G is positively invariant.* □

***Proof of Theorem 10.2.1*** Let

$$x = (S, I, R). \tag{10.7}$$

We will prove that $\{S \geq 0\}$ is positively invariant. So, let

$$L(x) = -S.$$

$L$ is differentiable, and

$$\nabla L(x) = (-1, 0, 0) \neq 0 \text{ for all } x \in L^{-1}(0) = \{x \in \mathbb{R}^3 / L(x) = 0\}.$$

The vector field on $\{S = 0\}$ is

$$X(x) = \begin{pmatrix} B \\ -(\mu_2 + \gamma)I \\ -\mu_3 R \end{pmatrix}. \tag{10.8}$$

Then $< X(x), \nabla L(x) >= -B < 0$. This proves that $\{S \geq 0\}$ is positively invariant. Similarly, we prove that $\{I \geq 0\}$, $\{R \geq 0\}$ are positively invariant.

Then $\{(S, I, R) \in \mathbb{R}^3 : S \geq 0, I \geq 0\ R \geq 0\}$ is positively invariant for system (10.5). □

Therefore, the model is mathematically well-posed and epidemiologically reasonable since all the variables remain non-negative for all $t > 0$. □

**Theorem 10.2.2** *Any solution $(S, I, R)$ of system (10.5)–(10.6) with the initial conditions satisfies*

$$\limsup_{t \to +\infty}(S(t) + I(t) + R(t)) \leq \frac{B}{\overline{\mu}}, \quad \textit{where } \overline{\mu} = \min\{\mu_1, \mu_2, \mu_3\}.$$

□

***Proof*** With $N(t) = S(t) + I(t) + R(t)$.

Adding the two equations of (10.5) and (10.6), we get

$$\begin{aligned} \dot{N}(t) &= B - \mu_1 S - \mu_2 I - \mu_3 R \\ &\leq B - \overline{\mu} N(t). \end{aligned} \tag{10.9}$$

According to [2], it follows that

$$N(t) \le \frac{B}{\overline{\mu}} + (N(0) - \frac{B}{\overline{\mu}})e^{-\overline{\mu}t}. \tag{10.10}$$

Thus, as $t \to +\infty$, $N(t) \le \frac{B}{\overline{\mu}}$. □

## 10.3 Basic Reproduction Number and Equilibria

The disease-free equilibrium is given by

$$E_0 = (S^0, I^0, R^0) = \left(\frac{B}{\mu_1}, 0, 0\right). \tag{10.11}$$

**Proposition 10.3.1** *The basic reproduction number for model system (10.5) is defined by*

$$R_0 = \frac{f_2(S^0, 0)}{\mu_2 + \gamma}.$$

*The basic reproduction number $R_0$ represents the average number of new cases generated by a single infected individual in a completely susceptible population.*

***Proof*** Note that in this case the disease-free equilibrium

$$E_0 = (S^0, I^0, R^0) = \left(\frac{B}{\mu_1}, 0, 0\right)$$

and

$$A = f_2(S^0, 0) - (\mu_2 + \gamma).$$

Hence, $M = f_2(S^0, 0)$, $D = \mu_2 + \gamma$, and

$$R_0 = MD^{-1} = \frac{f_2(S^0, 0)}{\mu_2 + \gamma}.$$

□

Now, let us study the behaviour of system (10.5) with respect to $R_0$.

**Theorem 10.3.3**

*(i) If $R_0 \le 1$, then model (10.5) has a disease-free equilibrium $E_0$.*
*(ii) If $R_0 > 1$, then model (10.5) has an endemic equilibrium.*

□

***Proof*** Let $E = (S, I, R)$ be an equilibrium point of system (10.5).
Using the second equations of (10.5), we have

$$f(S, I) = (\mu_2 + \gamma)I;$$

therefore, we have

$$\frac{f(S, I)}{I} = \mu_2 + \gamma.$$

Let

$$\Phi(I) = \frac{f(S^0 - \frac{(\mu_2 + \gamma)I}{\mu_1}, I)}{I} - (\mu_2 + \gamma)$$

$$\lim_{I \to 0^+} \Phi(I) = f_2 - (\mu_2 + \gamma),$$

$$\lim_{I \to 0^+} \Phi(I) = (\mu_2 + \gamma)(R_0 - 1),$$

and we also have $\Phi(\bar{I}) = -(\mu_2 + \gamma)$ with $\bar{I} = \dfrac{S^0 \mu_1}{\mu_2 + \gamma}$. When $R_0 \le 1$, we have $\lim_{I \to 0^+} \Phi(I) \le 0$; thus, there is not any $I^* > 0$ such that $\Phi(I^*) = 0$. So system (10.5) has a free-disease equilibrium $E_0$.

When $R_0 > 1$, we have $\lim_{I \to 0^+} \Phi(I) \ge 0$, so there exists $I^* \in ]0, \bar{I}[$. This implies that system (10.5) has an endemic equilibrium point $E^*$. □

## 10.4 Stability of Equilibria

### *10.4.1 Stability of the Disease-Free Equilibrium*

In this section, we study the local and global behaviours of the disease-free equilibrium.

**Theorem 10.4.4** *Disease-free equilibrium $E_0$ is locally asymptotically stable if $R_0 \le 1$.* □

***Proof*** The characteristic equation of linear system of (10.5) at $E_0$ gives the following equation:

$$(-\mu_3 - \lambda)\Big[(-\mu_1 - f_1(S^0, 0) - \lambda)(f_2(S^0, 0) - (\mu_2 + \gamma) - \lambda)$$
$$+ f_1(S^0, 0) f_2(S^0, 0)\Big] = 0. \tag{10.12}$$

We can see that any solution $\lambda$ of equation (10.12) is negative.

Indeed, the equation (10.12) has negative root $\lambda = -\mu_3$, and other roots are given by

$$(-\mu_1 - f_1(S^0, 0) - \lambda)(f_2(S^0, 0) - (\mu_2 + \gamma) - \lambda) + f_1(S^0, 0) f_2(S^0, 0) = 0. \tag{10.13}$$

By developing (10.13), we get

$$\lambda^2 + (\mu_1 + f_1(S^0, 0) - f_2(S^0, 0) + \mu_2 + \gamma)\lambda - \mu_1 f_2(S^0, 0) + \mu_1(\mu_2 + \gamma)$$
$$+ f_1(S^0, 0)(\mu_2 + \gamma) = 0. \tag{10.14}$$

Since $R_0 \leq 1$, we obtain

$$\mu_1 + f_1(S^0, 0) - f_2(S^0, 0) + \mu_2 + \gamma > 0.$$

Therefore, by the Routh–Hurwitz criterion, all the roots of equation (10.14) have the negative real parts. This shows that equilibrium $E_0$ is locally asymptotically stable. This completes the proof. □

**H3** For all $(S, I) \in \mathbb{R}^2$, $f(S, I_\tau) \leq f_2(S^0, 0)I$.

**Theorem 10.4.5** *Disease-free equilibrium is globally asymptotically stable if* $R_0 \leq 1$. □

***Proof*** The proof is based on comparison theorem [13]. Note that the equations of infected components in system (10.5) can be expressed as

$$\dot{I} \leq \Big(f_2(S^0, 0) - (\mu_2 + \gamma)\Big)I. \tag{10.15}$$

So, we deduce that the constant $f_2(S^0, 0) - (\mu_2 + \gamma)$ is negative since $R_0 \leq 1$.

Thus, $I(t) \to 0$ as $t \to \infty$ for the system (10.15). Consequently, by a standard comparison theorem [13], $I(t) \to 0$ as $t \to \infty$, and substituting $I = 0$ into system (10.5), $S \to S^0$ as $t \to \infty$.

Thus,$(S, I, R) \to (S^0, 0, 0)$ as $t \to \infty$ for $R_0 \leq 1$. Therefore, $E_0$ is globally asymptotically stable if $R_0 \leq 1$. □

### 10.4.2 Global Stability of the Endemic Equilibrium

In this section, we study the global dynamics for $R_0 > 1$ by using some technical conditions. We recall that the endemic equilibrium $E^*$ exists if and only if $R_0 > 1$. So let:

**H4**: $\forall (S, I) \in \mathbb{R}^2_+, \dfrac{I}{I^*} \leq \dfrac{S}{S^*} \leq \dfrac{f(S, I_\tau)}{f(S^*, I^*)}$.

*Remark 10.4.2* Assumption **H4** can be seen as a technical assumption to have $\dfrac{dV}{dt} \leq 0$ and biologically correct because at the endemic equilibrium $S^* > 0$, $I^* > 0$, and $f(S^*, I^*) > 0$. □

**Theorem 10.4.6** *If $R_0 > 1$, the endemic equilibrium $E^*$ is globally asymptotically stable.* □

***Proof*** Evaluating both sides of (10.5) at $E^*$, we have

$$B = \mu_1 S^* + f(S^*, I^*) \tag{10.16}$$

and

$$f(S^*, I^*) = (\mu_2 + \gamma)I^*, \tag{10.17}$$

which will be used as substitutions in the calculations below.

Let

$$g(x) = x - 1 - \ln x$$

and

$$\begin{aligned} V_s(t) &= g\left(\frac{S(t)}{S^*}\right) \\ V_I(t) &= g\left(\frac{I(t)}{I^*}\right) \\ V_+(t) &= \int_0^\tau g\left(\frac{I(t-s)}{I^*}\right) ds. \end{aligned} \tag{10.18}$$

We study the behaviour of the Lyapunov functional

$$V(t) = \frac{S^*}{f(S^*, I^*)} V_s(t) + \frac{I^*}{f(S^*, I^*)} V_I(t) + V_+(t). \tag{10.19}$$

We note that $g : \mathbb{R}_{>0} \to \mathbb{R}_{\geq 0}$ has the strict global minimum $g(1) = 0$. Thus, $V(t) \geq 0 \ \forall t \geq 0$ with equality if and only if $\dfrac{S(t)}{S^*} = 1$, $\dfrac{I(t)}{I^*} = 1$, and $\dfrac{I(t-s)}{I^*} = 1$ for all $s \in [0, \tau]$.

By Theorem 10.2.2, solutions are bounded above and bounded away from zero for large time. Without loss of generality, we may assume that the solution in question satisfies these bounds for all $t \geq 0$. Thus, $V(t)$ is defined for all $t \geq 0$.

For clarity, the derivatives of $V_s$, $V_I$, and $V_+$ will be calculated separately and then combined to obtain $\frac{dV}{dt}$.

$$\begin{aligned}\frac{dV_s}{dt} &= \frac{1}{S^*}\left(1 - \frac{S^*}{S}\right)\frac{dS}{dt}\\ &= \frac{1}{S^*}\left(1 - \frac{S^*}{S}\right)(B - \mu_1 S - f(S, I_\tau)).\end{aligned}$$

Using (10.16) to replace $B$, we have

$$\begin{aligned}\frac{dV_s}{dt} &= \frac{1}{S^*}\left(1 - \frac{S^*}{S}\right)(\mu_1(S^* - S) + (f(S^*, I^*) - f(S, I_\tau)))\\ &= -\frac{\mu_1}{SS^*}(S - S^*)^2 + \frac{f(S^*, I^*)}{S^*}\left(1 - \frac{S^*}{S}\right)\left(1 - \frac{f(S, I_\tau)}{f(S^*, I^*)}\right).\end{aligned}$$

Let

$$x = \frac{S}{S^*}, \quad y = \frac{I}{I^*} \quad \text{and} \quad z = \frac{I_\tau}{I^*}.$$

Additionally, let

$$F(z) = \frac{f(S, I^* z)}{f(S^*, I^*)} = \frac{f(S, I_\tau)}{f(S^*, I^*)}.$$

Then we may write

$$\frac{dV_s}{dt} = -\mu_1 \frac{(S - S^*)^2}{SS^*} + \frac{f(S^*, I^*)}{S^*}\left(1 - \frac{1}{x} - F(z) + \frac{F(z)}{x}\right). \tag{10.20}$$

Next, we calculate $\frac{dV_I}{dt}$.

$$\begin{aligned}\frac{dV_I}{dt} &= \frac{1}{I^*}\left(1 - \frac{I^*}{I}\right)\frac{dI}{dt}\\ &= \frac{1}{I^*}\left(1 - \frac{I^*}{I}\right)(f(S, I_\tau) - (\mu_2 + \gamma)I)\\ &= \frac{1}{I^*}\left(1 - \frac{I^*}{I}\right)\left(f(S^*, I^*)\frac{f(S, I_\tau)}{f(S^*, I^*)} - (\mu_2 + \gamma)I^* \frac{I}{I^*}\right).\end{aligned}$$

Using (10.17) to replace $(\mu_2 + \gamma)I^*$, we have

$$\begin{aligned}\frac{dV_I}{dt} &= \frac{f(S^*, I^*)}{I^*}\left(1 - \frac{I^*}{I}\right)\left(\frac{f(S, I_\tau)}{f(S^*, I^*)} - \frac{I}{I^*}\right)\\ &= \frac{f(S^*, I^*)}{I^*}\left(1 - y - \frac{F(z)}{y} + F(z)\right). \end{aligned} \tag{10.21}$$

The derivative of $V_+(t)$ is calculated as follows:

$$\begin{aligned}\frac{dV_+}{dt} &= \frac{d}{dt}\int_0^\tau g\left(\frac{I(t-s)}{I^*}\right)ds\\ &= \int_0^\tau \frac{d}{dt} g\left(\frac{I(t-s)}{I^*}\right)ds\\ &= \int_0^\tau -\frac{d}{ds} g\left(\frac{I(t-s)}{I^*}\right)ds\\ &= g\left(\frac{I(t)}{I^*}\right) - g\left(\frac{I(t-\tau)}{I^*}\right)\\ &= g(y) - g(z)\\ &= y - z + \ln(z) - \ln(y). \end{aligned} \tag{10.22}$$

Combining Eqs. (10.20)–(10.22), multiplying appropriately by coefficients determined by (10.19), we obtain

$$\frac{dV}{dt} = -\mu_1 \frac{(S - S^*)^2}{SS^*} + 2 - \frac{1}{x} + \frac{F(z)}{x} - \frac{F(z)}{y} - z + \ln(z) - \ln(y).$$

By adding and subtracting the quantity $\ln x$, $\ln(\frac{F(z)}{y})$, and $\ln(\frac{F(z)}{x})$, we obtain

$$\frac{dV}{dt} = -\mu_1 \frac{(S - S^*)^2}{SS^*} - g\left(\frac{1}{x}\right) - g\left(\frac{F(z)}{y}\right) - g(z) + g\left(\frac{F(z)}{x}\right).$$

By using **H4** and the fact that the function $g$ is monotone (decreasing and increasing) on each side of point 1 and minimized at this point 1, we get

$$g\left(\frac{F(z)}{x}\right) \leq g\left(\frac{F(z)}{y}\right).$$

Thus, $\frac{dV}{dt} \leq 0$. By Theorem 5.3.1 of [12], solutions limit to $M$, the largest invariant subset of $\{\frac{dV}{dt} = 0\}$.

We note that $\frac{dV}{dt} = 0$ if and only if $x = y = z = 1$. In particular, this requires that for any solution in $M$ we have $S(t) = S^*$ and $I(t) = I^*$ for all $t$, and so $M$ consists of the single point $E^*$. Thus we see that all solutions limit to the endemic equilibrium. $E^*$ is globally asymptotically stable. □

## 10.5 Observer Design

This section is devoted to observers construction, so we consider a dynamical model described by

$$\begin{cases} \dot{x}_1(t) = B - \mu_1 x_1(t) - f\Big(x_1(t), X_2(t-\tau)\Big), \\ \dot{X}_2(t) = f\Big(x_1(t), X_2(t-\tau)\Big) - (\mu_2 + \gamma) X_2(t), \\ \dot{x}_3(t) = \gamma X_2(t) - \mu_3 x_3(t), \end{cases} \tag{10.23}$$

where:

$x_1(t) = S =$ the susceptible
$X_2(t) = I =$ the infected
$x_3(t) = R =$ the recovered

and $f(x_1, X_2) = x_1 X_2$, the mass action. We construct two observers, a simple one and the high-gain one, and we compare them with respect to their convergence.

### 10.5.1 A Simple Observer for an SIR Epidemic Model

Let us consider system (10.23). The compact set

$$D = \{(S, I, R) \in \mathbb{R}^3 : S \geq 0, I \geq 0, R \geq 0\}$$

is positively invariant set under the flow of the system (10.23).

Let $y(t) = X_2(t) = I$ be the measurable variable that is the output. A simple candidate observer for system (10.23) on the set $D$ is given by

$$\begin{cases} \dot{\hat{x}}_1(t) = B - \mu_1 \hat{x}_1(t) - f\Big(\hat{x}_1(t), \hat{X}_2(t-\tau)\Big), \\ \dot{\hat{X}}_2(t) = f\Big(\hat{x}_1(t), \hat{X}_2(t-\tau)\Big) - (\mu_2 + \gamma)\hat{X}_2(t) + L_1(y - \hat{X}_2), \\ \dot{\hat{x}}_3(t) = \gamma \hat{X}_2(t) - \mu_3 \hat{x}_3(t). \end{cases} \tag{10.24}$$

This system (10.24) is simply a copy of system (10.23) plus a corrective term given by $L_1(y - \hat{X}_2)$. The parameter $L_1$ is a positive constant that will be chosen in order to ensure the convergence of estimation error.

We will denote $x(t) = \Big(x_1(t), X_2(t)\Big)$ the state vector of the system (10.23) and $\hat{x}(t) = \Big(\hat{x}_1(t), \hat{X}_2(t)\Big)$ the state vector of the candidate observer (10.24). The estimation error is $e(t) = \Big(e_1(t), e_2(t)\Big) = x(t) - \hat{x}(t)$.

Let us make the following assumptions:

**H5:**

$$\int_0^1 f_1(se(t) + \hat{x})ds > 0 \text{ and } \int_0^1 f_2(se(t) + \hat{x})ds > 0.$$

**Proposition 10.5.2** *The system governed by (10.24) is an exponential observer for system (10.23) for $L_1$ satisfying*

$$L_1 > \max\left(\int_0^1 f_2(se(t) + \hat{x})ds - (\mu_2 + \gamma); 0\right),$$

*i.e., there exists a positive real number $\lambda$ such that for all initial conditions $\Big(\hat{x}(0), x(0)\Big) \in D \times D$, one has $|\hat{x}(t) - x(t)| \le e^{-\lambda t}|\hat{x}(0) - x(0)|$.* □

***Proof*** The estimation error $e(t) = \Big(e_1(t), e_2(t)\Big) = x(t) - \hat{x}(t)$ obeys the following differential equation:

$$\dot{e} = A_d e + F(x) - F(\hat{x}), \tag{10.25}$$

where

$$A_d = \begin{pmatrix} -\mu_1 & 0 \\ 0 & -\mu_2 - \gamma - L_1 \end{pmatrix}, \quad F(x) = \begin{pmatrix} -f\Big(x_1(t), X_2(t-\tau)\Big) \\ f\Big(x_1(t), X_2(t-\tau)\Big) \end{pmatrix}.$$

Let us consider the following candidate Lyapunov function for the error equation(10.25):

$$V(e) = e^T Pe \text{ where } P = \begin{pmatrix} \dfrac{1}{2\mu_1} & 0 \\ 0 & \dfrac{1}{2(\mu_2 + \gamma + L_1)} \end{pmatrix};$$

we can write

$$F(x) - F(\hat{x}) = \int_0^1 \frac{\partial F}{\partial x}(sx + (1-s)\hat{x})dse = R(\hat{x})e.$$

The explicit expression of the matrix $R(\hat{x})$ is

$$R(e, \hat{x}) = \begin{pmatrix} r_{11} & r_{12} \\ r_{21} & r_{22} \end{pmatrix},$$

where

$$r_{11} = -\int_0^1 f_1(se(t) + \hat{x})ds, \qquad r_{12} = -\int_0^1 f_2(se(t) + \hat{x})ds,$$

$$r_{21} = \int_0^1 f_1(se(t) + \hat{x})ds, \qquad r_{22} = \int_0^1 f_2(se(t) + \hat{x})ds.$$

Therefore, $\dot{e} = (A_d + R(\hat{x}))e$, and then the derivative of $V(e)$ with respect to time along the solutions of the estimation error equation is

$$\dot{V}(e) = e^T\left(PA_d + A_d^T P + PR(\hat{x}) + R(\hat{x})^T P\right)e.$$

$$\dot{V}(e) = e^T \begin{pmatrix} \Lambda_{11} & \Lambda_{12} \\ \Lambda_{21} & \Lambda_{22} \end{pmatrix} e,$$

where:

$$\Lambda_{11} = -1 - \frac{1}{\mu_1}\int_0^1 f_1(se(t) + \hat{x})ds$$

$$\Lambda_{12} = -\frac{1}{2\mu_1}\int_0^1 f_2(se(t) + \hat{x})ds + \frac{1}{2(\mu_2 + \gamma + L_1)}\int_0^1 f_1(se(t) + \hat{x})ds$$

$$\Lambda_{21} = \frac{1}{2(\mu_2 + \gamma + L_1)}\int_0^1 f_1(se(t) + \hat{x})ds - \frac{1}{2\mu_1}\int_0^1 f_2(se(t) + \hat{x})ds$$

$$\Lambda_{22} = -1 + \frac{1}{\mu_2 + \gamma + L_1}\int_0^1 f_2(se(t) + \hat{x})ds$$

so

$$\dot{V}(e) = (\Lambda_{11}e_1 + \Lambda_{21}e_2)e_1 + (\Lambda_{12}e_1 + \Lambda_{22}e_2)e_2$$

$$= \left(-1 - \frac{1}{\mu_1}\int_0^1 f_1(se(t) + \hat{x})ds\right)e_1^2$$

$$+\left(\frac{1}{2(\mu_2+\gamma+L_1)}\int_0^1 f_1(se(t)+\hat{x})ds-\frac{1}{2\mu_1}\int_0^1 f_2(se(t)+\hat{x})ds\right)e_1e_2$$

$$+\left(-\frac{1}{2\mu_1}\int_0^1 f_2(se(t)+\hat{x})ds+\frac{1}{2(\mu_2+\gamma+L_1)}\int_0^1 f_1(se(t)+\hat{x})ds\right)e_1e_2$$

$$+\left(-1+\frac{1}{\mu_2+\gamma+L_1}\int_0^1 f_2(se(t)+\hat{x})ds\right)e_2^2$$

$$=\left(-1-\frac{1}{\mu_1}\int_0^1 f_1(se(t)+\hat{x})ds\right)e_1^2$$

$$+\left(\frac{1}{(\mu_2+\gamma+L_1)}\int_0^1 f_1(se(t)+\hat{x})ds-\frac{1}{\mu_1}\int_0^1 f_2(se(t)+\hat{x})ds\right)e_1e_2$$

$$+\left(-1+\frac{1}{\mu_2+\gamma+L_1}\int_0^1 f_2(se(t)+\hat{x})ds\right)e_2^2$$

$$-\dot{V}(e)=\left(1+\frac{1}{\mu_1}\int_0^1 f_1(se(t)+\hat{x})ds\right)$$

$$\times\left(e_1^2+\frac{-\frac{1}{(\mu_2+\gamma+L_1)}\int_0^1 f_1(se(t)+\hat{x})ds+\frac{1}{\mu_1}\int_0^1 f_2(se(t)+\hat{x})ds}{1+\frac{1}{\mu_1}\int_0^1 f_1(se(t)+\hat{x})ds}e_1e_2\right.$$

$$\left.+\frac{1-\frac{1}{\mu_2+\gamma+L_1}\int_0^1 f_2(se(t)+\hat{x})ds}{1+\frac{1}{\mu_1}\int_0^1 f_1(se(t)+\hat{x})ds}e_2^2\right).$$

Let

$$l_1=(1+\frac{1}{\mu_1}\int_0^1 f_1(se(t)+\hat{x})ds),$$

$$l_2=\frac{-\frac{1}{(\mu_2+\gamma+L_1)}\int_0^1 f_1(se(t)+\hat{x})ds+\frac{1}{\mu_1}\int_0^1 f_2(se(t)+\hat{x})ds}{1+\frac{1}{\mu_1}\int_0^1 f_1(se(t)+\hat{x})ds},$$

and

$$l_3=\frac{1-\frac{1}{\mu_2+\gamma+L_1}\int_0^1 f_2(se(t)+\hat{x})ds}{1+\frac{1}{\mu_1}\int_0^1 f_1(se(t)+\hat{x})ds}.$$

The derivative of $V(e)$ can be seen as a quadratic form in $e_i$. Applying the Gauss–Lagrange reduction to this quadratic form leads to

$$
\begin{aligned}
-\dot{V}(e) &= l_1\left(e_1^2 + l_2 e_1 e_2 + l_3 e_2^2\right) \\
&= l_1\left((e_1 + \frac{1}{2} l_2 e_2)^2 - (\frac{1}{2} l_2 e_2)^2 + l_3 e_2^2\right) \\
&= l_1\left((e_1 + \frac{1}{2} l_2 e_2)^2 + (-\frac{1}{4} l_2^2 + l_3) e_2^2\right),
\end{aligned}
$$

where $l_1$, $l_2$, and $l_3$ are functions of the parameters.

Taking into account the conditions on $L_1$, we can argue that the $\dot{V}$ is negative definite, and this ends the proof. □

### 10.5.2 *High-Gain Observer for an SIR Epidemic Model*

Here, we construct a high-gain observer for system (10.23) using the techniques developed in [8, 9]. We denote by $x(t) = \left(x_1(t), X_2(t)\right)$ the state vector of the system (10.23). Let $g$ be the vector field defining the dynamics of the system (10.23) and $h$ be the output function, that is, $y(t) = h(x(t)) = X_2(t)$, and

$$
g = \begin{pmatrix} B - \mu_1 x_1(t) - f(x_1(t), X_2(t-\tau)) \\ f(x_1(t), X_2(t-\tau)) - (\mu_2 + \gamma) X_2(t) \end{pmatrix}.
$$

To construct a high-gain observer for (10.23), one has to perform a change of coordinates in order to write the system in a simpler form. Usually, this is done by using the output function together with its time derivative.

Let $\Phi$ be the function $\Phi : \mathcal{D} \to \mathbb{R}^3$ defined as follows:

$$
\Phi(x) = \begin{pmatrix} h(x) \\ L_g h(x) \end{pmatrix},
$$

where $L_g$ denotes the Lie derivative operator with respect to the vector field $g$. Thus

$$
\Phi(x) = \begin{pmatrix} X_2(t) \\ f(x_1(t), X_2(t-\tau)) - (\mu_2 + \gamma) X_2(t). \end{pmatrix}.
$$

The Jacobian of $\Phi$ can be written:

$$\frac{d\Phi}{dx} = \begin{pmatrix} 0 & 1 \\ f_1(x_1, X_2(t-\tau) & f_2(x_1, X_2(t-\tau)) - (\mu_2+\gamma) \end{pmatrix}.$$

The determinant of $\frac{d\Phi}{dx}$ can be expressed by

$$\Gamma(x_1; X_2) = -f_1(x_1; X_2(t-\tau)).$$

The Jacobian $\frac{d\Phi}{dx}$ is nonsingular in the region $\overset{\circ}{D}$, and moreover, $\Phi(x)$is one-to-one from $\overset{\circ}{D}$ in $\Phi(\overset{\circ}{D})$. So the map $\Phi$ is a diffeomorphism from $\overset{\circ}{D}$ to $\Phi(\overset{\circ}{D})$. This implies that the system (10.23) with the output $y(t) = X_2(t)$ is observable. In the news coordinates defined by $(z_1, z_2)^T = z = \Phi(x) = (h(x), L_g h(x))^T$, our system can be written in the canonical form as follows:

$$\begin{cases} \dot{z}(t) = \begin{pmatrix} 0 & 1 \\ 0 & 0 \end{pmatrix} z(t) + \begin{pmatrix} 0 \\ \Psi(z(t)) \end{pmatrix} , \\ y(t) = z_1(t) = (0, 1)z(t) \end{cases} \tag{10.26}$$

where: $A = \begin{pmatrix} 0 & 1 \\ 0 & 0 \end{pmatrix}$; $C = (0, 1)$ and

$$\Psi(z) = L_g^2 h(\Phi^{-1}(z)) = L_g^2 h(x) = \psi(x).$$

The function $\psi(x)$ is smooth (it is a polynomial function of $x = (x_1; X_2)$ on the compact set $D$). Hence, it is globally Lipschitz on $D$. Therefore, it can be extended by $\tilde{\psi}$, a Lipschitz function on $\mathbb{R}^2$ that satisfies $\tilde{\psi}(x) = \psi(x)$, for all $x \in D$. Doing as before, we define $\tilde{\Psi}$ the Lipschitz prolongation of the function $\Psi$. So we have the following system (10.27) defined on the whole space $\mathbb{R}^2$. The restriction of system (10.27) to the domain $D$ is the system (10.26):

$$\begin{cases} \dot{z} = Az + \begin{pmatrix} 0 \\ \tilde{\Psi}(z) \end{pmatrix}, \\ y = Cz. \end{cases} \tag{10.27}$$

According to [8], an exponential (high-gain) observer for system (10.27) is given by

$$\dot{\tilde{z}} = A\tilde{z} + \begin{pmatrix} 0 \\ \tilde{\Psi}(z) \end{pmatrix} - S_\theta^{-1} C^T (y - C\tilde{z}), \tag{10.28}$$

where $S(\theta)$ is the solution of $-\theta S_\theta - A^T S_\theta - S_\theta A^T + C^T C = 0$ and $\theta$ is large enough.

Here,

$$S(\theta) = \begin{pmatrix} \frac{1}{\theta} & -\frac{1}{\theta^2} \\ -\frac{1}{\theta^2} & \frac{2}{\theta^3} \end{pmatrix}.$$

This observer is particularly simple since it is only a copy of system (10.27), together with a corrective term depending on $\theta$. For more details of the proof, we refer to [8, 9].

An observer for the original system (10.23) can then be written by

$$\begin{cases} \dot{\tilde{z}} = A\tilde{z} + \begin{pmatrix} 0 \\ \tilde{\Psi}(\tilde{z}) \end{pmatrix} - S_\theta^{-1} C^T (y - C\tilde{z}) \\ \\ \hat{x}(t) = \Phi^{-1}(z(t)). \end{cases} \tag{10.29}$$

Or more simply, a high-gain observer for the original system (10.23) can be given by

$$\dot{\hat{x}} = \tilde{f}(\hat{x}) + \left[\frac{d\Phi}{dx}\right]^{-1}_{x=\hat{x}} \times S_\theta^{-1} C^T (y - h(\hat{x})). \tag{10.30}$$

The expression of observer system is

$$\begin{cases} \dot{\hat{x}}_1 = B - \mu_1 \hat{x}_1(t) - f(\hat{x}_1(t), \hat{X}_2(t-\tau)) - \dfrac{1}{f_1(\hat{x}_1(t), \hat{X}_2(t-\tau))} \times \\ [\theta^2(f_2(\hat{x}_1(t), \hat{X}_2(t-\tau)) - \mu_2 - \gamma) - \theta^3](X_2 - \hat{X}_2) \\ \dot{\hat{X}}_2 = f(\hat{x}_1(t), \hat{X}_2(t-\tau)) - (\mu_2 + \gamma) X_2(t) + \theta^2 (X_2 - \hat{X}_2). \end{cases} \tag{10.31}$$

However, the set $D$ that is positively invariant for system (10.23) is not necessary positively invariant for the observer (10.30), and $\Phi(D)$ is not positively invariant for the observer (10.28).

Therefore, the expressions $[\frac{d\Phi}{dx}]^{-1}_{x=\hat{x}}$ and $\Phi^{-1}(z(t))$ are not well-defined in general.

If there exists $\tilde{\Phi}$ a prolongation of the diffeomorphism $\Phi$ to the whole space $\mathbb{R}^2$, that is, $\tilde{\Phi}$ is a diffeomorphism from $\mathbb{R}^2$ to $\mathbb{R}^2$ whose restriction to $\mathring{D}$ is $\Phi$, then it will be sufficient to replace $\Phi$ by $\tilde{\Phi}$ in (10.29) and (10.30) and so all the expressions will be well-defined. □

## 10.6 Numerical Simulation and Comments

In this section, we expose the computation work that supports our study. In this computation, the function $f$ is chosen as follows: $f(x_1, X_2) = x_1 X_2$ (mass action). $x_1$ represents the state of the susceptible state and $X_2$ the infectious one. In this part, we have simulated systems (10.5), (10.24), (10.31) using the parameters given in the table below. The results of the simulations are presented in Figs. 10.2 and 10.3 and illustrate the evolutions of the original state variables and the estimated states given by the high-gain observer and simple observer (Table 10.1).

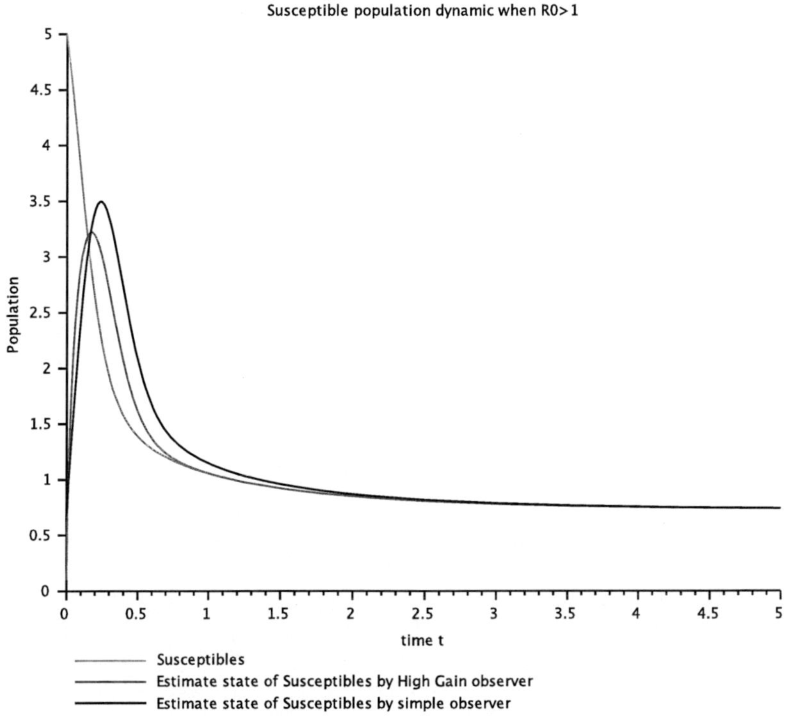


**Fig. 10.2** The temporal evolution of the number of susceptible persons (red line) given by (10.4) its estimate (blue line) delivered by the high-gain observer given by (10.31) and its estimate (black line) delivered by the simple observer given by (10.24)

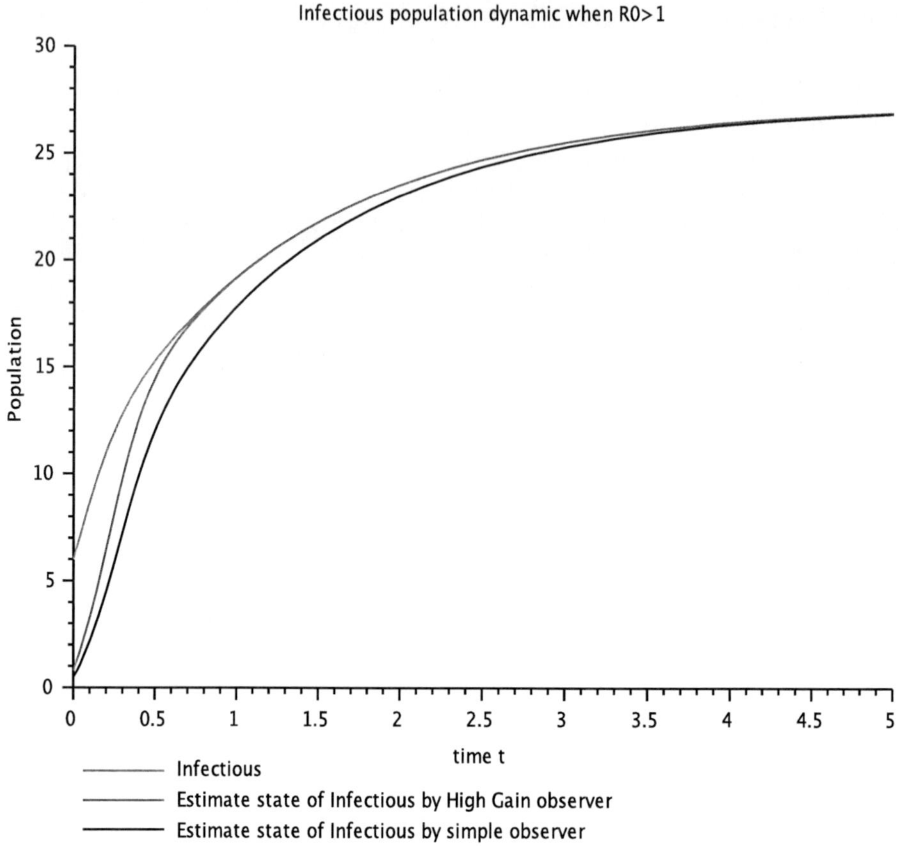


**Fig. 10.3** The temporal evolution of the number of infectious persons (red line) given by (10.4) its estimate (blue line) delivered by the high-gain observer given by (10.31) and its estimate (black line) delivered by the simple observer given by (10.24)

**Table 10.1** Parameters values of the model

| Symbols | Values | Sources |
|---|---|---|
| $B$ | 20 | Estimated |
| $\mu_1$ | 0.1 | Estimated |
| $\mu_2$ | 0.003 | Estimated |
| $\gamma$ | 0.0000027 | Estimated |
| $\theta$ | 2 | Estimated |

## 10.7 Conclusion

In this chapter, an SIR epidemic model with delay in the general incidence function is derived. In one hand, the global behaviour of the model system was studied. We proved that, if $R_0 \leq 1$ holds, then the disease-free equilibrium is globally asymptotically stable, which implies that the disease fades out from the population. If $R_0 > 1$, then there exists a unique endemic equilibrium that is globally asymptotically stable, and this implies that the disease will persist in the population. In the second part of this chapter, we deal with state identification, which is called nonlinear observer design. We just supposed that the infectious population is measured and gave an algorithm that allows to estimate the unmeasured states ($S(t)$ and $R(t)$) that are the susceptibles and the recovered. We construct two kinds of observers, a simple one and the so-called high-gain observer. With both observers, we reconstruct the unmeasured states. We corroborate the convergence of our observers with numerical simulation. For that, we present the curves when $R_0 \leq 1$ and $R_0 > 1$.

**Acknowledgments** The authors want to thank the anonymous referee for his valuable comments on the paper.

**Competing Interests** The authors declare that they have no competing interests.

**Authors' Contribution** Aboudramane Guiro provided the subject, wrote the introduction and the conclusion, and verified some calculation. Dramane Ouedraogo conceived the study and computed the equilibria and their local stabilities. Harouna Ouedraogo wrote the mathematical formula, brought up the Lyapunov functional, and did all the calculus with the first author. All the authors read and approved the final manuscript.

## References

1. D.M. Bichara, A. Guiro, D. Ngom, A. Iggidr, State and parameter estimation for a class of schistosomiasis models. Math. Biosci. **315**, 108–226 (2019)
2. G. Birkhoff, G.C. Rota, *Ordinary Differential Equations* (Ginn, Boston, 1982)
3. J.M. Bony, Principe du maximum, inégalité de Harnack et unicité du problème de Cauchy pour les operateurs elliptiques dégénérés . Ann. Inst. Fourier (Grenoble) **19**(1), 277–304 (1969)
4. C. Connell McCuskey, Global stability for an SIR epidemic model with delay and nonlinear incidence. Nonlinear Anal. RWA **11**, 3106–3109 (2010)
5. C. Connell McCuskey, Complete global stability for an SIR epidemic model with delay-distributed or discrete. Nonlinear Anal. RWA **11**, 55–59 (2010)
6. C. Connell McCuskey, Global stability for an SEIR epidemiological model with varying infectivity and infinite delay. Math. Biosci. Eng. **6**, 603–610 (2009)
7. L. Esteva, C. Vargas, Analysis of a dengue disease transmission model. Math. Biosci. **150**(2), 131–151 (1998)
8. J.P. Gauthier, H. Hammouri, S. Othman, A simple observer for nonlinear systems applications to bioreactors. IEEE Trans. Autom. Control **37**(6), 875–880 (1992)
9. J.P. Gauthier, I.A. Kupka, Observability and observers for nonlinear systems. SIAM J. Control Optim. **32**(4), 975–994 (1994)
10. A. Guiro, Iggidr A, Ngom D, Touré H, A non linear observer for a fishery model, in *Proceedings of 17th Triennial IFAC World Congress*, Seoul, Korea, 2008
11. A. Guiro, D. Ngom, D. Ouedraogo, *Global Stability for a Class of Discrete Schistosomiasis Models with General Incidence.* Advances in Difference Equations (2017), pp. 1–16
12. J. Hale, S. Verduyn Lunel, *Introduction to Functional Differential Equation* (Springer, Berlin, 1993)

13. V. Lakshmikantham, S. Leela, A.A. Martynyuk, *Stability Analysis of Nonlinear Systems* (Marcel Dekker, New York, 1989)
14. D. Moulay, M.A. Aziz-Alaoui, M. Cadivel, The chikungunya disease: modeling, vector and transmission global dynamics. Math. Biosci. **229**, 50–63 (2011)
15. D. Ngom, A. Iggidr, A. Guiro, A. Ouahbi, An observer for a nonlinear age-structured model of a harvested fish population. Math. Biosci. Eng. **5**(2), 337–354 (2008)
16. R. Xu, Z. Ma, Global stability of a SIR epidemic model with nonlinear incidence rate and time delay. Nonlinear Anal. RWA **10**, 3175–3189 (2009)

# Chapter 11
# Threshold Parameters of Stochastic SIR and SIRS Epidemic Models with Delay and Nonlinear Incidence

**Ali Traoré**

**Abstract** In this chapter, we study stochastic SIR and SIRS epidemic models with delay. A nonlinear incidence function that includes some special incidence rates is also considered. Two thresholds $\mathcal{R}_0^S$ and $\tilde{\mathcal{R}}_0^S$ of the two models are derived by using the nonnegative semimartingale convergence theorem. The disease goes extinct when the value of $\mathcal{R}_0^S$ is below 1, and it prevails when $\tilde{\mathcal{R}}_0^S$ value is above 1 for any size of the white noise. The comparison between the two thresholds is made.

**Keywords** Delays · Stochastic SIR model · Nonlinear incidence · Extinction · Persistence in mean

## 11.1 Introduction

The use of mathematical model for understanding the infectious disease dynamics is well-established. An SIR (Susceptible, Infected, Removed) epidemic model is often used to describe the prevalence of the disease in a population. The deterministic and stochastic models are applied to capture the propagation of the epidemic depending on the appropriate circumstances [1–3]. The transformation of a deterministic model into a stochastic model has been analysed by many authors [4–7]. The approach of that modelling random fluctuation consists of introducing parameter perturbations in the ordinary differential equations. The noise is introduced by replacing the model parameters by the fixed parameters plus an amplitude randomly fluctuation. In general, the parameters have great variability depending on the errors in the observed and measured data. The noise is hence introduced when some variables cannot be measured, and there is a lack of knowledge to illustrate the presence of random environment. Recently, Yang et al. [8] introduced a stochastic

A. Traoré (✉)
LAboratoire de Mathématiques et Informatique (LAMI), Université Joseph KI-ZERBO, Ouagadougou, Burkina Faso



T. Diagana et al. (eds.), *Partial Differential Equations and Applications*, Springer Proceedings in Mathematics & Statistics 420,
https://doi.org/10.1007/978-3-031-27661-3_11

perturbation into an SIR epidemic model with saturated incidence and investigated their dynamics according to the basic reproduction number. In [9], Zhao extended the work of Yang et al. by establishing a method to obtain the threshold values of the system in [8, 10]. Liu et al. [11] studied the equilibria of the following model:

$$\begin{cases} dS(t) = \big(\Lambda - \mu S(t) - \beta e^{-\mu\tau} S(t)G(I(t-\tau))\big)dt + \sigma_1 S(t)dB_1(t), \\ \\ dI(t) = \big(\beta e^{-\mu\tau} S(t)G(I(t-\tau)) - (\mu+\gamma+\alpha)I(t)\big)dt + \sigma_2 I(t)dB_2(t), \end{cases} \tag{11.1}$$

where $S(t)$ and $I(t)$ denote the number of susceptible individuals to the disease and the number of infective individuals, respectively. $\Lambda$ is the recruitment rate of the population, $\mu$ represents the natural death rate of the population, $\beta$ is the transmission rate between compartments $S$ and $I$, $\gamma$ is the recovered rate of infectious individuals, $\alpha$ is the disease-caused death rate of infectious individuals, $\tau \geq 0$ is the incubation time, and $\beta e^{-\mu\tau} S(t)G(I(t-\tau))$ is the force of infection. The term $e^{-\mu\tau}$ denotes the survival of vector population in which the time taken to become infectious is $\tau$. $B_1(t)$ and $B_2(t)$ are mutually independent standard Brownian motions defined on the probability space $\big(\Omega, \mathcal{F}, P\big)$ with a filtration $\{\mathcal{F}_t\}_{t\geq 0}$, and $\sigma_1$ and $\sigma_2$ denote the intensities of the white noise. The parameters are all supposed to be positive. The initial conditions of system (11.1) are set as follows:

$$\begin{cases} S(\theta) = \varphi_1(\theta), \ I(\theta) = \varphi_2(\theta), \\ \varphi_i(\theta) \geq 0, \theta \in [-\tau, 0], i = 1, 2, \\ (\varphi_1, \varphi_2) \in C, \end{cases} \tag{11.2}$$

where $C$ denotes the Banach space $C([-\tau, 0]; \mathbf{R}_+^2)$ of continuous functions mapping $[-\tau, 0]$ into $\mathbf{R}_+^2$. The threshold value of the epidemic is an important concept in mathematical epidemiology and is also important when studying properties of the extinction time [12]. However, in [11], the authors did not accurately point out the threshold whose value can completely determine the dynamics of the considered model. In this chapter, we derive the threshold parameters of system (11.1) and the threshold parameters of its corresponding SIRS epidemic model.

We organize the remainder of this chapter as follows. We establish the threshold parameter of model (11.1) that will allow the disease to fade out exponentially in Sect. 11.2. In Sect. 11.3, we derive the threshold parameter of model (11.1) for the disease being persistent in mean. In Sect. 11.4, we extend the model (11.1) to a stochastic SIRS epidemic model, and by using the method stated in previous sections, we establish the threshold parameters. In Sect. 11.5, this chapter ends with a conclusion.

## 11.2 Extinction of the Epidemic Model (11.1)

Liu et al. [11] have shown that the model (11.1) admits a unique positive solution $(S(t), I(t))$ on $t > 0$ and that this solution remains in $\mathbf{R}^2_+$ with probability one. We now focus on establishing the threshold of (11.1). We assume that the function $G$ is continuous on $[0, \infty)$ and is a twice differentiable function satisfying the following hypotheses:

**(H1)** $G(I) \geq 0$ with equality if and only if $I = 0$.
**(H2)** $G'(I) \geq 0$.
**(H3)** $G''(I) \leq 0$.

*Remark 11.2.1* From biological view, the three hypotheses are reasonable:

- Hypothesis **(H1)** means that if there are no infectives, then obviously there is no disease transmission.
- Hypothesis **(H2)** expresses the fact that increasing the number of the infective hosts increases the chance for the occurrence of new infections.
- Hypothesis **(H3)** describes the fact that susceptible individuals take measures to reduce contagion if the epidemics breaks out.

We first start by preparing some previous results.

**Lemma 11.2.1 (See [13])** *Let $U(t)$ and $W(t)$ be two continuous adapted increasing processes on $t \geq 0$ with $U(0) = W(0) = 0$ a.s. Let $M(t)$ be a real-value continuous local martingale with $M(0) = 0$ a.s. Let $X_0$ be a nonnegative $F_0$-measurable random variable such that $EX_0 < \infty$. Define $X(t) = X_0 + U(t) - W(t) + M(t)$ for all $t \geq 0$. If $X(t)$ is nonnegative, then $\lim_{t\to\infty} U(t) < \infty$ implies $\lim_{t\to\infty} W(t) < \infty$, $\lim_{t\to\infty} X(t) < \infty$, and $-\infty < \lim_{t\to\infty} M(t) < \infty$ hold with probability one.* □

**Lemma 11.2.2 (See [14])** *Let $M(t), t \geq 0$, be a local martingale vanishing at time 0 and define*

$$\rho_M(t) := \int_0^t \frac{d\langle M, M\rangle(s)}{(1+s)^2}, \quad t \geq 0,$$

*where $\langle M, M\rangle(t)$ is Meyer's angle bracket process. Then $\lim_{t\to\infty} \frac{M(t)}{t} = 0$ a.s. provided that $\lim_{t\to\infty} \rho_M(t) < \infty$.* □

We now start the study of the model (11.1).

**Lemma 11.2.3** *Assume that $(S(t), I(t))$ be the solution of system (11.1) with initial value given by (11.2), and then*

$$\limsup_{t\to\infty}(S(t) + I(t)) < \infty, \quad a.s.$$

***Proof*** From (11.1), we get

$$d(S(t)+I(t)) = (\Lambda - \mu(S(t)+I(t)) - (\gamma+\alpha)I + \sigma_1 S dB_1(t) + \sigma_2 I dB_2(t). \tag{11.3}$$

The solution of Eq. (11.3) satisfies the following inequality:

$$\begin{aligned} S(t)+I(t) &= \frac{\Lambda}{\mu} + \left(S(0)+I(0) - \frac{\Lambda}{\mu}\right)e^{-\mu t} \\ &\quad - (\alpha+\gamma)\int_0^t e^{-\mu(t-s)} I(s)ds + M(t), \\ &\leq \frac{\Lambda}{\mu} + \left(S(0)+I(0) - \frac{\Lambda}{\mu}\right)e^{-\mu t} + M(t), \end{aligned}$$

where $M(t) = \sigma_1 \int_0^t e^{-\mu(t-k)} S(k) dB_1(k) + \sigma_2 \int_0^t e^{-\mu(t-k)} I(k) dB_2(k)$ is a continuous local martingale with $M(0) = 0$ a.s.

Define $X(t) = X(0) + U(t) - W(t) + M(t)$, with $X(0) = S(0) + I(0)$, $U(t) = \frac{\Lambda}{\mu}(1 - e^{-\mu t})$, and $W(t) = (S(0) + I(0))(1 - e^{-\mu t})$ for all $t \geq 0$.

It follows that $S(t) + I(t) \leq X(t)$ a.s. Moreover, $U(t)$ and $W(t)$ are continuous adapted increasing processes on $t \geq 0$ with $U(0) = W(0) = 0$. In addition, $X(t)$ is clearly nonnegative and $\lim_{t\to\infty} U(t) = \frac{\Lambda}{\mu} < \infty$. Then, from Lemma 11.2.1, we deduce that $\lim_{t\to\infty} X(t) < \infty$, which implies $\limsup_{t\to\infty}(S(t) + I(t)) < \infty$. □

**Lemma 11.2.4** *Assume that $(S(t), I(t))$ be the solution of system (11.1) with initial value given by (11.2); then*

$$\lim_{t\to\infty} \frac{1}{t}\int_0^t \sigma_1 S(\xi) dB_1(\xi) = 0 \ \ a.s; \quad \lim_{t\to\infty} \frac{1}{t}\int_0^t \sigma_2 I(\xi) dB_2(\xi) = 0 \ a.s.$$

□

***Proof*** Let denote

$$M_1(t) = \sigma_1 \int_0^t S(\xi) dB_1(\xi); \quad M_2(t) = \sigma_2 \int_0^t I(\xi) dB_2(\xi).$$

Compute that $\langle M_1, M_1\rangle(t) = \sigma_1^2 \int_0^t S^2(\xi) d\xi$ and $\langle M_2, M_2\rangle(t) = \sigma_2^2 \int_0^t I^2(\xi) d\xi$.

Then, $\lim_{t\to\infty} \rho_{M_1}(t) = \lim_{t\to\infty} \sigma_1^2 \int_0^t \frac{S^2(\xi)d\xi}{(1+\xi)^2} \leq \sigma_1^2 \sup_{t\geq 0}\{S^2(t)\}$, and from Lemma 11.2.3, we get $\sigma_1^2 \sup_{t\geq 0}\{S^2(t)\} < \infty$. Thus, $\lim_{t\to\infty} \rho_{M_1}(t) < \infty$, and by

Lemma 11.2.2, it follows that $\lim_{t\to\infty} \frac{M_1(t)}{t} = 0$ a.s. By the same technique, we prove that $\lim_{t\to\infty} \frac{M_2(t)}{t} = 0$ a.s.

In the remaining part of this chapter, we set $\langle y(t)\rangle = \frac{1}{t}\int_0^t y(s)ds$.

**Theorem 11.2.1** *Let $(S(t), I(t))$ be the solution of system (11.1) with initial value given by (11.2). Let define*

$$\mathcal{R}_0^S = \frac{1}{\mu+\gamma+\alpha}\Big(\beta\Lambda G'(0)\frac{e^{-\mu\tau}}{\mu} - \frac{\sigma_2^2}{2}\Big).$$

*If $\mathcal{R}_0^S < 1$, then*

$$\limsup_{t\to\infty} \frac{\ln I(t)}{t} \le (\mu+\gamma+\alpha)(\mathcal{R}_0^S - 1) < 0 \ a.s.$$

□

*In addition,*

$$\lim_{t\to\infty} \langle S(t)\rangle = \frac{\Lambda}{\mu}. \tag{11.4}$$

***Proof*** By summing the two equations of (11.1) and integrating, we get

$$\frac{S(t)+I(t)-S(0)-I(0)}{t} - \frac{M_1(t)}{t} - \frac{M_2(t)}{t}$$
$$= \Lambda - \mu\langle S(t)\rangle - (\mu+\gamma+\alpha)\langle I(t)\rangle.$$

Therefore,

$$\langle S(t)\rangle = \frac{1}{\mu}\Big[\Lambda + \frac{M_1(t)}{t} + \frac{M_2(t)}{t} - \frac{S(t)+I(t)-S(0)-I(0)}{t} - (\mu+\gamma+\alpha)\langle I(t)\rangle\Big]. \tag{11.5}$$

On the other hand, applying Itô's formula to the second equation of (11.1) yields

$$d\ln(I(t)) = \Big[\beta e^{-\mu\tau}S(t)\frac{G(I(t-\tau))}{I(t)} - \Big(\mu+\gamma+\alpha+\frac{\sigma_2^2}{2}\Big)\Big]dt + \sigma_2 dB_2(t). \tag{11.6}$$

If assumption **H2** holds, then

$$d\ln(I(t)) \le [\beta G'(0)e^{-\mu\tau}S(t) - \Big(\mu+\gamma+\alpha+\frac{\sigma_2^2}{2}\Big)]dt + \sigma_2 dB_2(t),$$

and by integration, we get

$$\frac{\ln(I(t)) - \ln(I(0))}{t} \le \beta G^{'}(0)e^{-\mu\tau}\langle S(t)\rangle - \left(\mu+\gamma+\alpha+\frac{\sigma_2^2}{2}\right) + \frac{\sigma_2 B_2(t)}{t}. \tag{11.7}$$

Substituting (11.5) into (11.7), we obtain

$$\begin{aligned}\frac{\ln I(t)}{t} &\le \beta\Lambda G^{'}(0)\frac{e^{-\mu\tau}}{\mu} - \frac{\sigma_2^2}{2} - (\mu+\gamma+\alpha) + \beta G^{'}(0)\frac{e^{-\mu\tau}}{\mu}\Big[\frac{M_1(t)}{t} + \frac{M_2(t)}{t} \\ &\quad -(\mu+\gamma+\alpha)\langle I(t)\rangle - \frac{S(t)+I(t)-S(0)-I(0)}{t}\Big] \\ &\quad + \frac{\sigma_2 B_2(t)}{t} + \frac{\ln I(0)}{t}.\end{aligned} \tag{11.8}$$

From inequality (11.8), we derive

$$\begin{aligned}\frac{\ln I(t)}{t} &\le (\mu+\gamma+\alpha)(\mathcal{R}_0^S - 1) + \beta G^{'}(0)\frac{e^{-\mu\tau}}{\mu}\Big[\frac{M_1(t)}{t} + \frac{M_2(t)}{t} \\ &\quad - (\mu+\gamma+\alpha)\langle I(t)\rangle - \frac{S(t)+I(t)-S(0)-I(0)}{t}\Big] \\ &\quad + \frac{\sigma_2 B_2(t)}{t} + \frac{\ln I(0)}{t}.\end{aligned} \tag{11.9}$$

Further, from the law of large number, we have

$$\lim_{t\to\infty} \frac{B_2(t)}{t} = 0. \tag{11.10}$$

Moreover,

$$\lim_{t\to\infty}\Big[\frac{M_1(t)}{t} + \frac{M_2(t)}{t} - \frac{S(t)+I(t)-S(0)-I(0)}{t}\Big] = 0, \tag{11.11}$$

due to Lemmas 11.2.3 and 11.2.4.

In view of (11.10) and (11.11), taking the limit superior on both sides of (11.8), if $\mathcal{R}_0^S < 1$ and by the fact that $I(t) > 0$, we get

$$\limsup_{t\to\infty} \frac{\ln I(t)}{t} \le (\mu+\gamma+\alpha)(\mathcal{R}_0^S - 1) < 0 \ a.s.,$$

which implies that

$$\lim_{t\to\infty} I(t) = 0 \ a.s. \tag{11.12}$$

We now verify (11.4). From (11.5), (11.11), and (11.12), we obtain

$$\lim_{t\to\infty}\langle S(t)\rangle = \frac{\Lambda}{\mu} \quad a.s.$$

This completes the proof. □

## 11.3 Persistence in Mean of the Epidemic Model (11.1)

In this section, we derive a sufficient condition for the persistence in mean of the epidemic model (11.1). For that, we start by defining the notion of persistence in mean.

**Definition 11.3.1 ([10])** System (11.1) is said to be persistence in the mean if

$$\liminf_{t\to\infty}\langle I(t)\rangle > 0 \; a.s.$$

□

The following previous results will be used to attend our goal.

**Proposition 11.3.1 ([15])** $G'(I) \leq \dfrac{G(I)}{I}$. □

**Lemma 11.3.1 (See [16])** *Let* $f \in C([0,\infty),(0,\infty))$. *If there exist positive constants* $\lambda_0$ *and* $\lambda$ *such that*

$$\ln f(t) \geq \lambda t - \lambda_0 \int_0^t f(s)ds + F(t), a.s.$$

*for all* $t \geq 0$, *where* $F \in C[[0,\infty),(-\infty,\infty)]$ *and* $\lim_{t\to\infty}\dfrac{F(t)}{t} = 0$ *a.s.*, *then*

$$\liminf_{t\to\infty}\langle f(t)\rangle \geq \frac{\lambda}{\lambda_0} \quad a.s.$$

□

**Lemma 11.3.2 (See [16])** *Let* $f \in C([0,\infty),(0,\infty))$. *If there exist positive constants* $\lambda_0$, $\lambda$ *such that*

$$\ln f(t) \leq \lambda t - \lambda_0 \int_0^t f(s)ds + F(t), a.s.$$

*for all* $t \geq 0$, *where* $F \in C([0,\infty),(-\infty,\infty))$ *and* $\lim_{t\to\infty}\dfrac{F(t)}{t} = 0$ *a.s.*, *then*

$$\limsup_{t\to\infty}\langle f(t)\rangle \le \frac{\lambda}{\lambda_0}\ a.s.$$

□

We now state the result of this section as follows.

**Theorem 11.3.1** *Let $(S(t), I(t))$ be the solution of system (11.1) with initial value given by (11.2). Define*

$$\tilde{\mathcal{R}}_0^S = \frac{1}{\mu+\gamma+\alpha}\Big(\beta\Lambda G'\Big(\frac{\Lambda}{\mu}\Big)\frac{e^{-\mu\tau}}{\mu} - \frac{\sigma_2^2}{2}\Big).$$

*If $\tilde{\mathcal{R}}_0^S > 1$, then*

$$\liminf_{t\to\infty}\langle I(t)\rangle \ge \frac{\mu e^{\mu\tau}}{\beta G'\Big(\dfrac{\Lambda}{\mu}\Big)}(\tilde{\mathcal{R}}_0^S - 1)\ a.s,\quad \limsup_{t\to\infty}\langle I(t)\rangle \le \frac{\mu e^{\mu\tau}}{\beta G'(0)}(\mathcal{R}_0^S - 1)\ a.s.$$

□

***Proof*** From (11.6) and by using Proposition 11.3.1, we obtain

$$d\ln(I(t)) \ge \Big[\beta e^{-\mu\tau}S(t)G'(I) - \Big(\mu+\gamma+\alpha+\frac{\sigma_2^2}{2}\Big)\Big]dt + \sigma_2 dB_2(t).$$

Note that $S \le \dfrac{\Lambda}{\mu}$, and if condition (**H3**) is satisfied, then

$$d\ln(I(t)) \ge \Big[\beta G'\Big(\frac{\Lambda}{\mu}\Big)e^{-\mu\tau}S(t) - \Big(\mu+\gamma+\alpha+\frac{\sigma_2^2}{2}\Big)\Big]dt + \sigma_2 dB_2(t),$$

which yields after integration

$$\frac{\ln(I(t))}{t} \ge \beta G'\big(\frac{\Lambda}{\mu}\big)e^{-\mu\tau}\langle S(t)\rangle - \big(\mu+\gamma+\alpha+\frac{\sigma_2^2}{2}\big) + \frac{\ln(I(0))}{t} + \frac{\sigma_2 B_2(t)}{t}. \tag{11.13}$$

By substituting (11.5) into (11.13), we get

$$\begin{aligned}\frac{\ln I(t)}{t} \ge{}& \beta\Lambda G'\big(\frac{\Lambda}{\mu}\big)\frac{e^{-\mu\tau}}{\mu} - \frac{\sigma_2^2}{2} - (\mu+\gamma+\alpha) + \beta G'\Big(\frac{\Lambda}{\mu}\Big)\frac{e^{-\mu\tau}}{\mu}\Big[\frac{M_1(t)}{t}\\ &+\frac{M_2(t)}{t} - (\mu+\gamma+\alpha)\langle I(t)\rangle - \frac{S(t)+I(t)-S(0)-I(0)}{t}\Big]\\ &+\frac{\sigma_2 B_2(t)}{t} + \frac{\ln I(0)}{t}.\end{aligned}$$

This inequality can be rewritten as

$$\frac{\ln I(t)}{t} \geq (\mu+\gamma+\alpha)(\tilde{\mathcal{R}}_0^S - 1) - \beta G'\left(\frac{\Lambda}{\mu}\right)\frac{e^{-\mu\tau}}{\mu}(\mu+\gamma+\alpha)\langle I(t)\rangle + \frac{F(t)}{t},$$

where

$$F(t) = \beta G'\left(\frac{\Lambda}{\mu}\right)\frac{e^{-\mu\tau}}{\mu}\Big[M_1(t) + M_2(t) - S(t) - I(t) + S(0) + I(0)\Big]$$
$$+ \sigma_2 B_2(t) + \ln I(0).$$

From Lemma 11.3.1, we have

$$\liminf_{t\to\infty}\langle I(t)\rangle \geq \frac{\mu e^{\mu\tau}}{\beta G'(\frac{\Lambda}{\mu})}(\tilde{\mathcal{R}}_0^S - 1).$$

On the other hand, inequality (11.9) can be rewritten as

$$\frac{\ln I(t)}{t} \leq (\mu+\gamma+\alpha)(\mathcal{R}_0^S - 1) - \beta G'(0)\frac{e^{-\mu\tau}}{\mu}(\mu+\gamma+\alpha)\langle I(t)\rangle + \frac{F(t)}{t}.$$

Therefore, from Lemma 11.3.2, we derive

$$\limsup_{t\to\infty}\langle I(t)\rangle \leq \frac{\mu e^{\mu\tau}}{\beta G'(0)}(\mathcal{R}_0^S - 1).$$

This completed the proof. □

*Remark 11.3.1* The value $\mathcal{R}_0^S < 1$ will lead to the extinction of the epidemic, while the value $\tilde{\mathcal{R}}_0^S > 1$ will lead to the disease prevailing. We have $\mathcal{R}_0^S \geq \tilde{\mathcal{R}}_0^S$ with equality if the incidence function $G(I) = I$, that is, when the mass action incidence function is considered. □

## 11.4 The Threshold of the Stochastic SIRS Epidemic Model

Now, we will study an extension of the model (11.1). We are not interested on the existence of the unique positive solution for the considered model since it can be proved by the standard process (see [5, 11]). We consider that the recovered individuals lose immunity and return to the susceptible class at rate $\epsilon$, and then (11.1) takes the form

$$\begin{cases} dS(t) = \big(\Lambda - \mu S(t) - \beta e^{-\mu\tau} S(t)G(I(t-\tau)) + \epsilon R(t)\big)dt + \sigma_1 S(t)dB_1(t), \\ dI(t) = \big(\beta e^{-\mu\tau} S(t)G(I(t-\tau)) - (\mu+\gamma+\alpha)I(t)\big)dt + \sigma_2 I(t)dB_2(t), \\ dR(t) = \big(\gamma I(t) - (\mu+\epsilon)R(t)\big)dt + \sigma_3 R(t)dB_3(t). \end{cases} \tag{11.14}$$

By summing the three equations of (11.14) and after integration, we get

$$\frac{S(t)+I(t)+R(t)-S(0)-I(0)-R(0)}{t} - \frac{M_1(t)}{t} - \frac{M_2(t)}{t} - \frac{M_3(t)}{t} = \Lambda$$
$$-\mu\langle S(t)\rangle - (\mu+\gamma+\alpha)\langle I(t)\rangle - \mu\langle R(t)\rangle, \tag{11.15}$$

where $M_3(t) = \sigma_3 \int_0^t R(\xi)dB_3(\xi)$.

Moreover, from the last equation of (11.14), we obtain

$$\langle R(t)\rangle = \frac{1}{\mu+\epsilon}\Big[\gamma\langle I(t)\rangle - \frac{R(t)-R(0)}{t} + \frac{M_3(t)}{t}\Big]. \tag{11.16}$$

We state the following result.

**Theorem 11.4.1** *Let $(S(t), I(t), R(t))$ be the solution of system (11.14) with positive initial value.*

(1) *If $\mathcal{R}_0^S < 1$, then* $\limsup_{t\to\infty} \frac{\ln I(t)}{t} \le (\mu+\gamma+\alpha)(\mathcal{R}_0^S - 1) < 0$ *a.s.*

(2) *If $\tilde{\mathcal{R}}_0^S > 1$, then*

$$\limsup_{t\to\infty}\langle I(t)\rangle \le \frac{(\mu+\gamma+\alpha)e^{\mu\tau}}{\beta G'(0)\Big(\frac{\mu+\gamma+\alpha}{\mu} + \frac{\gamma}{\mu+\epsilon}\Big)}(\mathcal{R}_0^S - 1) \ \ a.s,$$

$$\textit{and } \liminf_{t\to\infty}\langle I(t)\rangle \ge \frac{(\mu+\gamma+\alpha)e^{\mu\tau}}{\beta G'\big(\frac{\Lambda}{\mu}\big)\Big(\frac{\mu+\gamma+\alpha}{\mu} + \frac{\gamma}{\mu+\epsilon}\Big)}(\tilde{\mathcal{R}}_0^S - 1) \ \ a.s.$$

□

***Proof*** Equation (11.15) shows that

$$\langle S(t)\rangle = \frac{1}{\mu}\Big[\Lambda + \frac{M_1(t)}{t} + \frac{M_2(t)}{t} + \frac{M_3(t)}{t}$$
$$- \frac{S(t)+I(t)+R(t)-S(0)-I(0)-R(0)}{t}$$
$$-(\mu+\gamma+\alpha)\langle I(t)\rangle - \mu\langle R(t)\rangle\Big]. \tag{11.17}$$

Plugging (11.16) into (11.17) gives

$$
\begin{aligned}
\langle S(t)\rangle = \frac{1}{\mu}\Big[\Lambda + \frac{M_1(t)}{t} + \frac{M_2(t)}{t} + \frac{\epsilon}{\mu+\epsilon}\frac{M_3(t)}{t} \\
- \frac{S(t)+I(t)-S(0)-I(0)}{t} - (\mu+\gamma+\alpha+\frac{\gamma\mu}{\mu+\epsilon})\langle I(t)\rangle \\
- \frac{\epsilon}{\mu+\epsilon}\frac{R(t)-R(0)}{t}\Big].
\end{aligned}
\tag{11.18}
$$

Since the second equations of systems (11.1) and (11.14) are the same, then equations (11.7) and (11.13) are also satisfied for system (11.14).

Replacing $\langle S(t)\rangle$ by its expression in (11.7) gives

$$
\begin{aligned}
\frac{\ln I(t)}{t} \leq \beta\Lambda G^{'}(0)\frac{e^{-\mu\tau}}{\mu} - \frac{\sigma_2^2}{2} - (\mu+\gamma+\alpha) + \beta G^{'}(0)\frac{e^{-\mu\tau}}{\mu}\Big[\frac{M_1(t)}{t} + \frac{M_2(t)}{t} \\
+ \frac{\epsilon}{\mu+\epsilon}\frac{M_3(t)}{t} - (\mu+\gamma+\alpha+\frac{\gamma\mu}{\mu+\epsilon})\langle I(t)\rangle \\
- \frac{S(t)+I(t)-S(0)-I(0)}{t} - \frac{\epsilon}{\mu+\epsilon}\frac{R(t)-R(0)}{t}\Big] \\
+ \frac{\sigma_2 B_2(t)}{t} + \frac{\ln I(0)}{t}.
\end{aligned}
$$

That is

$$
\begin{aligned}
\frac{\ln I(t)}{t} \leq (\mu+\gamma+\alpha)(\mathcal{R}_0^S - 1) + \beta G^{'}(0)\frac{e^{-\mu\tau}}{\mu}\Big[\frac{M_1(t)}{t} + \frac{M_2(t)}{t} + \frac{\epsilon}{\mu+\epsilon}\frac{M_3(t)}{t} \\
-(\mu+\gamma+\alpha+\frac{\gamma\mu}{\mu+\epsilon})\langle I(t)\rangle - \frac{S(t)+I(t)-S(0)-I(0)}{t} \\
- \frac{\epsilon}{\mu+\epsilon}\frac{R(t)-R(0)}{t}\Big] + \frac{\sigma_2 B_2(t)}{t} + \frac{\ln I(0)}{t}.
\end{aligned}
\tag{11.19}
$$

As in Sect. 11.2, taking the limit superior on both sides of (11.19) yields

$$
\limsup_{t\to\infty} \frac{\ln I(t)}{t} \leq (\mu+\gamma+\alpha)(\mathcal{R}_0^S - 1) < 0 \; a.s.,
$$

which completes the result 1).

Replacing $\langle S(t)\rangle$ in inequality (11.13) gives

$$\frac{\ln I(t)}{t} \geq \beta\Lambda G'(\frac{\Lambda}{\mu})\frac{e^{-\mu\tau}}{\mu} - \frac{\sigma_2^2}{2} - (\mu+\gamma+\alpha) + \beta G'(0)\frac{e^{-\mu\tau}}{\mu}\Big[\frac{M_1(t)}{t} + \frac{M_2(t)}{t}$$
$$+\frac{\epsilon}{\mu+\epsilon}\frac{M_3(t)}{t} - (\mu+\gamma+\alpha+\frac{\gamma\mu}{\mu+\epsilon})\langle I(t)\rangle$$
$$-\frac{S(t)+I(t)-S(0)-I(0)}{t} - \frac{\epsilon}{\mu+\epsilon}\frac{R(t)-R(0)}{t}\Big]$$
$$+\frac{\sigma_2 B_2(t)}{t} + \frac{\ln I(0)}{t},$$

which is equivalent to

$$\frac{\ln I(t)}{t} \geq (\mu+\gamma+\alpha)(\tilde{\mathcal{R}}_0^S - 1) - \beta G'\left(\frac{\Lambda}{\mu}\right)\frac{e^{-\mu\tau}}{\mu}\left(\mu+\gamma+\alpha+\frac{\gamma\mu}{\mu+\epsilon}\right)\langle I(t)\rangle$$
$$+\frac{E(t)}{t},$$

where

$$E(t) = \beta G'\left(\frac{\Lambda}{\mu}\right)\frac{e^{-\mu\tau}}{\mu}\Big[M_1(t) + M_2(t) + \frac{\epsilon}{\mu+\epsilon}M_3(t) - S(t) - I(t)$$
$$+S(0) + I(0) - \frac{\epsilon}{\mu+\epsilon}(R(t) - R(0))\Big] + \sigma_2 B_2(t) + \ln I(0).$$

From Lemma 11.3.1, we obtain

$$\liminf_{t\to\infty}\langle I(t)\rangle \geq \frac{(\mu+\gamma+\alpha)e^{\mu\tau}}{\beta G'\left(\frac{\Lambda}{\mu}\right)\left(\frac{\mu+\gamma+\alpha}{\mu} + \frac{\gamma}{\mu+\epsilon}\right)}(\tilde{\mathcal{R}}_0^S - 1).$$

Inequality (11.19) shows that

$$\frac{\ln I(t)}{t} \leq (\mu+\gamma+\alpha)(\mathcal{R}_0^S - 1) - \beta G'(0)\frac{e^{-\mu\tau}}{\mu}\left(\mu+\gamma+\alpha+\frac{\gamma\mu}{\mu+\epsilon}\right)\langle I(t)\rangle$$
$$+\frac{E(t)}{t}.$$

Using Lemma 11.3.2, we get

$$\limsup_{t\to\infty}\langle I(t)\rangle \leq \frac{(\mu+\gamma+\alpha)e^{\mu\tau}}{\beta G'(0)\left(\frac{\mu+\gamma+\alpha}{\mu} + \frac{\gamma}{\mu+\epsilon}\right)}(\mathcal{R}_0^S - 1) \;\; a.s.$$

This completed the proof. □

## 11.5 Conclusion

In this chapter, we considered two stochastic delayed SIR and SIRS epidemic models. For both models, a nonlinear incidence function that includes some special incidence rates is considered. We established sufficient conditions for extinction and persistence in the mean of the epidemic for each model. The thresholds that allow extinction and persistence are obtained. We found that, for each epidemic model, these thresholds are equal when the mass action incidence function is considered, that is when $G(I) = I$.

## References

1. T. Britton, D. Lindenstrand, Epidemic modelling: aspects where stochasticity matters. Math. Biosci. **222**(2), 109–116 (2009)
2. T. Britton, Stochastic epidemic models: a survey. Math. Biosci. **225**(1), 24–35 (2010)
3. M. Roberts, V. Andreasen, A. Lloyd, L. Pellis, Nine challenges for deterministic epidemic models. Epidemics **10**, 49–53 (2015)
4. A. Gray, D. Greenhalgh, L. Hu, X. Mao, J. Pan, A stochastic differential equation SIS epidemic model. SIAM J. Appl. Math. **71**(3), 876–902 (2011)
5. X. Mao, G. Marion, E. Renshaw, Environmental Brownian noise suppresses explosions in population dynamics. Stochastic Proc. Appl. **97**(1), 95–110 (2002)
6. B. Spagnolo, D. Valenti, A. Fiasconaro, Noise in ecosystems: a short review (2004). arXiv preprint q-bio/0403004
7. D. Stanescu, B.M. Chen-Charpentier, Random coefficient differential equation models for bacterial growth. Math. Comput. Modell. **50**(5–6), 885–895 (2009)
8. Q. Yang, D. Jiang, N. Shi, C. Ji, The ergodicity and extinction of stochastically perturbed SIR and SEIR epidemic models with saturated incidence. J. Math. Analy. Appl. **388**(1), 248–271 (2012)
9. D. Zhao, Study on the threshold of a stochastic SIR epidemic model and its extensions. Commun. Nonlinear Sci. Numer. Simul. **38**, 172–177 (2016)
10. Y. Zhao, D. Jiang, The threshold of a stochastic SIS epidemic model with vaccination. Appl. Math. Comput. **243**, 718–727 (2014)
11. Q. Liu, D. Jiang, N. Shi, T. Hayat, A. Alsaedi, Asymptotic behaviors of a stochastic delayed sir epidemic model with nonlinear incidence. Commun. Nonlinear Sci. Numer. Simul. **40**, 89–99 (2016)
12. T. Britton, A. Traoré, A stochastic vector-borne epidemic model: quasistationarity and extinction. Math. Biosci. **289**, 89–95 (2017)
13. X. Mao, *Stochastic Differential Equations and Applications* (Elsevier, Amsterdam, 2007)
14. R.S. Liptser, A strong law of large numbers for local martingales. Stochastics **3**(1–4), 217–228 (1980)
15. R.P. Sigdel, C.C. McCluskey, Global stability for an SEI model of infectious disease with immigration. Appl. Math. Comput. **243**, 684–689 (2014)
16. C. Ji, D. Jiang, Threshold behaviour of a stochastic SIR model. Appl. Math. Modell. **38**(21–22), 5067–5079 (2014)

# Chapter 12
# Weak Solutions for Nonlinear Boltzmann–Poisson System Modelling Electron–Electron Interactions

**Mohamed Lazhar Tayeb**

**Abstract** The existence of weak solutions of an initial boundary-value problem of a Boltzmann–Poisson model is studied. A dynamics describing electron–electron and electron–impurity interactions is considered. A fixed-point procedure is used to construct a weak solution for a regularized system, using the compactness properties of dynamics. Useful uniform estimates are established and used to carry out the proof of existence of the unregularized system.

**Keywords** Kinetic transport equations · Semiconductors · Entropy dissipation · Velocity-averaging lemma · Free energy

## 12.1 Introduction

Our aim is to analyse the existence of solutions for a nonlinear Boltzmann–Poisson system. Let $f \equiv f(t, x, v)$ be a distribution function depending on the time variable $t$ and the phase variable, $(x, v)$, belonging to a domain $\Omega = \omega \times \mathbb{R}^d$, where the dimension $d \in \{1, 2, 3\}$ and $\omega$ is a bounded subset of $\mathbb{R}^d$. The dynamics of collisions we considered in the present analysis takes into account the electron–impurity and electron–electron interactions [16]. Electron–impurity interactions are elastic, given by

$$Q_0(f)(v) = \int_{S^{d-1}} \sigma_0(v, |v|w)(f(|v|w) - f(v))dw, \tag{12.1}$$

where $S^{d-1}$ is the unit sphere of $\mathbb{R}^d$ and $\sigma_0$ is the cross-section of electron–impurity collisions, assumed to satisfy the detailed balance principle detailed later

M. L. Tayeb (✉)
Department of Mathematics, Faculty of Sciences of Tunis, University of Tunis El-Manar, Tunis, Tunisia
e-mail: mohamedlazhr.tayeb@fst.utm.tn

 
T. Diagana et al. (eds.), *Partial Differential Equations and Applications*,
Springer Proceedings in Mathematics & Statistics 420,
https://doi.org/10.1007/978-3-031-27661-3_12

on. Electron–electron interactions are trilinear, preserve mass, momentum, and kinetic energy, and satisfy the Pauli exclusion principle:

$$Q_1(f)(v) = \int_{\mathbb{R}^{3d}} \sigma_1(v, v_1, v', v_1')\delta(v + v_1 - v' - v_1')\delta(|v|^2 + |v_1|^2 - |v'|^2 - |v_1'|^2) \times \left[f' f_1'(1 - f)(1 - f_1) - f f_1(1 - f')(1 - f_1')\right] dv_1 dv' dv_1', \tag{12.2}$$

where

$$f = f(v), \quad f_1 = f(v_1), \quad f' = f(v'), \quad f_1' = f(v_1'),$$

and $\delta$ is Dirac distribution. We notice that the presence of the product term $\delta(v + v_1 - v' - v_1')\delta(|v|^2 + |v_1|^2 - |v'|^2 - |v_1'|^2)$ means that the first and second momenta (in velocity) are conserved during collisions. Indeed, if $v$ and $v_1$ are the velocities of two particles before collisions and $v'$ and $v_1'$ their post-collisional velocities, then the conservation can be expressed by the following relation:

$$\begin{cases} v + v_1 = v' + v_1', \\ |v|^2 + |v_1|^2 = |v'|^2 + |v_1'|^2 \end{cases}. \tag{12.3}$$

Using (12.3), we have

$$v' = v - (v - v_1 \cdot w)w, \qquad v_1' = v_1 + (v - v_1 \cdot w)w, \tag{12.4}$$

where $w$ belongs to $S^{d-1}$ and $(v - v_1) \cdot w)$ is the inner product between $v - v_1$ and $w$. The function $\sigma_1$ is the cross-section associated with electron–electron collisions, depending on $|v - v_1|$ and $(v - v_1 \cdot w)$ [31]. By writing

$$\sigma_1(v, v_1, v', v_1') := B(v - v_1, w),$$

we can rewrite $Q_1$ as follows:

$$Q_1(f)(v) = \int\int_{\mathbb{R}^d \times S^{d-1}} B(v - v_1, w)\left[f' f_1'(1 - f)(1 - f_1) - f f_1(1 - f')(1 - f_1')\right] dv_1 dw,$$

where $v, \ v_1, \ v'$, and $v_1'$ satisfy (12.4) [31, 45].

The transport equation satisfied by the distribution function $f$ is the following Boltzmann equation:

$$\partial_t f + v \cdot \nabla_x f + E \cdot \nabla_v f = (Q_0 + Q_1)(f) = Q(f). \tag{12.5}$$

The electrostatic field is self-consistent solving the Poisson equation

$$\begin{cases} E = -\nabla_x \Phi, \\ -\Delta_x \Phi = \varrho = \int_{\mathbb{R}^d} f dv \end{cases} \tag{12.6}$$

We assume that the initial value of $f$ is given, its boundary satisfies the condition of specular reflection on the boundary $\partial\Omega = \partial\omega \times \mathbb{R}^d$, and the potential $\Phi$ satisfies the Dirichlet condition:

$$f(t = 0, x, v) = f_0(x, v) \tag{12.7}$$

$$f(t, x, \bar{v}) = f(t, x, v), \quad (x, v) \in \partial\Omega, \tag{12.8}$$

$$\Phi(t, x) = \Phi_0(t, x), \quad x \in \partial\omega, \tag{12.9}$$

where $\bar{v} = v - 2(v \cdot n(x))n(x)$ and $n(x)$ is the unit normal vector to $\partial\omega$ at the position $x$.

### 12.1.1 Assumptions and Notations

We assume that the cross-section $\sigma_0$ of electron–impurity collisions satisfies the detailed balance principle [13, 14, 16]:

$$\text{(H1)} \qquad \begin{cases} 0 \le \sigma_0(v, |v|w) = \sigma_0(|v|w, v), \quad \forall (v, w) \in \mathbb{R}^d \times S^{d-1}, \\ \int_{S^{d-1}} \sigma_0(v, |v|w) dw \le C_0, \quad \forall v \in \mathbb{R}^d, \end{cases}$$

and the cross-section $B$ of electron–electron collisions satisfies

$$\text{(H2)} \qquad \begin{cases} B \in L^1(\mathbb{R}^d \times S^{d-1}), \\ B(z, w) \textit{ depends only on } |z| \textit{ and } |(z \cdot w)|, \\ \lim_{|z| \to 0} B(z, w) = 0, \qquad \lim_{|(z \cdot w)| \to 0} B(z, w) = 0, \\ \forall R > 0, \ |z| \gg 1, \int_{|v-z| \le R} \int_{S^{d-1}} B(v, w) dw dv = o(1 + |z|^2). \end{cases}$$

The initial data $f_0$ and the boundary data $\Phi_0$ satisfy

$$\text{(H3)} \qquad \begin{cases} 0 \le f_0 \le 1, \quad \displaystyle\int_\Omega (1+|v|^2) f_0(x,v)dxdv < +\infty, \\ \Phi_0 \ge 0, \ \Phi_0 \in L^\infty_{loc}(\mathbb{R}^+;\ W^{2,\infty}(\omega)) \ and \ \partial_t \Phi_0 \in L^\infty_{loc}(\mathbb{R}^+;\ L^\infty(\omega)). \end{cases}$$

In all the sequel, $f, \ f_1, \ f'$, and $f'_1$ denote, respectively, the functions $f(v), \ f(v_1), \ f(v')$, and $f(v'_1)$.

The phase space and the incoming and outgoing parts are denoted by

$\Omega = \omega \times \mathbb{R}^d, \quad \Gamma = \partial\omega \times \mathbb{R}^d,$
$\Gamma^\pm = \{(x,v) \in \Omega / \pm v \cdot n(x) > 0\},$
$d\sigma_x$ is the elementary measure on the surface $\partial\omega$.

The charge and current densities and the kinetic and potential energies stand as

$$\begin{aligned} \varrho(t,x) &= \int_{\mathbb{R}^d} f dv, \quad j(t,x) = \int_{\mathbb{R}^d} vf(t,x,v)dv, \\ K(t) &= \int_\Omega |v|^2 f(t,x,v)dxdv, \quad V(t) = \int_\omega \varrho(t,x)\Phi(t,x)dx. \end{aligned} \tag{12.10}$$

By extending the boundary data $\Phi_0$ in a harmonic way on $\bar{\omega}$ (denoted also $\Phi_0$):

$$E_0 = -\nabla_x \Phi_0, \qquad -\Delta\Phi_0 = 0,$$

we can rewrite our Boltzmann–Poisson system as follows:

$$(BP) \qquad \begin{cases} \partial_t f + v \cdot \nabla_x f + (E + E_0) \cdot \nabla_v f = (Q_0 + Q_1)(f) = Q(f), \\ E = -\nabla_x \Phi, \\ -\Delta_x \Phi = \varrho = \displaystyle\int_{\mathbb{R}^d} f dv, \\ f(t=0,x,v) = f_0(x,v), \\ f(t,x,\bar{v}) = f(t,x,v), \quad (x,v) \in \partial\Omega, \\ \Phi(t,x) = 0, \quad x \in \partial\omega, \end{cases}$$

where

$$\bar{v} = v - 2(v \cdot n(x))n(x),$$

$$Q_0(f)(v) = \int_{S^{d-1}} \sigma_0(v, |v|w)(f(|v|w) - f(v))dw,$$

$$Q_1(f)(v) = \int_{\mathbb{R}^d \times S^{d-1}} B(v - v_1, w)[f' f_1'(1 - f)(1 - f_1)$$
$$- f f_1(1 - f')(1 - f_1')]dv_1 dw.$$

Our main result is the following:

**Theorem 12.1.1** *Assume that $d \leq 3$ and (H1)–(H3) are satisfied. Then, the Boltzmann–Poisson system (BP) has a weak solution $(f, E)$ satisfying*

$$f \in L^\infty(\mathbb{R}^+;\ L^1 \cap L^\infty(\Omega)), \quad 0 \leq f \leq 1, \quad \|f(t)\|_{L^1(\Omega)} = \|f_0\|_{L^1(\Omega)}, a.e.$$

$$E \in L^\infty_{loc}(\mathbb{R}^+;\ [W^{1,\frac{d+2}{d}}(\omega)]^d),$$

$$t \mapsto \int_\Omega |v|^2 f(t, x, v)dxdv + \int_\omega |E(t, x)|^2 dx + 2\int_\omega \varrho(t, x)\varphi_0(x, t)dx \in L^\infty_{loc}(\mathbb{R}^+).$$

The analysis of the existence of solution is detailed as follows. The next section is devoted to the properties of the collision operators $Q_0$ and $Q_1$ (Sect. 12.2). Then, we recall some basic properties of the Vlasov equation posed on the free space with a given and regular potential. These properties are useful to apply to the penalization method due to S. Mischler [49] constructs a solution satisfying the specular reflection boundary condition, giving a solution of our Boltzmann equation with prescribed potential. In Section 12.4.2, useful uniform estimates on $\varrho$, $j$ and $K$ are proved. These estimates are enough to prove the stability results obtained by P.-L. Lions [45], based on the fact that $Q_1$ is a pseudo-differential operator and a velocity-averaging lemma (Sect. 12.4). Section 12.6 is devoted to the proof of the existence for the coupled setting.

## 12.2 Properties of the Dynamics

### *12.2.1 Continuity of $Q_0$*

**Lemma 12.2.1** *The operator $Q_0$ is continuous on $L^p(\mathbb{R}^d)$ for all $p \in [1, \infty]$.* □

***Proof of Lemma 12.2.1*** This result is based on the co-area formula [32]. Let us consider the function $\varepsilon : \mathbb{R}^d \mapsto \mathbb{R}^+ \in C^2(\mathbb{R}^d)$ with a finite number of critical points. By denoting $dS_r$ the unit surface:

$$S_r := \{v \in \mathbb{R}^d / \varepsilon(v) = r\}$$

and $N(r) := \{v \in \mathbb{R}^d/\varepsilon(v) = r\}$, then

$$N(r) = \int_{\{v/\varepsilon(v)=r\}} dN_r(v) \quad and \quad dN_r(v) = \frac{dS_r(v)}{|\nabla_v \varepsilon(v)|},$$

and if $f \in L^1(\mathbb{R}^d)$, then

$$\int_{\mathbb{R}^d} f dv = \int_0^{+\infty} \left\{ \int_{\{v\in\mathbb{R}^d \ / \ \varepsilon(v)=r\}} f(v) dN_r \right\} dr. \tag{12.11}$$

As a consequence, $f \in L^\infty(\mathbb{R}^d)$ implies for almost every $R > 0$, $f_{|S_R} \in L^\infty(S_R, dN_R)$ and

$$\|f_{|S_R}\|_{L^\infty(S_R)} \leq \|f\|_{L^\infty(\mathbb{R}^d)}. \tag{12.12}$$

By denoting

$$\chi_0(v) = \int_{S^{d-1}} \sigma_0(v, |v|w) dw, \tag{12.13}$$

we can rewrite $Q_0$ as follows:

$$Q_0(f)(v) = \int_{S^{d-1}} \sigma_0(v, |v|w) f(|v|w) dw - \chi_0(v) f(v).$$

Using the assumption (H1) and the fact that $\chi_0$ is bounded, we get the continuity of $Q_0$ on $L^\infty$. Indeed,

$$0 \leq \chi_0(v) \leq C_0 \quad and \quad \|Q_0(f)\|_{L^\infty(\mathbb{R}^d)} \leq 2C_0 \|f\|_{L^\infty(\mathbb{R}^d)}.$$

By the same argument, using the co-area formula (12.11), we can prove the continuity of $Q_0$ on $L^1(\mathbb{R}^d)$ :

$$\int_{\mathbb{R}^d} |Q_0(f)| dv \leq 2\|\chi_0\|_{L^\infty} \|f\|_{L^1}.$$

The continuity of $Q_0$ on $L^p$ is deduced using interpolation argument due to M. Riesz-Thorin and Marcinkiewicz (see [18], page 77).

### 12.2.2 Entropy Inequalities and Invariants of Collisions

**Lemma 12.2.2**

$$\int_{\mathbb{R}^d} Q_0(f)g dv = -\frac{1}{2}\int_{\mathbb{R}^d\times S^{d-1}} \sigma_0(v,|v|w)(f(|v|w)-f(v))(g(|v|w)-g(v))dvdw,$$

$$\int_{\mathbb{R}^d} Q_1(f)g dv = -\frac{1}{4}\int_{\mathbb{R}^{2d}\times S^{d-1}} B(v-v_1,w)[g'+g_1'-g-g_1]$$

$$\{f'f_1'(1-f)(1-f_1)-ff_1(1-f')(1-f_1')\}dvdv_1dw,$$

where $v' = v-(v-v_1.w)w$ *and* $v_1' = v_1+(v-v_1.w)w$. In particular:

**Lemma 12.2.3** *H-theorem*
*For all increasing function $H$,*

$$\int_{\mathbb{R}^d} Q_0(f)H(f)dv \le 0,$$

$$\int_{\mathbb{R}^d} Q_1(f)Log\frac{f}{1-f}dv \le 0, \quad \forall f \in ]0,1[.$$

**Lemma 12.2.4** *Invariants of collision and equilibrium state*

$$\int_{\mathbb{R}^d} Q_0(f)\begin{pmatrix}1\\ g(|v|^2)\end{pmatrix}dv = 0 \tag{12.14}$$

*and*

$$\int_{\mathbb{R}^d} Q_1(f)\begin{pmatrix}1\\ v\\ |v|^2\end{pmatrix}dv = 0. \tag{12.15}$$

*Moreover,*

$$(Q_0+Q_1)(f) = 0 \Leftrightarrow \exists\, \mu, T\ /\ f(v) = \frac{1}{1+\exp\frac{|v|^2/2-\mu(x,t)}{T(x,t)}}. \tag{12.16}$$

## 12.3 Free-Space Vlasov Equation

Some basic results related to the Vlasov equation for $(x,v)\in\mathbb{R}^{2d}$ (12.17) are used to construct a solution satisfying the reflexion boundary condition. To do this, we expect some estimates on the trace of the solution of the Vlasov equation to give a sense of all solution satisfying (12.8).

Let $E \in L^1_{loc}(\mathbb{R}^d \times \mathbb{R}^+)$, $\lambda \in L^\infty_{loc}(\mathbb{R}^{2d} \times \mathbb{R}^+)$, and $G \in L^\infty_{loc}(\mathbb{R}^{2d} \times \mathbb{R}^+)$. We consider

$$\Lambda_E f + \lambda f = \partial_t f + v.\nabla_x f + E.\nabla_v f + \lambda(x, v, t) f = G, \quad (x, v) \in \mathbb{R}^{2d}. \tag{12.17}$$

**Definition 12.3.1** We say that a function $f \in L^1_{loc}(\mathbb{R}^{2d} \times \mathbb{R}^+)$ is a weak solution of (12.17)–(12.7) if for all $\psi \in C^1_c(\mathbb{R}^{2d} \times \mathbb{R}^+)$,

$$\int_{\mathbb{R}^+ \times \mathbb{R}^{2d}} \{f \Lambda_E \psi - \lambda \psi) + G\psi\} + \int_{\mathbb{R}^{2d}} f_0 \psi(0, x, v) = 0. \tag{12.18}$$

We notice that the concept of weak solution is related to the sense we give to its trace on the boundary, which depends on the regularity of its coefficients. Such a problem was studied for example of the case of free transport ($E = 0$), for the netronics ($E = 0$ and $v \in S^{d-1}$) by V.I. Agoshkov [2], M. Cessenat [22], L. Arkeryd, C. Cercignani [4] and few years ago by S. Mischler in the context of the Boltzmann equation. The case of Lipschitz force field was analysed by C. Bardos [7] and N. B. Abdallah [12] using the characteristics: $z = (x, v)$ and $Z = (X, V)$ satisfying

$$\begin{cases} \dfrac{dZ}{ds}(s; t, z) = (V(s; t, z), E(t, X(s; t, z))), \\ Z(t; t, z) = z. \end{cases} \tag{12.19}$$

This implies that the solution of (12.17)–(12.7) is given by:

**Theorem 12.3.2 ((Existence) [7, 49])** *Let $p \in [1, \infty]$, $f_0 \in L^p(\mathbb{R}^{2d})$, $E \in W^{1,\infty}(\mathbb{R}^d \times [0, T])$, $G \in L^1_{loc}(\mathbb{R}^+; L^p(\mathbb{R}^{2d}))$, and $\lambda \in L^p_{loc}(\mathbb{R}^{2d} \times \mathbb{R}^+)$ . Then the solution of (12.17)–(12.7) reads as*

$$\begin{aligned} f(t, z) &= f_0(Z(0; t, z)) exp\left\{-\int_0^t \lambda(s, Z(s; t, z)) ds\right\} \\ &\quad + \int_0^t G(s, Z(s; t, z)) exp\left\{-\int_s^t \lambda(s', Z(s'; t, z)) ds'\right\} ds \end{aligned} \tag{12.20}$$

*and satisfies the weak maximum principle*

$$G \geq 0 \ \& \ f_0 \geq 0, \quad which\ implies \quad f(t, x, v) \geq 0. \tag{12.21}$$

*Moreover, $\lambda \in L^\infty(\mathbb{R}^d \times [0, T])$ and $\lambda \geq \lambda_0$; then $f \in L^\infty(0, T; L^p(\mathbb{R}^{2d}))$ and*

$$\sup_{s \in [0,t]} \|f(s)\|_{L^p(\mathbb{R}^{2d})} \leq \|f_0\|_{L^p(\mathbb{R}^{2d})} e^{-\lambda_0 t} + \int_0^t \|G(s)\|_{L^p(\mathbb{R}^{2d})} e^{-\lambda_0 (t-s)} ds. \tag{12.22}$$

### 12.3.1 Renormalized Solution

This concept of solution also called weak–weak solution is introduced to give a sense for PDEs presenting coefficients without enough regularity [24, 25]. It consists in considering an equation satisfied only by some $\beta(f)$ for a class of functions $\beta$, generally, in $W^{1,\infty}_{loc}$. The definition of renormalized solution for the Vlasov equation is given by:

**Definition 12.3.2** Let $f \in L^1_{loc}(\mathbb{R}^{2d} \times \mathbb{R}^+)$. We say $f$ is a renormalized solution for the initial value problem (12.17)–(12.7) if for all $\beta \in W^{1,\infty}(\mathbb{R})$, the function is a weak solution of

$$\begin{cases} \Lambda_E \beta(f) = \beta'(f)(G - \lambda f), \\ \beta(f)(t = 0) = \beta(f_0). \end{cases}$$

We notice that we need to deal with this concept of solution if the regularity of $f$ and $E$ is not enough to deal with $E \cdot \nabla_v \psi f$ in $L^1_{loc}$. The functions $\beta \in L^\infty_{loc}$ and $E \in L^1_{loc}$. This is enough to define $E \cdot \nabla_v \psi \beta(f)$. S. Mischler explained twenty years ago the relation with weak and weak–weak solution for a Boltzmann equation associated with a quadratic collision operator [50].

**Theorem 12.3.3 ([49])**

*1. Let $p \in [1, \infty[$.*
*Let $f \in L^\infty_{loc}(\mathbb{R}^+;\ L^p_{loc}(\mathbb{R}^{2d}))$ be a solution of (12.17)–(12.7). Then, for all $t \in [0, T]$, $f(t, .) \in L^p_{loc}(\mathbb{R}^{2d})$ and*

$$f \in C(\mathbb{R}^+;\ L^1_{loc}(\mathbb{R}^{2d}))$$

*and for all open and regular subsets $O$ of $\mathbb{R}^{2d}$, the trace of $\gamma_f$ is the unique function*

$$\gamma_f \in L^1_{loc}(\partial O, (v.n(x))^2 d\sigma_x dv ds)$$

*satisfying the following Green formula:*

$$\int_{t_0}^{t_1} \int_O \{f(\Lambda_E \psi - \lambda \psi) + G\psi\} dx dv dt$$
$$= \left[\int_O f(t, .)\psi\right]_{t_0}^{t_1} + \int_{t_0}^{t_1}\int_{\partial O} \gamma_f \psi (v.n(x)) d\sigma_x dv dt \tag{12.23}$$

*for all $t_0$, $t_1$, for all $\psi \in \mathcal{D}(\mathbb{R}^{2d} \times \mathbb{R}^+)$ such that $\psi = 0$ on $\Gamma_0 \times \mathbb{R}^+$, ($\Gamma_0 = \Gamma \cap \{v \cdot n(x) = 0\}$).*

2. *Let $p = +\infty$.*
   *Let $f \in L^\infty_{loc}(\mathbb{R}^{2d} \times \mathbb{R}^+)$, a weak solution of (12.17)–(12.7). Then $f(t,.)$ is defined for all $t$, and $\gamma_f$ is given by the formula (12.23) and satisfies*

$$f \in C(\mathbb{R}^+;\ L^a_{loc})\ \forall a < \infty \quad and \quad \gamma_f \in L^\infty_{loc}(\partial\Omega \times \mathbb{R}^+, d\sigma_x dv dt). \tag{12.24}$$

   *Moreover, (12.23) is satisfied for all $\psi \in \mathcal{D}(\bar{\Omega} \times \mathbb{R}^+)$.*

### 12.3.2 Penalization Method

We consider a force field $E$ defined on $\omega \times [0, T]$. We will denote, in the sequel, by $\bar{E}$ its extension by zero outside $\bar{\omega} \times [0, T]$. Now, we shall explain how we construct a solution for the Vlasov equation satisfying the specular reflection boundary condition (12.8) in the sense of the following definition:

**Definition 12.3.3** A function $f \in L^1_{loc}(\Omega \times \mathbb{R}^+)$ is said to be a weak solution of (12.17)–(12.7)–(12.8) if

$$\int_{\mathbb{R}^+\times\Omega} \{f(\Lambda_E\psi - \lambda\psi) + G\psi\} dx dv dt + \int_\Omega f_0 \psi(0, x, v) dx dv = 0, \tag{12.25}$$

for all $\psi \in C^1_c(\bar{\Omega} \times \mathbb{R}^+)$ and $\psi(t, x, \bar{v}) = \psi(t, x, v)$ sur $\Gamma^- \times \mathbb{R}^+$. □

The method of construction of a weak solution, in the sense of the previous definition, consists in using the free-space equation with a "strong force field" tending to confine in $\Omega^c$. To do this, we define a function $\delta \in W^{2,\infty}(\mathbb{R}^d)$, which is equal, on a neighbourhood $\mathcal{V}$ of $\partial\Omega$, to the distance $d(x) = dist(x, \partial\Omega)$ and that $d(x) \geq \delta_0 > 0$ outside $\mathcal{V}$. Let $\delta(x) = d(x)\chi_{\{x\in\Omega^c\}}$; $\delta(x) := dist(x, \bar{\Omega})$ sur $\mathcal{V}$. The vector $n(x) = \nabla_x d(x)$ does not vanish on a neighbourhood $\mathcal{W}$ of $\partial\Omega$. We define, for all $x \in \mathcal{W}$, the projection $\Pi_x$ on $\langle n(x)\rangle^\perp$ by

$$\forall v \in \mathbb{R}^d, \quad v = (n(x)\cdot v)n(x) + \Pi_x v \quad et \quad n(x)\cdot \Pi_x v = 0,$$

and we extend $\Pi_x$ arbitrarily outside $\mathcal{W}$.

Let $\varphi \in \mathcal{D}(\mathbb{R}^d \times \mathbb{R}^+)$, $supp(\varphi) \subset \bar{\Omega} \times \mathbb{R}^+$, $\theta \in \mathcal{D}(\mathbb{R}^+)$, $\theta(0) = 0$, and $\Psi \in \mathcal{D}(\mathbb{R}^d)$. We consider the set of functions $\psi \in \mathcal{D}(\mathbb{R}^{2d} \times \mathbb{R}^+)$ such that:

$$\psi(x, v, t) = \varphi(x, t)\theta((v.n(x))^2)\Psi(\Pi_x v). \tag{12.26}$$

Using density argument 12.3.3, the previous definition is equivalent with:

**Lemma 12.3.5 ([49])** *A function $f \in L^1_{loc}(\Omega\times\mathbb{R}^+)$ is a solution of (12.17)–(12.7)–(12.8) if*

$$\int_{\mathbb{R}^+\times\Omega}\{f(\Lambda_E\psi-\lambda\psi)+G\psi\}dxdvdt+\int_{\Omega}f_0\psi(t,x,v)dxdv=0$$

*for all function satisfying (12.26).* □

The existence result of weak solution is given by:

**Theorem 12.3.4** *Let $p\in[1,\infty]$, $f_0\in L^p(\Omega)$, $G\in L^1(0,T;\ L^p(\Omega))$, $E\in L^1(0,T;\ W^{1,p'}(\omega))$, and $\lambda\in L^{p'}(\Omega)$. Then, (12.17)–(12.7):*

$$f\in L^\infty(\mathbb{R}^+;\ L^\infty(\Omega))\cap C(\mathbb{R}^+;\ L^p(\Omega)),\forall p\in[1,\infty[.$$

*If $\lambda\geq\lambda_0,\ a.e.$, then*

$$\sup_{s\in[0,t]}\|f(s)\|_{L^p(\Omega)}\leq\|f_0\|_{L^p(\Omega)}e^{-\lambda_0 t}+\int_0^t\|G(s)\|_{L^p(\Omega)}e^{-\lambda_0(t-s)}ds \tag{12.27}$$

*for $(p,q,r)$ such that $1/p+1/q=1/s\leq1$, $r=p(1-1/q)$, $E\in L^1(0,T;\ L^q(\omega))$, and for all compact subset $K\subset\partial\Omega\times[0,T]$, the function $f$ has a trace*

$$\gamma_f\in L^r(K;\ |v.n(x)|d\sigma_x dvds).$$

*Moreover, there exists a constant $C(K,r,\|E\|_{L^1(0,T;\ L^q(\omega))})$ such that*

$$\|\gamma_f\|_{L^r(K;\ |v.n(x)|d\sigma_x dvds)}\leq C(K,r,\|E\|_{L^1(0,T;\ L^q(\omega))}).$$

***Proof*** The proof is based on the idea introduced in [25] and used in [49].

We define the force field $E$ by

$$E^\varepsilon=\bar{E}-\frac{\delta(x)}{\varepsilon}n(x). \tag{12.28}$$

$E^\varepsilon\in L^\infty_{loc}(\mathbb{R}^+;\ W^{1,p'}_{loc}(\mathbb{R}^d))$. An application of the previous results (Theorem 12.3.2) implies

$$\sup_{s\in[0,T]}\|f^\varepsilon(t)\|_{L^\infty(\mathbb{R}^{2d})}\leq C(T,\|f_0\|_{L^\infty(\mathbb{R}^{2d})},\|G\|_{L^1(0,T;\ L^\infty(\mathbb{R}^{2d})}).$$

By passing to the limit in the Green formula satisfied by the solution of the free-space Vlasov equation with the following family test functions: $\psi\in\mathcal{D}(\Omega\times\mathbb{R}^+)$, satisfying

$$\psi_\varepsilon=\varphi\theta_\varepsilon\Psi=\varphi(x,t)\theta\left((v\cdot n(x))^2+\frac{\delta^2(x)}{\varepsilon}\right)\Psi(\Pi_x v),$$

where $\varphi$, $\theta$, and $\Psi$ are defined in (12.26), we get

$$\int_{\mathbb{R}^{2d}\times\mathbb{R}^+} f^\varepsilon \left\{\theta_\varepsilon \Psi(\partial_t\varphi + v.\nabla_x\varphi - \lambda\varphi) + \varphi\Psi(v\cdot\nabla_x\theta_\varepsilon + E^\varepsilon\cdot\nabla_v\theta_\varepsilon)\right\}$$

$$+\int_{\mathbb{R}^{2d}\times\mathbb{R}^+} \left\{\varphi\theta_\varepsilon(v\cdot\nabla_x\Psi + E^\varepsilon\cdot\nabla_v\Psi) + G\varphi\theta_\varepsilon\Psi\right\} dxdvdt = 0.$$

By remarking that $\delta(x) = 0$ on $\omega$, we infer that

$$f^\varepsilon\theta_\varepsilon = f^\varepsilon\theta\chi_\omega(x) \overset{*}{\rightharpoonup} f\theta\chi_\omega(x) \quad in \quad L^\infty,$$

$$v\cdot\nabla_x\theta_\varepsilon + E^\varepsilon\cdot\nabla_v\theta_\varepsilon = v\cdot\nabla_x\left[(v\cdot n(x))^2\right]\theta'\left((v\cdot n(x))^2 + \frac{\delta^2(x)}{\varepsilon}\right) + E\cdot\nabla_v\theta_\varepsilon,$$

and

$$E^\varepsilon\cdot\nabla_v\Psi = (E - \frac{\delta(x)}{\varepsilon}n(x))\cdot\nabla_v\left(\Psi(\Pi_x v)\right) = E\cdot[\Pi_x[\nabla\Psi](\Pi_x v)],$$

leading to

$$\int_{\mathbb{R}^{2d}\times\mathbb{R}^+} f\left\{\theta\Psi(\partial_t\varphi + v\cdot\nabla_x\varphi - \lambda\varphi) + \varphi\Psi(v\cdot\nabla_x\theta + E\cdot\nabla_v\theta)\right\}$$

$$+\int_{\mathbb{R}^{2d}\times\mathbb{R}^+} \{\varphi\theta(v\cdot\nabla_x\Psi + E\cdot\nabla_v\Psi) + G\varphi\theta\Psi\}\, dxdvdt = 0,$$

which is equivalent to

$$\int_{\mathbb{R}^{2d}\times\mathbb{R}^+} \{f(\partial_t\psi + v\cdot\nabla_x\psi + E\cdot\nabla_v\psi - \lambda\psi) + G\psi\}\, dxdvdt = 0$$

for all $\psi$ satisfying (12.26). Then, $f$ is a weak solution of

$$\begin{cases} \Lambda_E f + \lambda f = G, & (x, v) \in \Omega, \\ f(t=0) = f_0, & \\ f(t, x, \bar{v}) = f(t, x, v), & (x, v) \in \Gamma^-, \end{cases}$$

where $\bar{v} = v - 2(v\cdot n(x))n(x)$.

## 12.4 Equation of Boltzmann with Specular Reflection Boundary Condition

This section is devoted to the existence and uniqueness of a weak solution of the Boltzmann equation satisfying the reflection boundary condition (12.8), given by this formulation

$$\int_{\Omega\times\mathbb{R}^+}\{f\Lambda_E\psi + Q(f)\psi\}dxdvdt + \int_{\Omega} f_0(x,v)\psi(0,x,v)dxdv = 0 \tag{12.29}$$

for all $\psi \in C_c^1(\mathbb{R}^{2d}\times\mathbb{R}^+)$ and satisfying (12.26).

### 12.4.1 Existence of a Weak Solution

**Theorem 12.4.5** *Let $E \in L^1_{loc}(\mathbb{R}^+;\ W^{1,\infty}(\omega))$. Then (12.17) has a weak solution satisfying*

$$\begin{cases} f \in L^\infty(\mathbb{R}^+;\ L^1\cap L^\infty(\Omega))\cap C(\mathbb{R}^+;\ L^p(\Omega)), \quad \forall p \in ]1,+\infty[, \\ 0\le f\le 1, \\ \|f(t)\|_{L^1(\Omega)} = \|f_0\|_{L^1(\Omega)}. \end{cases} \tag{12.30}$$

*Furthermore, its trace $\gamma_f$ on $\Gamma\times]0,T[$ satisfies:*

*For all $p, q \in ]1,\infty[,\ 1/p + 1/q = 1/s \le 1,\ r = p(1-1/q)\ge 2$ and for all compact subset $K\subset\Gamma\times[0,T]$, we have*

$$\gamma_f \in L^\infty(\Gamma\times\mathbb{R}^+, d\sigma_x dvds)\cap L^r(K;\ |v.n(x)|d\sigma_x dvds)$$

$$\|\gamma_f\|_{L^r(K;\ |v.n(x)|d\sigma_x dvds)} \le C(K, r, \|E\|_{L^1(0,T;\ L^q(\omega))}).$$

***Proof of Theorem 12.4.5*** We define, for all $f$, its extension $\bar{f}$ by

$$\bar{f}(t,x,v) = \begin{cases} 0 & si \quad f<0, \\ f(t,x,v) & si \quad 0\le f\le 1, \\ 1 & si \quad f>1. \end{cases}$$

We also define the function $F$ on $[0,1]^4$ as

$$F(x_1,x_2,x_3,x_4) = x_3x_4(1-x_1)(1-x_2) - x_1x_2(1-x_3)(1-x_4).$$

We have

$$\sup_i \|\partial_{x_i} F\|_{L^\infty(0,1)} = 2.$$

For all $f$ and $g$,

$$|\bar{f} - \bar{g}| \leq |f - g|$$

and

$$|Q(\bar{f}) - Q(\bar{g})| \leq \left(8\|B\|_{L^1(\mathbb{R}^d \times S^{d-1})}\right) |\bar{f} - \bar{g}|,$$

where $C_0$ depends only on (H1).

Now assume that $E \in W^{1,\infty}(\omega \times [0, T])$, and for all $f$, we define $\tau(f)$ is the weak solution of the following transport equation:

$$\begin{cases} \Lambda_E \tau(f) = (\partial_t + v \cdot \nabla_x + E \cdot \nabla_v)\tau(f) = Q(\bar{f}), \quad (x, v) \in \Omega, \\ \tau(f)(t, x, \bar{v}) = \tau(f)(t, x, v), \quad (x, v) \in \Gamma^-, \\ \tau(f)(t = 0) = \bar{f}_0 \quad (\equiv f_0). \end{cases} \tag{12.31}$$

The proof consists in proving that $\tau$ has a fixed point. Indeed, let $\tau(f)$ and $\tau(g)$ be two weak solutions of (12.31) with the same initial data $f_0$. The difference $h = \tau(f) - \tau(g)$ is a solution of

$$\begin{cases} \Lambda_E h = Q(\bar{f}) - Q(\bar{g}), \quad (x, v) \in \Omega \\ h(x, v, 0) = 0, \\ h(x, \bar{v}, t) = h(x, v, t), \quad (x, v) \in \Gamma^-, \end{cases}$$

and

$$\sup_{s\in[0,t]} \|h(s)\|_{L^p(\Omega)} \leq \int_0^t \|Q(\bar{f}) - Q(\bar{g})(s)\|_{L^p(\Omega)} ds \leq Ct \sup_{s\in[0,t]} \|(f-g)(s)\|_{L^p(\Omega)},$$

where $C$ depends only on the cross-section $B$.

As a consequence, $\tau$ has a fixed point in $L^\infty(0, t_0; \; L^1(\Omega))$ for $t_0 < 1/C$. The constant $C$ is independent of $t$. Then, we can construct a fixed solution on a sequence of intervals $[t_n, t_{n+1}[$ with $t_{n+1} - t_n < 1/C$ with an initial data $f(t = t_n) = f(t_n)$ (in a weak sense). With this procedure, we can construct a weak solution on $[0, T]$ for all $T > 0$.

We also remark that the solution of the Vlasov equation satisfying the specular reflection boundary condition on $\partial\Omega$ is a weak limit (in $L^p(]0,T[\times\Omega))$ of $g^\varepsilon$, the weak solution of the same equation defined on hole space associated with the confinement force field (12.28), which satisfies the weak maximum principle.

The operator $Q$ satisfies

$$-C\max(f,0) \le -C\bar{f} \le Q(\bar{f});$$

then, $\varphi(f) = f - \max(f,0)$; thus

$$\begin{cases} \Lambda_E(\varphi(f)) & = \varphi'(f)\Lambda_E(f) = \Lambda_E(f)(1 - \chi_{\{f\ge 0\}}) \\ & \ge -C\max(f,0)(1-\chi_{\{f\ge 0\}}) \equiv 0, \\ \varphi(f)(x,\bar{v},t) = \varphi(f)(x,v,t), & (x,v)\in\Gamma^-, \\ \varphi(f)(t=0) & = 0, \end{cases}$$

Implying that $\varphi(f)\ge 0$, and $f\ge 0$. Furthermore,

$$Q(\bar{f}) \le C(1-\bar{f}) \le C(1-\min(1,f))$$

so

$$\begin{cases} \Lambda_E(\min(1,f)-f) & = \Lambda_E(f)(\chi_{\{f\ge 1\}}-1) \\ & \ge C(1-\min(1,f))(\chi_{\{f\ge 1\}}-1) \equiv 0, \\ \varphi(f)(x,\bar{v},t) & = \varphi(f)(x,v,t), \quad (x,v)\in\Gamma^-, \\ \min(1,f) - f(t=0) = 0. \end{cases}$$

So,

$$\max(f,0) \le f \le \min(1,f) \quad a.e. \quad 0\le f\le 1.$$

To prove that the solution of the Boltzmann equation belongs to $C([0,T];\ L^p(\mathbb{R}^{2d}))$.

We can proceed as in [49]. Indeed, $f$ $(=\tau(f))$ is a weak solution of a Vlasov equation with a force field $E\in L^1_{loc}(\mathbb{R}^+;\ W^{1,\infty}(\omega))$ and a source term $G = Q(f)\in L^\infty(\Omega\times\mathbb{R}^+)$. This implies using [49] (Theorem 4) that

$$f\in C(\mathbb{R}^+;\ L^a_{loc}(\Omega)) \qquad \forall a<+\infty,$$

and its trace $\gamma_f$ is well defined by the Green formula and satisfies

$$\gamma_f \in L^\infty(\Gamma \times \mathbb{R}^+;\ d\sigma_x dv ds).$$

If $O$ is a smooth open subset of $\mathbb{R}^{2d}$ and $K$ is a compact subset of $\partial O \times \mathbb{R}_t^+$, if we denote $L_{loc}^{a,b} := L_{loc}^a(\mathbb{R}^+;\ L_{loc}^b(\Omega))$, then for all $p,\ q \in [1, +\infty[$ such that $1/p + 1/q = 1/s \leq 1$ and $r = p(1 - 1/q)$ we have

$$\|f\|_{L^r(K;\ (v.n(x))^2 d\sigma_x dv ds)} \leq C_K(1 + \|E\|_{L^1_{loc}(\mathbb{R}^+;\ L^q(\omega))}\|f\|_{L^{\infty,p}_{loc}}$$
$$+C_K\|Q(f)\|^{1/r}_{L^{1,s}_{loc}}\|f\|^{1-1/r}_{L^{\infty,p}_{loc}}.$$

The uniform bound $0 \leq f \leq 1$ and the $L^1$-estimate (verified later on) lead for all $p$ and $p'$ such that $1/p + 1/p' = 1$,

$$\|f\|_{L^{\infty,p}_{loc}} \leq \|f\|^{1/p}_{L^{\infty,1}_{loc}}\|f\|^{1/p'}_{L^{\infty,\infty}_{loc}} \leq \|f_0\|^{1/p}_{L^1},$$

$$\|Q(f)\|_{L^s_{loc}} \leq C(s)\|Q(f)\|_{L^\infty_{loc}} \leq C(\|B\|_{L^1}, s),$$

$$\|\gamma_f\|_{L^r(K;\ (v.n(x))^2 d\sigma_x dv ds)} \leq C_K(1 + \|E\|_{L^{1,q}_{loc}})\|f_0\|^{1/p}_{L^1} + C_{K,r,s}\|f_0\|^{1/p(1-1/r)}_{L^1},$$

and if $r \geq 2$, $L^r_{loc}(K;\ (v.n(x))^2 d\sigma_x dv ds) \hookrightarrow L^{r/2}_{loc}(K;\ |v.n(x)| d\sigma_x dv ds)$, then

$$\begin{cases} \gamma_f \in L^r(K;\ |v.n(x)| d\sigma_x dv ds) \qquad \forall r < +\infty \\ \text{and} \\ \|\gamma_f\|_{L^r(K;\ |v.n(x)| d\sigma_x dv ds)} \leq C(K, r, \|E\|_{L^{1,q}_{loc}}, s, p, q). \end{cases}$$

The $L^1$-estimate is the well-known consequence of the mass conservation property.

**Lemma 12.4.6** *For all $\xi \in C^1_c(\bar{\omega} \times \mathbb{R}^+)$, the charge and current densities $\varrho$ and $j$ satisfy*

$$\int_{\omega\times\mathbb{R}^+} [\varrho(t,x)\partial_t\xi(t,x) + j(t,x)\cdot\nabla_x\xi(t,x)]dxdt + \int_\Omega f_0(x,v)\xi(0,x)dxdv = 0. \tag{12.32}$$

*As a consequence,*

$$\partial_t\varrho + \nabla_x \cdot j = 0, \quad in \quad \mathcal{D}'(\mathbb{R}_+^* \times \omega). \tag{12.33}$$

□

***Proof of Lemma 12.4.6*** The proof is a consequence of the mass conservation property of the collision operator:

$$\int Q(f)dv = 0.$$

## 12.4.2 Energy Estimate

We shall establish a control on the charge and current densities and the kinetic energies depending on the force field $E$.

**Lemma 12.4.7** *Let $E \in L^1_{loc}(\mathbb{R}^+;\ L^{d+2}(\omega))$. The Boltzmann equation has a weak solution satisfying*

$$\|\varrho(t)\|_{L^{\frac{d+2}{d}}(\omega)} \le C_T K(t)^{\frac{d}{d+2}}, \tag{12.34}$$

$$\|j(t)\|_{L^{\frac{d+2}{d+1}}(\omega)} \le C_T K(t)^{\frac{d+1}{d+2}}, \tag{12.35}$$

$$\sup_{t\in[0,T]} \left\{ \|\varrho(t)\|_{L^{\frac{d+2}{d}}(\omega)} + K(t) \right\} \le C_T(1 + \int_0^T \|E(s)\|^{d+2}_{L^{d+2}(\omega)} ds). \tag{12.36}$$

*Moreover,* $\sup_{t\in[0,T]} f(t)$ *belongs to a weakly compact subset of* $L^1(\Omega)$. □

***Proof of Lemma 12.4.7*** The proof is based on truncation idea used by Horst [40]. We detail this idea for the (12.34). We start by writing the density as follows:

$$\varrho(x,t) = \int_{|v|\le R} f dv + \int_{|v|\ge R} f dv.$$

This implies

$$|\varrho(x,t)| \le C R^d \|f\|_{L^\infty} + \frac{1}{R^2} \int_{\mathbb{R}^d} |v|^2 f dv.$$

By taking $R = \left[ \int_{\mathbb{R}^d} |v|^2 f(t) dv / C \|f\|_{L^\infty} \right]^{1/d+2}$, we get

$$\|\varrho(t)\|_{L^{\frac{d+2}{d}}(\omega)} \le C_T K(t)^{\frac{d}{d+2}}.$$

To prove (12.36), we remark that if we multiply the Boltzmann equation by $|v|^2$ and integrate by parts using the properties of the collision operator, we obtain

$$[K(s)]_0^t = \int_0^t \int_\omega j(x,s).E(x,s)dxds.$$

If $E \in L^\infty(\omega)$,

$$\begin{aligned} K(t) &\le K(0) + \int_0^t \|j(s)\|_{L^1(\omega)} \|E(s)\|_{L^\infty(\omega)} ds, \\ &\le K(0) + \int_0^t (\|f_0\|_{L^1} + K(s)) \|E(s)\|_{L^\infty(\omega)} ds \\ &\le C \exp\left(\int_0^t \|E(s)\|_{L^\infty(\omega)} ds\right), \end{aligned}$$

and if $E \in L^{d+2}(\omega)$, using the Hölder's inequality, we infer that

$$K(t) \le K(0) + \int_0^t \|j\|_{L^{\frac{d+2}{d+1}}(\omega)} \|E(s)\|_{L^{d+2}(\omega)} ds.$$

$$K(t) \le C \left\{ 1 + \int_0^t K(s)^{\frac{d+1}{d+2}} \|E(s)\|_{L^{d+2}(\omega)} ds \right\}.$$

Using the Young's inequality ($ab \le a^p/p + b^q/q, \quad a, b \ge 0, 1/p + 1/q = 1$) with $a = K(s)^{\frac{d+1}{d+2}}$ and $b = \|E(s)\|_{L^{d+2}}$, we get

$$K(t) \le C \left( 1 + \frac{d+1}{d+2} \int_0^t K(s) ds + \frac{1}{d+2} \int_0^t \|E(s)\|_{L^{d+2}(\omega)}^{d+2} ds \right).$$

The Gronwall lemma gives

$$\sup_{s \in [0,t]} K(s) \le C \left( 1 + \int_0^t \|E(s)\|_{L^{d+2}(\omega)}^{d+2} ds \right) e^t,$$

which ends the proof of (12.36). □

As a consequence of this lemma, we have $\sup_{t \in [0,T]} f(t)$ belongs to a weak compact subset $L^1(\Omega)$. Indeed,

$$\forall x \in ]0,1[, \quad x|\log x| \le 2\sqrt{x}$$

implies

$$\int_\Omega f|Logf|(t)dxdv = \int_{f<e^{-|v|^2}} f|Logf| + \int_{e^{-|v|^2}\le f\le 1} f|Logf|$$
$$\le \sup_{t\le T}\left\{\int_\Omega [2e^{-|v|^2/2} + |v|^2 f(t)]dxdv\right\} \le C_T. \tag{12.37}$$

Moreover,

$$\sup_{t\le T}\int_\omega\int_{|v|^2>R} f dxdv \le \frac{1}{R}\left\{\sup_{t\le T}\int_\Omega |v|^2 f(t)dxdv\right\} \le \frac{C_T}{R}.$$

With these two inequalities, we verify that $\sup_{t\le T} f(t)$ satisfies the assumptions of Dunford–Pettis theorem [18], which completes the proof of the Lemma. □

## 12.5 Stability Results

Let

$$\mathcal{T} := \partial_t + v\cdot\nabla_x$$

and the set

$$\mathcal{F} = \{f \in L^\infty(\mathbb{R}^+;\ L^1(\Omega)\ /\ 0\le f\le 1,\ K\in L^\infty_{loc},$$
$$\mathcal{T}f \in \textit{weakly compact subset of } L^1\}.$$

We would like to prove that $Q(F^\alpha) := \{Q_0(f) + Q_1(f),\ f\in F^\alpha\}$ belongs to a weakly compact subset of $L^1(\Omega\times]0,T[)$. This property is trivial for the operator $Q_0$. Let us detail for the nonlinear part.

### 12.5.1 $L^1$-weak Precompacity $Q_1(F^\alpha)$

We rewrite $Q_1(f)$ as follows:

$$Q_1(f) = \int_{\mathbb{R}^d\times S^{d-1}} B(v-v_1,w)\big\{(1-f)f'f_1' - f'f_1'f_1$$
$$+ ff_1f' + ff_1f_1' - ff_1\big\}dv_1dw$$
$$= (1-f)L_1(f,f) - L_2(f) + fL_3(f,f) + fL_4(f,f) - fL_5(f),$$

where

$$L_1(f,g)=\int_{\mathbb{R}^d\times S^{d-1}} B(v-v_1,w)f'g_1'dv_1dw,\quad L_2(f)=\int_{\mathbb{R}^d\times S^{d-1}} B(v-v_1,w)f'f_1'f_1dv_1dw,$$

$$L_3(f,g)=\int_{\mathbb{R}^d\times S^{d-1}} B(v-v_1,w)f_1g'dv_1dw,\quad L_4(f,g)=\int_{\mathbb{R}^d\times S^{d-1}} B(v-v_1,w)f_1g_1'dv_1dw,$$

$$L_5(f)=\int_{\mathbb{R}^d\times S^{d-1}} B(v-v_1,w)f_1dv_1dw=(A*_{v_1}f)(v),$$

and

$$A(z)=\int_{S^{d-1}} B(z,w)dw.$$

To simplify the notations, we denote by $a:=\|A\|_{L^1(\mathbb{R}^d)}$, and we define

$$\begin{cases} F_i^\alpha=\{L_i(f,f),\ f\in F^\alpha\}, & i\in\{1,3,4,5\}\\ F_2^\alpha=\{L_2(f),\ f\in F^\alpha\}. \end{cases} \tag{12.38}$$

The property of stability of $Q_1$ is given by:

**Theorem 12.5.6** *The set $\{Q_1(f),\quad f\in F^\alpha\}$ is weakly relatively compact in $L^1(]0,T[\times\Omega)$.*

The proof of this theorem consists in proving that the sets $F_i^\alpha$ satisfy the assumptions of the Dunford–Pettis theorem [18]. This will be the subject of the following lemmata. We begin by $L_5$.

**Lemma 12.5.8** *The set $\{fL_5(f),\quad f\in F^\alpha\}$ is weakly relatively compact in $L^1(]0,T[\times\Omega)$.* □

***Proof of Lemma** 12.5.8* Let $f\in F^\alpha$. We have

$$\begin{cases} 0\le L_5(f)\le a,\\ \|L_5(f)(t)\|_{L^1(\Omega)}=a\|f(t)\|_{L^1(\Omega)}. \end{cases} \tag{12.39}$$

The set $F_5^\alpha$ is bounded in $L^\infty(0,T;\ L^1\cap L^\infty(\Omega))$. We define the function

$$\varphi(t):=t(Logt)^+=t\sup(0,Logt). \tag{12.40}$$

The function $\varphi$ is increasing and satisfies

$$\varphi(t)\ge 0,\qquad \lim_{t\to+\infty}\frac{\varphi(t)}{t}=+\infty,$$

$$\forall t > 0, \quad \Phi(t) \leq a\varphi\left(\frac{t}{a}\right) + t|Loga|,$$

and

$$\int_\Omega \varphi(L_5(f)) \leq a \int_\Omega \Phi(L_5(f)/a) + \|L_5(f)\|_{L^1(\Omega)}|Loga|.$$

Using the convexity of $\varphi$ and the Jensen inequality with $d\mu = \dfrac{B(v - v_1, w)dwdv_1}{a}$, we get

$$\int_\Omega \varphi(L_5(f)) \leq \int_\Omega L_5(\varphi(f)) + a|Loga|\|f\|_{L^1(\Omega)}.$$

We deduce, thanks to (12.37) and (12.39), that

$$\forall t \leq T, \quad \int_\Omega \varphi(L_5(f))(t) \leq a\{\|\varphi(f)\|_{L^1} + |Loga|\|f\|_{L^1}\} \leq C_T, \tag{12.41}$$

leading to: $\varphi(L_5(f))$ is bounded in $L^\infty(0, T;\ L^1(\Omega))$, and to conclude that $L_5(f)$ is in a weakly compact subset, we shall prove

$$\lim_{R\to+\infty} \left\{ \sup_{0\leq t\leq T} \int_\omega \int_{|v|\geq R} L_5(f)(t)dxdv \right\} = 0. \tag{12.42}$$

Indeed,

$$\begin{aligned}
\int_\omega \int_{|v|\geq R} L_5(f)(t) &\leq \int_\omega dx \int_{|v|\geq R} dv \int_{|v_1|\geq R/2} A(v - v_1)f(v_1)dv_1 \\
&\quad + \int_\omega dx \int_{|v|\geq R} dv \int_{|v_1|<R/2} A(v - v_1)f(v_1)dv_1 \\
&\leq \frac{4aK(t)}{R^2} + \int_\Omega \int_{|v-v_1|\geq R/2} A(v - v_1)f(v_1)dxdvdv_1 \\
&\leq \frac{4aK(t)}{R^2} + \|f(t)\|_{L^1} \int_{|z|\geq R/2} A(z)dz,
\end{aligned}$$

which implies (12.42) due to the fact that $A \in L^1(\mathbb{R}^d)$.

We proceed in the same manner for $fL_5(f)$ using the uniform bound $f \in [0, 1]$. We have

$$0 \leq fL_5(f) \leq L_5(f) \leq a. \tag{12.43}$$

Lemma 12.4.7 implies

$$\int_{\Omega} [\varphi(fL_5(f)) + (1+|v|^2)fL_5(f)]dxdv \leq C_T\left(1+\int_0^T \|E(s)\|^{d+2}_{L^{d+2}(\omega)}ds\right), \tag{12.44}$$

which leads to

$$\lim_{R\to+\infty}\left\{\sup_{0\leq t\leq T}\int_{\omega}\int_{|v|\geq R} fL_5(f)dxdv\right\} = 0. \tag{12.45}$$

Moreover, the set $\{fL_5(f),\ f\in F^{\alpha}\}$ is bounded in $L^{\infty}(0,T;\ L^1(\Omega))$ and weakly relatively compact in $L^1(]0,T[\times\Omega)$ faible. □

**Lemma 12.5.9** *The set* $\{fL_1(f,f),\ f\in F^{\alpha}\}$ *belongs to a weakly compact subset of* $L^1(\Omega\times]0,T[)$. □

***Proof of Lemma** 12.5.9* Let $f\in F^{\alpha}$; we have $\forall t\in]0,T[$,

$$0\leq L_1(f,f)(t)\leq a,$$

$$\int_{\Omega} L_1(f,f)(t)dxdv \leq \int_{\Omega} A(v-v_1)f_1dxdvdv_1 \leq a\|f(t)\|_{L^1(\Omega)},$$

$$\int_{\Omega}|v|^2L_1(f,f)dxdv = \int_{\Omega}\int_{\mathbb{R}^d\times S^{d-1}} B(v-v_1,w)|v|^2f'f_1'dwdv_1dxdv.$$

Using the new coordinates: $(v,v_1)\mapsto(v',v_1')$, we get

$$\begin{aligned}\int_{\Omega}|v|^2L_1(f,f)dxdv &= \int_{\Omega}\int_{\mathbb{R}^d\times S^{d-1}} B(v-v_1,w)|v'|^2ff_1dwdv_1dxdv\\ &\leq \int_{\Omega}\int_{\mathbb{R}^d} A(v-v_1)(|v|^2+|v_1|^2)ff_1dv_1dxdv\\ &\leq 2\int_{\Omega}\int_{\mathbb{R}^d} A(v-v_1)|v|^2ff_1dv_1dxdv \leq 2aK(t).\end{aligned}$$

The properties of the function $\varphi$ given in (12.40) lead to

$$\int_{\Omega}\varphi(L_1(f,f)) \leq a\int_{\Omega}\varphi\left(\frac{L_1(f,f)}{a}\right) + \|L_1(f,f)\|_{L^1(\Omega)}|Loga|. \tag{12.46}$$

The Jensen inequality (with $d\mu = \frac{B(v-v_1,w)dwdv_1}{a}$) gives

$$\varphi\left(\frac{L_1(f,f)}{a}\right) \leq \int_{\mathbb{R}^d\times S^{d-1}}\varphi(f'f_1')d\mu. \tag{12.47}$$

Using again the coordinates ( $(v, v_1) \mapsto (v', v_1')$, we obtain

$$a\int_{\Omega}\varphi\left(\frac{L_1(f,f)}{a}\right) \leq a\int_{\Omega}\int_{\mathbb{R}^d\times S^{d-1}}\varphi(ff_1)d\mu dvdx$$

$$\leq \int_{\Omega}\int_{\mathbb{R}^d}A(v-v_1)\varphi(ff_1)dv_1dvdx.$$

The monotony of $\varphi$ implies that

$$a\int_{\Omega}\varphi\left(\frac{L_1(f,f)}{a}\right) \leq \int_{\Omega}\int_{\mathbb{R}^d}A(v-v_1)\varphi(f)dv_1dvdx \leq a\|\varphi(f)\|_{L^1(\Omega)},$$

and (12.46) becomes

$$\int_{\Omega}\varphi(L_1(f,f)) \leq a\{\|\varphi(f)\|_{L^1} + |Loga|\|f\|_{L^1}\} \leq C_T(1+\int_0^T \|E(t)\|^{d+2}_{L^{d+2}(\omega)}dt)$$

and

$$\sup_{t\in[0,T]}\int_{\Omega}\{(1+|v|^2)L_1(f,f)+\varphi(L_1(f,f))\}(t) \leq C_T(1+\int_0^T\|E(t)\|^{d+2}_{L^{d+2}(\omega)}dt). \tag{12.48}$$

This implies that the set $\{L_1(f,f),\ f \in F^{\alpha}\}$ is a weakly relatively compact in $L^1(\Omega\times]0,T[)$.

Also, this inequality is satisfied by f $fL_1(f,f)$, because $f \in [0,1]$ and $\varphi$ is increasing, then the set $\{fL_1(f,f),\ f \in F^{\alpha}\}$ is a weakly relatively compact of $L^1(\Omega\times]0,T[)$, and it satisfies

$$\sup_{t\in[0,T]}\int_{\Omega}\{(1+|v|^2)fL_1(f,f)+\varphi(fL_1(f,f))\}(t) \leq C_T(1+\int_0^T\|E(t)\|^{d+2}_{L^{d+2}(\omega)}dt),$$

where the constant $C_T$ depends only on the time $T$. □

***Proof of Theorem 12.5.6*** *Conclusion*: The sets $F_1^{\alpha}$ and $F_5^{\alpha}$ are relatively compact in $L^1(\Omega\times]0,T[)$. To extend this property to $F_2^{\alpha}$, $F_3^{\alpha}$, and $F_4^{\alpha}$, we remark that, for all $f \in F^{\alpha}$,

$$0 \leq L_2(f) \leq L_5(f), \quad 0 \leq L_i(f,f) \leq L_5(f), \quad i = 3, 4.$$

This implies that these quantities are bounded in $L^{\infty}(0,T;\ L^1 \cap L^{\infty}(\Omega))$. The function $\varphi$ given in (12.40) is increasing, and the inequalities (12.41) and (12.45) are satisfied by $L_2(f)$, $L_3(f,f)$, and $L_4(f,f)$. Then, the weak compacity of $F_2^{\alpha}$, $F_3^{\alpha}$, and $F_4^{\alpha}$ is in $L^1(\Omega\times]0,T[)$). The compacity of $fL_3(f,f)$ and $fL_4(f,f)$ is immediate because $f \in [0,1]$ and $\varphi$ is increasing. □

It remains to prove that $Q_1(F^\alpha)$ is closed for topology $\sigma(L^1, L^\infty)$ ; it is equivalent to proving that if $(f_n)_n$ is a sequence of $F^\alpha$ weakly converge towards $f$, then for all $\psi \in \mathcal{D}(\Omega \times \mathbb{R}^+)$

$$\lim_{n\to\infty} \int_{\Omega\times\mathbb{R}^+} Q_1(f_n)\psi dxdvdt = \int_{\Omega\times\mathbb{R}^+} Q_1(f)\psi dxdvdt. \tag{12.49}$$

To do this, we need a velocity-averaging argument.

### 12.5.2 Velocity-Averaging Lemma

We consider the operator $\mathcal{T} = \partial_t + v \cdot \nabla_x$.

**Theorem 12.5.7 ([28])** *Let $f \in C([0,T], \mathcal{D}'(\mathbb{R}^{2d}))$, $f \in L^p_{loc}([0,T] \times \mathbb{R}^{2d})$, and $\mathcal{T}(f) \in W^{\alpha,p}(0,T;\ W^{\alpha,p}_{loc}(\mathbb{R}^d_x;\ W^{\beta,p}_{loc}(\mathbb{R}^d_v)))$, with $p \in ]1,\infty[$, $\alpha > -1$ and $\beta \in \mathbb{R}$. Assume that $f(0) \in L^p_{loc}(\mathbb{R}^{2d})$. Then, there exists $s(p,\alpha,\beta,d) > 0$ such that, for all $\psi \in \mathcal{D}(\mathbb{R}^d_v)$,*

$$M_\psi(f) := \int_{\mathbb{R}^d} f\psi dv \in W^{s,p}(0,T;\ W^{s,p}_{loc}(\mathbb{R}^d)).$$

*Therefore, for $R > 0$, there exists $R' > 0$ such that*

$$\begin{aligned} \|M_\psi(f)\|_{W^{s,p}(]0,T[\times B_R)} \le C\{\psi, \|f\|_{L^p([0,T]\times B_{R'}\times B_{R'})},\\ \|f(0)\|_{L^p(B_{R'}\times B_{R'})}, \|\mathcal{T}f\|_{W^{\alpha,p}([0,T]\times B_{R'};\ W^{\beta,p}(B_{R'})}\}. \end{aligned} \tag{12.50}$$

The idea of the proof of this theorem in the case ($\alpha = 0$, $\beta = -1$, and $p = 2$) is given in [17]. We can prove a compacity $L^2(\omega\times]0,T[)$ of the sequence $\int_{\mathbb{R}^d} f^n\psi(t,x,v)dv$, where $f^n$ is a weak solution of

$$\partial_t f^n + v \cdot \nabla_x f^n = Q(f^n) - E^n \cdot \nabla_v f^n, \quad (t,x,v) \in \mathbb{R}^*_+ \times \omega. \tag{12.51}$$

We assume that

$$\begin{cases} \|E^n\|_{L^2([0,T]\times\omega)} \le C_T. \\ Q(f^n) \in L^\infty(\mathbb{R}^+;\ L^1 \cap L^\infty(\Omega)). \end{cases} \tag{12.52}$$

The compactness of $f_n$ is described by:

**Lemma 12.5.10** *Let $\psi \in \mathcal{D}(\bar{\Omega} \times [0,T[)$, and $f^n$ is a weak solution of (12.51). We assume that (12.52) is satisfied; then, $M^n_\psi = \int_{\mathbb{R}^d} f^n\psi(x,v,t)dv$ is relatively compact in $L^2(\omega\times]0,T[)$.* □

***Proof of Lemma 12.5.10*** This lemma is a consequence of the result of regularity on the evolutionary equations. The first part of the proof is given in [17] for the case of a function $\psi := \psi(v)$. We resume the same analysis for a function $\psi := \psi(x, v, t)$, and we satisfy that the hypothesis (12.52) is sufficient to deduce (12.53).

Let $g^n = f^n \psi$. Then $g^n$ is a weak solution of

$$\begin{cases} \mathcal{T}(g^n) = S^n = \mathcal{Q}(f^n)\psi + \bar{E^n}.\nabla_v \psi f^n - f^n \mathcal{T}(\psi) - \nabla_v.(\psi f^n \bar{E^n}) \\ \qquad := S_1^n + \nabla_v.S_2^n, \quad in \quad \mathcal{D}'(]0, T[\times\omega), \\ g^n(0) = f_0 \psi. \end{cases}$$

The term $S^n \in L^2(0, T;\ H^{-1}(\Omega))$. Under the assumption of Theorem 12.5.7 with $\alpha = 0, \beta = -1$, and $p = 2$.

We denote $F_x^\alpha(g^n)$ the Fourier transform of $x$ of $g^n$; then

$$F_x^\alpha(g^n)(t, \xi, v) = F_x^\alpha(f_0\psi)(\xi, v)e^{-i(v\cdot\xi)t} + \int_0^t F_x^\alpha(S_1^n)(s, \xi, v)e^{-i(v\cdot\xi)(t-s)}ds$$
$$+ \int_0^t \nabla_v \cdot [F_x^\alpha(S_2^n)](s, \xi, v)e^{-i(v.\cdot\xi)(t-s)}ds.$$

We integrate this equation with respect to $v$, we denote by $F_{x,v}^\alpha$ the transformation with respect to $(x, v)$ and $M_\psi^n := \int_{\mathbb{R}^d} g^n dv$, and we have

$$F_x^\alpha(M_\psi^n)(t, \xi) = F_{x,v}^\alpha(f_0\psi)(\xi, t\xi) + \int_0^t F_{x,v}^\alpha(S^n)(\xi, (t-s)\xi, s)ds.$$

By using the Holder inequality,

$$\int_0^T |F_x^\alpha(M_\psi^n)(t, \xi)|^2 dt \leq 2\int_0^T |F_{x,v}^\alpha(f_0\psi)(\xi, t\xi)|^2 dt$$
$$+2T \int_0^T\int_0^T |F_{x,v}^\alpha(S^n)(\xi, t\xi, s)|^2 dt ds,$$

and with the change $(t \mapsto t|\xi|)$, the previous inequality becomes

$$\int_0^T |F_x^\alpha(M_\psi^n)(t, \xi)|^2 dt$$
$$\leq 2\int_0^T |F_{x,v}^\alpha(f_0\psi)(\xi, t\frac{\xi}{|\xi|})|^2 \frac{dt}{|\xi|} + 2T\int_0^T\int_0^T |F_{x,v}^\alpha(S^n)\left(\xi, t\frac{\xi}{|\xi|}, s\right)|^2 \frac{dt}{|\xi|}ds.$$

Then

$$\int_0^T \int_{\mathbb{R}^d} |\xi||F_x^\alpha(M_\psi^n)(\xi,t)|^2 \le 2\int_{\mathbb{R}^{2d}} |(I-\Delta_\eta)^{d/2} F_{x,v}^\alpha(f_0\psi)(\xi,\eta)|^2 d\eta d\xi$$

$$+2T\int_0^T \int_{\mathbb{R}^{2d}} |(I-\Delta_\eta)^{d/2} F_{x,v}^\alpha(S^n)(\xi,\eta,s)|^2 d\eta d\xi ds$$

$$\le 2(\|f_0\psi\|_{L^2}^2 + T\|S^n\|_{L^2(0,T;\ L^2(\omega;\ H^{-1}(\mathbb{R}^d)))})\int |\psi(v)|^2(1+|v|^2)^d dv$$

$$\le C_\psi(1+\|E^n\|_{L^2(]0,T[\times\omega)}).$$

So,

$$\|M_\psi^n\|_{L^2(0,T;\ H^{1/2}(\omega))} \le C_T(1+\|E^n\|_{L^2(\omega)}) \le C_T. \tag{12.53}$$

Otherwise,

$$\frac{d}{dt}M_\psi^n = \frac{d}{dt}\left(\int_{\mathbb{R}^d} f^n\psi dv\right)$$

$$= \int_{\mathbb{R}^d}(Q(f^n) - E^n\nabla_v f^n - v.\nabla_x f^n)\psi dv + \int_{\mathbb{R}^d} f^n\partial_t\psi dv$$

$$= \int_{\mathbb{R}^d} Q(f^n)\psi dv + E^n.\int_{\mathbb{R}^d} f^n(\nabla_v\psi)dv$$

$$+\int_{\mathbb{R}^d} f^n(v.\nabla_x\psi)dv - \nabla_x.\int_{\mathbb{R}^d} f^n\psi v dv + \int_{\mathbb{R}^d} f^n\partial_t\psi dv.$$

Taking into account the uniform bound $f^n \in [0,1]$ and the estimate (12.53), we obtain

$$\left\|\frac{d}{dt}M_\psi^n\right\|_{L^2(0,T;\ H^{-1/2}(\omega))} \le C_T, \tag{12.54}$$

where $C_T$ depends only on $T$, $\|E^n\|_{L^2(0,T;\ L^2(\omega))}$, and $\psi$.

Estimates, (12.53) and (12.54), imply that $(M_\psi^n)_n$ is relatively compact in $L^2(0,T;\ L^2(\omega))$.

*Remark 12.5.1* The regularity of the function $\psi$ introduced in the previous theorem is not optimal to get the compactness of $M_\psi^n$ in $L^p(\omega\times]0,T[)$, $(p < +\infty)$. Indeed, by the same proof, we use the transformation into $(t,x)$, and we can prove that $M_\psi^n \in H^{1/4}(\omega\times]0,T[)$ for all $\psi \in C_c^1(\bar{\Omega}\times[0,T[)$ and that $f^n$ satisfy (12.51). Therefore,

$$\|M_\psi^n\|_{H^{1/4}(\omega\times]0,T[)} \le C_T(1+\|E^n\|_{L^2(0,T;\ L^2(\omega))}) \le C_T,$$

where the constant $C_T$ depends only on $T$ and $\psi$.

We also remark that if in the previous proof $\psi := \psi(v)$ and it has a compact support, then the $L^\infty(\mathbb{R}^d)$ regularity is enough to obtain the result of the lemma. Indeed, several other forms of this lemma are given in [28], where the function $\psi$ is chosen in different spaces. We quote other lemma who writes the average compacity in $L^1$. □

**Lemma 12.5.11 ([24])** *Let $g_n$ and $G_n$ be two sequences weakly compact in $L^1_{loc}(\mathbb{R}^+;\ L^1(\mathbb{R}^{2d}))$ and satisfying*

$$\mathcal{T} g_n = G_n \quad in \quad \mathcal{D}'(\mathbb{R}^+_* \times \mathbb{R}^{2d}).$$

*We assume that $supp(g_n) \subset [\alpha, T-\alpha] \times \mathbb{R}^d_x \times B_R$, $\alpha > 0$. Then, for all function $\psi \in L^\infty(\mathbb{R}^+_*;\ \mathbb{R}^{2d})$, we have $\int_{\mathbb{R}^d} \psi g_n dv$ belongs to a compact of $L^1(]0,T[\times\mathbb{R}^{2d})$. Consequently:*

*(i) If $K$ a compact of $[0,T] \times \mathbb{R}^{2d}$, $g_n$ and $G_n$ belong to a weakly compact of L $L^1(K)$, for all $\psi \in L^\infty(]0,T[\times\mathbb{R}^{2d})$ with compact support, and the sequence $\int_{\mathbb{R}^d} g_n\psi dv$ is in a compact of $L^1_{loc}(]0,T[\times\mathbb{R}^d)$.*

*(ii) Therefore, if the sequence $g_n$ is in a weak compact of $L^1(]0,T[\times\mathbb{R}^{2d})$, then for all $\psi \in L^\infty(]0,T[\times\mathbb{R}^{2d})$ with compact support, the sequence $\int_{\mathbb{R}^d} g_n\psi dv$ is in a compact of $L^1(]0,T[\times\mathbb{R}^d)$.* □

These lemmas are used in the following subsection.

### 12.5.3 $L^1$-compactness of $L_5(f_n)$

**Lemma 12.5.12** *Let $(f_n)$ be a sequence of $F^\alpha$, such that $f_n \rightharpoonup f$ in $L^p_{loc}$ weak. Then*

$$L_5(f_n) \longrightarrow L_5(f) \quad L^1(0,T;\ L^1_{loc}(\Omega))\ strong$$

*and*

$$f_n L_5(f_n) \rightharpoonup f L_5(f) \quad L^1(\mathbb{R}^{2d}\times]0,T[)\ weak.$$

□

***Proof*** The sequence $(L_5(f_n))_n$ is bounded in $L^\infty(0,T;\ L^1(\Omega))$, and it is weakly relatively compact in $L^1(\Omega\times]0,T[)$. Let $\theta_R \in \mathcal{D}(B_R)$, a localizing function; $A_\varepsilon \in \mathcal{D}(\mathbb{R}^d)$ such that $\|A - A_\varepsilon\|_{L^1} \le \varepsilon$. Then,

$$\begin{aligned} L_5(f_n - f)(v) &= \int_{\mathbb{R}^d} (f_n - f)(v_1)(A - A_\varepsilon)(v - v_1)dv_1 \\ &\quad + \int_{\mathbb{R}^d} (f_n - f)(v_1)A_\varepsilon(v - v_1)dv_1 \\ &= [(f_n - f) *_v (A - A_\varepsilon)](v) + [(f_n - f) *_v A_\varepsilon](v). \end{aligned}$$

We have

$$\|(f_n - f) *_v (A - A_\varepsilon)\|_{L^1(\Omega\times]0,T[)} \leq \|f_n - f\|_{L^1}\|A - A_\varepsilon\|_{L^1(\mathbb{R}^d)} \leq C_T\varepsilon$$

and

$$\int_{\mathbb{R}^d} (f_n - f)(v_1)A_\varepsilon(v - v_1)dv_1\theta_R(v) = \int_{\mathbb{R}^d} (f_n - f)(v_1)\psi_R(v, v_1)dv_1$$

$$\in C^\infty(\mathbb{R}^d_v;\ H^{1/4}(\omega \times [0, T]))$$

or $\int_{\mathbb{R}^d}(f_n - f)(v_1)\psi_R(v, v_1)dv_1$ is relatively compact in $L^2(\Omega\times]0, T[)$ because $\psi_R \in \mathcal{D}(\mathbb{R}^{2d})$. This implies that for all $\varepsilon > 0$, we have

$$\lim_{n\to+\infty} \int_{\Omega\times]0,T[} |\int_{\mathbb{R}^d} (f_n - f)(v_1)A_\varepsilon(v - v_1)dv_1\theta_R(v)|dxdvdt = 0.$$

By passing to the limit on $\varepsilon$, we obtain $L_5(f_n)$ converge towards $L_5(f)$ in $L^1(0, T;\ L^1_{loc}(\Omega))$.

By using this lemma, we prove the stability of $f L_5(f)$. By using the parity of $A$, we write

$$\int_{\mathbb{R}^{2d}\times]0,T[} f_n L_5(f_n)\psi dxdvdt = \int_{supp(\psi)} f_n(v)g_n(v)dxdvdt,$$

where

$$g_n(v) = \int_{\mathbb{R}^d} (A - A_\varepsilon)(v - v_1) f_n(v_1)\psi(v_1)dv_1 + \int_{\mathbb{R}^d} A_\varepsilon(v - v_1) f_n(v_1)\psi(v_1)dv_1.$$

The first term converges immediately towards ($\|A - A_\varepsilon\|_{L^1} \leq \varepsilon$). The second term is bounded in $L^\infty(\Omega\times]0, T[)$, and by using the previous lemma, it is in a compact of $L^1_{loc}(\Omega \times [0, T]))$; its norm depends only on $\|E\|_{L^2(\omega\times[0,T])}$, $\|A\|_{L^1}$, and $\psi$. This is enough to deduce the weak convergence of $f_n L_5(f_n)$ to $f L_5(f)$. □

After that, we study $L_1$, $L_2$, $L_3$, and $L_4$. We define the following operators $T$ and $\tilde{T}$:

$$T\varphi(z) = \int_{S^{d-1}} B(z, w)\varphi(z - (z.w)w)dw,$$

$$\tilde{T}\varphi(z) = \int_{S^{d-1}} B(z, w)\varphi((z.w)w)dw.$$

We begin by analysing the operator $L_1$; using the averaging lemma, we prove the result of compacity of the sequence $L_1(f_n, f_n)$ for all $f_n \in F^\alpha$ (Lemma 12.5.13).

Then, we prove the $L^1$- compacity of $L_1(f_n, f_n)$, which will be enough to obtain the compactness of $L_3$ and $L_4$ (Lemma 12.5.15) and the weak compactness of $L_2$.

### 12.5.4 Compactness of $L_1(f_n, f_n)$

**Lemma 12.5.13** *Let $(f_n)_n$ be a sequence of functions of $F^\alpha$, and $f$ is weak limit. Then, for all function $\psi \in \mathcal{D}(\bar{\Omega} \times [0, T])$,*

$$\int_{\mathbb{R}^d} L_1(f_n, f_n)\psi dv \longrightarrow \int_{\mathbb{R}^d} L_1(f, f)\psi dv \quad in \quad L^1(\omega \times [0, T]). \tag{12.55}$$

***Proof of Lemma 12.5.13*** Let $supp(\psi) \subset K \times \bar{\omega} \times [0, T]$,

$$\int_{\mathbb{R}^d} L_1(f, f)\psi dv = \int_{K \times \mathbb{R}^d_{v_1}} f(v) f(v_1) a(v, v_1) dv dv_1,$$

where

$$a(x, v, v_1, t) = \int_{S^{d-1}} B(v - v_1, w)\psi(v_1 + (v - v_1.w)w)dw.$$

We assume that the cross-section is regular ($C^\infty$), and the function $a \in C^\infty(supp(\psi) \times \mathbb{R}^d_{v_1})$. We define

$$\mathcal{A}(x, v_1, t) = \int_K f(v)a(x, v, v_1, t)dv, \quad \mathcal{A}_n(x, v_1, t) = \int_K f_n(v)a(x, v, v_1, t)dv.$$

Under the assumption (H2), the function $\mathcal{A} \in L^\infty_{loc}(\mathbb{R}^+;\ L^1 \cap L^\infty(\Omega))$, and by using the previous lemma, the sequence $\mathcal{A}_n$ belongs to $C^\infty(\mathbb{R}^d_{v_1};\ H^{1/4}(\omega \times ]0, T[)$ bounded, and it depends on $\|E\|_{L^2(\omega \times [0,T])}$,

$$\begin{aligned}\int_K (L_1(f_n, f_n) - L_1(f, f))\psi dv = &\int_{\mathbb{R}^d} (\mathcal{A}_n - \mathcal{A})(x, v_1, t) f_n(v_1) dv_1 \\ &+ \int_{\mathbb{R}^d} (f_n - f)(v_1)\mathcal{A}(x, v_1, t)dv_1.\end{aligned} \tag{12.56}$$

Let $R > 0$, $B_R = B(0, R)$

$$\begin{aligned}\left|\int_{\mathbb{R}^d} (\mathcal{A}_n - \mathcal{A})(x, v_1, t) f_n(v_1) dv_1\right| \le &\|\mathcal{A}_n - \mathcal{A}\|_{L^1(B_R)} \\ &+2 \int_{\{|v_1| > R\} \times K} |a(x, v, v_1, t)| dv dv_1;\end{aligned}$$

then

$$\lim_{n\to\infty}\left\|\int_{\mathbb{R}^d}(\mathcal{A}_n-\mathcal{A})(v_1)f_n(v_1)dv_1\right\|_{L^1(\omega\times[0,T])}$$

$$\leq \lim_{n\to\infty}\|\mathcal{A}_n-\mathcal{A}\|_{L^1(B_R\times\omega\times[0,T])}+2\int_{\{|v_1|>R\}\times supp(\psi)}|a(x,v,v_1,t)|dvdv_1.$$

For fixed $R$, the sequence $\mathcal{A}_n$ is regular and uniformly bounded; then it converges strongly towards $\mathcal{A}$ in $L^1(B_R\times\omega\times[0,T])$. Otherwise,

$$\lim_{R\to\infty}\int_{|v_1|>R}\int_{supp(\psi)}|a(x,v,v_1,t)|dvdv_1\leq C_{K,\psi}\lim_{R\to\infty}\int_{\{|z|\geq R\}}A(z,w)dw=0. \tag{12.57}$$

For the last term (12.56), we take $\theta_R:=\theta_R(v)\in\mathcal{D}(B_R)$

$$\int_{\mathbb{R}^d}(f_n-f)(v_1)\mathcal{A}(x,v_1,t)dv_1=\int_{\mathbb{R}^d}(f_n-f)(v_1)\mathcal{A}(x,v_1,t)\theta_R(v_1)dv_1$$
$$+\int_{\mathbb{R}^d}(f_n-f)(v_1)\mathcal{A}(x,v_1,t)(1-\theta_R(v_1))dv_1.$$

The convergence in $L^1(\omega\times]0,T[)$ of $\int_{\mathbb{R}^d}(f_n-f)(v_1)\mathcal{A}(x,v_1,t)\theta_R(v_1)dv_1$ is a consequence of the averaging lemma (theorem 12.5.7), and the last term $\int_{\mathbb{R}^d}(f_n-f)(v_1)\mathcal{A}(x,v_1,t)(1-\theta_R(v_1))dv_1$ is the same as (12.57). □

### 12.5.5 $L^1$-Compactness of $L_1(f_n,f_n)$

**Theorem 12.5.8** *Let $(f_n)$ be a sequence in $\mathcal{F}$ and $f$ its weak limit. Then*

$$L_1(f_n,f_n)\longrightarrow L_1(f,f)\quad in\quad L^1(\Omega\times]0,T[). \tag{12.58}$$

□

The proof of this theorem is divided into three steps. In the first step, we regularize the sequence $f_n$ and the cross-section $B$. These two regularizations allow us to prove the result of the theorem by using the compactness lemmas.

**First Step: Regularization of the Sequence**
Let $M>0$ be a fixed real, $\varphi\in\mathcal{D}(\mathbb{R}^d)$, $0\leq\varphi\leq 1$, $\varphi\equiv 1$ in $B(0,1)$ and $\varphi_M=\varphi(\frac{\cdot}{M})$.

Let

$$f_n^M=f_n\varphi_M(v).$$

This step consists in showing that

$$L_1(f_n^M, f_n^M) \longrightarrow L_1(f^M, f^M) \quad in \quad L^1_{loc}(\Omega \times ]0, T[). \tag{12.59}$$

Indeed,

$$\begin{aligned} L_1(f_n, f_n) - L_1(f, f) &= (L_1(f_n, f_n) - L_1(f_n^M, f_n^M)) \\ &+ (L_1(f_n^M, f_n^M) - L_1(f^M, f^M)) + (L_1(f^M, f^M) - L_1(f, f)). \end{aligned} \tag{12.60}$$

The first term of the right-hand side of (12.60) satisfies

$$\begin{aligned} &\int_0^T \int_\Omega |L_1(f_n, f_n) - L_1(f_n^M, f_n^M)| dx dv dt \\ &= \int_0^T dt \int_\omega dx \int_{\mathbb{R}^{2d} \times S^{d-1}} dv dv_1 dw B(v - v_1, w) f_n(v') f_n(v_1')(1 - \varphi_M(v')\varphi_M(v_1')) \\ &\leq \int_0^T dt \int_\omega dx \int_{\mathbb{R}^{2d} \times S^{d-1}} dv dv_1 dw B(v - v_1, w) f_n(v) f_n(v_1)(1 - \varphi_M(v)\varphi_M(v_1)) \\ &\leq \frac{1}{M^2} \int_0^T dt \int_\omega dx \int_{\mathbb{R}^{2d}} dv dv_1 |v||v_1| A(v - v_1) f_n(v) f_n(v_1) \\ &\leq \frac{4}{M^2} \int_0^T dt \int_\omega dx \int_{\mathbb{R}^{2d}} dv dv_1 |v|^2 A(v - v_1) f_n(v) f_n(v_1) \leq \sup_{t \in [0,T]} \frac{4aK(t)}{M^2}. \end{aligned}$$

The third term is the remainder of a convergent integral:

$$\begin{aligned} &\int_0^T \int_\Omega |L_1(f^M, f^M) - L_1(f, f)| dx dv dt \\ &= \int_0^T dt \int_\omega dx \int_{\mathbb{R}^{2d} \times S^{d-1}} dw dv dv_1 B(v - v_1, w) f f_1 (1 - \varphi_M(v)\varphi_M(v_1)) \\ &\leq a \int_0^T \int_\omega \int_{|v|>M} f dx dv dt = O(1/M). \end{aligned}$$

**Corollary 1**

$$\int_{\mathbb{R}^d} f_n^M \psi dv \longrightarrow \int_{\mathbb{R}^d} f^M \psi dv \quad in \quad L^p(\omega \times ]0, T[)$$

*for all* $\psi \in L^\infty(\mathbb{R}^d_v)$, $p \in [1, \infty[$.

**Second Step: Regularization of the Cross-Section**
Now, we remark that we need only $B$ values for $v - v_1$ such that $|v'|,\ |v_1'| \leq CM$ ($C$ is a constant ). Then, we assume that $B = \varphi(|z|, |(z.w)|/|z|)$, where $\varphi(r,t)$is set to $[0, \infty[\times[0,1]$ and $\varphi(r,t) = 0$ for $r \gg 1$, uniformly for $t \in [0,1]$. Let $\varphi_\varepsilon \in \mathcal{D}^+(]0, \infty[\times]0,1[)$ and $B_\varepsilon = \varphi_\varepsilon(|z|, |(z.w)|/|z|)$. We have

$$B_\varepsilon \to B \quad in \quad L^1(\mathbb{R}^d \times S^{d-1}).$$

Let

$$L_1^\varepsilon(f,g) = \int_{\mathbb{R}^d \times S^{d-1}} B_\varepsilon(v - v_1, w) f' f_1' dw dv_1;$$

then (12.59) à :

$$L_1^\varepsilon(f_n^M, f_n^M) \longrightarrow L_1^\varepsilon(f^M, f^M) \quad in \quad L^1(\Omega\times]0,T[). \tag{12.61}$$

Indeed,

$$\begin{aligned} L_1(f_n^M, f_n^M) - L_1(f^M, f^M) &= (L_1(f_n^M, f_n^M) - L_1^\varepsilon(f_n^M, f_n^M)) \\ &+ (L_1^\varepsilon(f_n^M, f_n^M) - L_1^\varepsilon(f^M, f^M)) + (L_1^\varepsilon(f^M, f^M) - L_1(f^M, f^M)). \end{aligned}$$

The functions $f_n^M$ and $f^M$ have a compact support subset of $\bar{\omega} \times \bar{B}(0, M)$. Then, it is clear that $L_1(f_n^M, f_n^M) - L_1^\varepsilon(f_n^M, f_n^M)$ and $L_1^\varepsilon(f^M, f^M) - L_1(f^M, f^M)$ are bounded in $L^1(\Omega\times]0,T[)$ par $C_M \|B_\varepsilon - B\|_{L^1(\mathbb{R}^d \times S^{d-1})}$. This implies the equivalence between (12.59) and (12.61).

To simplify the notations, we omit the parameters $\varepsilon$ and $M$, and we assume that the sequence $(f_n) \in F^\alpha$ with a support subset of $\bar{\omega} \times \bar{B}(0, M)$) and satisfies the previous corollary. We also assume that the cross-section $B \in \mathcal{D}(\mathbb{R}^d \times S^{d-1})$.

**Third Step: Regularization in Velocity**
We regularize $L_1(f_n, f_n)$ by $\varrho_\delta = \dfrac{1}{\delta^d}\varrho(\dfrac{\cdot}{\delta})$, where $\varrho \in \mathcal{D}(\mathbb{R}^d)$, $\varrho \geq 0$, and $\int_{\mathbb{R}^d} \varrho dv = 1$.

Let

$$L_1^\delta(f_n, f_n) = L_1(f_n, f_n) *_v \varrho_\delta.$$

Explaining this formula, we have

$$\begin{aligned} L_1^\delta(f_n, f_n)(x, v_*, t) &= \int_{\mathbb{R}^{2d}} \int_{S^{d-1}} dw B(v - v_1, w) f_n(v') f_n(v_1') \varrho_\delta(v_* - v) dv dv_1 \\ &= \int_{\mathbb{R}^{2d}} \int_{S^{d-1}} dw B(v - v_1, w) f_n(v) f_n(v_1) \varrho_\delta(v_* - v') dv dv_1 \\ &= \int_{\mathbb{R}^{2d}} f_n(v) f_n(v_1) \varphi_\delta(v_*, v, v_1) dv dv_1 \end{aligned}$$

with

$$\varphi_\delta(v_*, v, v_1) = \int_{S^{d-1}} dw B(v - v_1, w)\varrho_\delta(v_* - v + (v - v_1.w)w) \in \mathcal{D}(\mathbb{R}^{3d}).$$

For $\delta > 0$, the sequences $f_n$ and $L_1^\delta(f_n, f_n)$ have a compact support subset of $\bar{\omega} \times \bar{B}(0, M)$; the function $\varphi_\delta$ is regular, and it has a compact support; then the averaging lemma implies that

$$L_1^\delta(f_n, f_n) \longrightarrow L_1^\delta(f, f) = L_1(f, f) *_v \varrho_\delta$$

in $L^p(\omega\times]0, T[)$ strongly for $p \in [1, \infty[$.

We write

$$L_1(f_n, f_n) - L_1(f, f) = (L_1(f_n, f_n) - L_1^\delta(f_n, f_n)) + (L_1^\delta(f_n, f_n) - L_1^\delta(f, f))$$
$$+(L_1^\delta(f, f) - L_1(f, f)).$$

We notice that it is sufficient to prove that: $L_1(f_n, f_n) - L_1^\delta(f_n, f_n)$ and $L_1^\delta(f, f) - L_1(f, f)$ converge towards zero in $L^1(0, T[\times\Omega)$ as $\delta$ goes to zero (uniformly on $n$).

***Proof of Theorem 12.5.8: Conclusion*** The functions $L_1(f_n, f_n)$ and $L_1(f, f)$ have supports subset of $\omega \times B(0, M)$. This is enough to prove the convergence in $L^2(]0, T[\times\Omega)$. Using the Plancherel formula, one can write

$$\left\| \left(L_1^\delta - L_1\right)(f_n, f_n) \right\|^2_{L^2(]0,T[\times\Omega)} = \frac{1}{(2\pi)^d} \int_0^T dt \int_\omega dx$$
$$\times \int_{\mathbb{R}^d} d\xi \left| F_v^\alpha \left[L_1(f_n, f_n)\right]\right|^2 (1 - \hat{\varrho}(\delta\xi))^2;$$

finally, we have $\hat{\varrho}$ is bounded, and it is sufficient to check that

$$\lim_{R\to\infty} \sup_{n\geq 1} \int_0^T dt \int_\omega dx \int_{|\xi|\geq R} d\xi \left| F_v^\alpha \left[L_1(f_n, f_n)\right]\right|^2 = 0.$$

This estimate is a consequence of the pseudo-differential regularity result of the operator $L_1$ given by:

**Theorem 12.5.9 (P.L. Lions [47])** *Let $B \in C^\infty\left(\mathbb{R}^d \times S^{d-1}\right)$, $f \in L_v^2$, and $g \in L_v^1$. We assume that B satisfies the assumption (H2). Then,*

$$L_1(f, g)(v) = \int_{\mathbb{R}^d \times S^{d-1}} B(v - v_1, w) f(v - (v - v_1 \cdot w) w)$$
$$\times g(v_1 + (v - v_1 \cdot w) w) dv_1 dw$$

*is in* $H^{(d-1)/2}\left(\mathbb{R}^d\right)$ *and satisfies*

$$\|L_1(f,g)\|_{H^{(d-1)/2}(\mathbb{R}^d_v)} \leq C\|f\|_{L^2(\mathbb{R}^d_v)}\|g\|_{L^1(\mathbb{R}^d_v)},$$

*where the constant* $C$ *is independent of* $f$ *and* $g$*. In particular, for all* $s \in \mathbb{R}$*, the operators* $T$ *and* $\hat{T}$ *are bounded in* $H^{s+\frac{d-1}{2}}\left(\mathbb{R}^d\right)$. □

*Remark 12.5.2*

1. The sequence $f_n^M$ is bounded in $L^\infty\left(0, T;; L^1 \cap L^\infty(\Omega)\right)$; the previous theorem allows us to deduce that

$$L_1\left(f_n^M, f_n^M\right) \textit{ is uniformly bounded with respect to } n \textit{ in}$$
$$L^\infty\left(]0, T[\times \omega_i;\ H^{(d-1)/2}\left(\mathbb{R}^d\right)\right).$$

   This implies the estimate (64).
2. The result of the previous theorem is based on an equivalent form of $L_1$. Indeed, with the relation $v = v' + \left(v_1' - v' - w\right) w$, we have

$$\int_{\mathbb{R}^d} L_1(f,f)\varphi(v)dv = \int\left\{\int_{S^{d-1}\times\mathbb{R}^d} f' B(v-v_1,w)\varphi \right.$$
$$\left. \times(v' + (v_1' - v'.w)w)dwdv'\right\} f_1' dv_1';$$

   replacing $(v, v_1) \mapsto (v', v_1')$, the previous expression becomes

$$\int_{\mathbb{R}^d} L_1(f,f)\varphi(v)dv = \int\left\{\int_{S^{d-1}\times\mathbb{R}^d} f(v)B(v-v_1,w)\varphi \right.$$
$$\left. \times(v - (v - v_1.w)w)dwdv\right\} f(v_1)dv_1,$$

   and if we denote by $\tau_h\psi := \psi(.-h)$, we get

$$\int_{\mathbb{R}^d} L_1(f,f)\varphi(v)dv = \int_{\mathbb{R}^d} f(v)\left\{\int_{\mathbb{R}^d} f(v_1)[(\tau_{-v_1}\circ\tilde{T}\circ\tau_{v_1})\varphi](v)dv_1\right\}dv,$$
$$= \int_{\mathbb{R}^d} f(v)\tilde{\mathcal{L}}_f(\varphi)(v)dv = \int_{\mathbb{R}^d} f(v)\tilde{\mathcal{L}}_\varphi(f)(v)dv$$

   with

$$(\tilde{\mathcal{L}}_\theta f)(x) = \int_{\mathbb{R}^d}\theta(z)[(\tau_{-z}\circ\tilde{T}\circ\tau_z)f](x)dz.$$

3. For all $\psi \in \mathcal{D}(\Omega \times [0, T])$, the sequences $L_1(\psi, f_n)$ and $L_1(f_n, \psi)$ converge, respectively, towards $L_1(\psi, f)$ and $L_1(f, \psi)$ in $L^1_{loc}(]0, T[\times\Omega)$. Therefore, for all $f \in F^\alpha$, the sequences $L_1(f, f_n)$ and $L_1(f_n, f)$ converge towards $L_1(f, f)$ in $L^1_{\text{loc}}(]0, T[\times\Omega)$ strongly. ■

**Lemma 12.5.14** *Let $\psi \in \mathcal{D}([0, T] \times \Omega)$, $\theta \in L^\infty(]0, T[\times\Omega)$. Then*

$$\int_{\mathbb{R}^d} L_1(\psi, f_n)\theta \to \int_{\mathbb{R}^d} L_1(\psi, f)\theta$$

*and*

$$\int_{\mathbb{R}^d} L_1(f_n, \psi)\theta \to \int_{\mathbb{R}^d} L_1(f, \psi)\theta$$

*in $L^1(]0, T[\times\omega)$.* □

***Proof of Lemma 12.5.14*** The proof is the same for both cases. We shall detail only the first case. We assume that supp( $\psi$) is a subset of a compact $[0, T] \times \bar{\omega} \times K$, then

$$\int_{\mathbb{R}^d} L_1(\psi, f_n - f)\theta dv = \int_{\mathbb{R}^d} (f - f_n)(v_1)g(v_1)dv_1$$

with

$$g(v_1) = \int_{K\times S^{d-1}} B(v - v_1, w)\psi(v)\theta(v - (v - v_1.w)w)dwdv,$$

the function $g$ is in $L^\infty(\Omega\times]0, T[)$, and then $\int_{\mathbb{R}^d}(f_n - f)gdv$ converges towards zero in $L^1(\Omega\times]0, T[)$ strongly. □

With the previous remark, we can deduce the stability of $L_2$, $L_3$, and $L_4$.

### 12.5.6 Analysis of $L_2$, $L_3$, and $L_4$

**Lemma 12.5.15** *For all $\psi \in \mathcal{D}(\bar{\Omega} \times [0, T])$, $i \in \{3, 4\}$, and $f_n \in F^\alpha$,*

$$\int_{\mathbb{R}^d} L_i(f_n, f_n)\psi dv \longrightarrow \int_{\mathbb{R}^d} L_i(f, f)\psi dv \quad in \quad L^1(\omega\times]0, T[).$$

*Then, for $i \in \{3, 4\}$, $f_n L_i(f_n, f_n)$ converges in $L^1(\Omega\times]0, T[)$ weakly towards $f L_i(f, f)$.* □

***Proof of Lemma 12.5.15*** By writing $L_3$ and $L_4$,

$$L_3(f,f)(v) = \int_{\mathbb{R}^d} f(v_1)[(\tau_{-v_1} \circ T \circ \tau_{v_1})f]dv_1,$$

$$L_4(f,f)(v) = \int_{\mathbb{R}^d} f(v_1)[(\tau_{-v_1} \circ \tilde{T} \circ \tau_{v_1})f](v)dv_1.$$

This is equivalent to $L_3(f,f) = \tilde{\mathcal{L}}_f f$ and $L_4(f,f) = \mathcal{L}_f f$ with

$$(\mathcal{L}_\theta f)(x) = \int_{\mathbb{R}^d} \theta(z)[(\tau_{-z} \circ T \circ \tau_z)f](x)dz,$$

$$(\tilde{\mathcal{L}}_\theta f)(x) = \int_{\mathbb{R}^d} \theta(z)[(\tau_{-z} \circ \tilde{T} \circ \tau_z)f](x)dz.$$

Let $\psi \in \mathcal{D}$, and by using the coordinates $(v, v) \mapsto (v', v_1')$, we check that

$$\int_{\mathbb{R}^d} L_3(f,g)\psi dv = \int_{\mathbb{R}^d} g(v)L_1(\psi, f)dv \tag{12.62}$$

and

$$\int_{\mathbb{R}^d} L_4(f,g)\psi dv = \int_{\mathbb{R}^d} g(v)L_1(f, \psi)dv. \tag{12.63}$$

Let us detail the proof for $L_3$. The proof of $L_4$ is exactly the same. Indeed, we deduce from (12.62) that

$$\int_{\mathbb{R}^d} [L_3(f_n, f_n) - L_3(f,f)]\psi dv = \int_{\mathbb{R}^d} L_1(\psi, f_n - f)f_n + \int_{\mathbb{R}^d} L_1(\psi, f)(f_n - f). \tag{12.64}$$

We insert the function $f_n^M$,

$$\begin{aligned}
&\int_0^T dt \int_\omega dx \left| \int_{\mathbb{R}^d} L_1(\psi, f_n - f) f_n dv \right| \\
&\quad \le \int_0^T dt \int_\Omega \left| L_1(\psi, f_n - f)(f_n - f_n^M) \right| dx dv + \int_0^T dt \int_\omega dx \\
&\qquad \times \int_{|v| \le M} dv |L_1(\psi, f_n - f)|
\end{aligned}$$

$$\leq C_\psi \int_0^T dt \int_\omega dx \int_{|v|\geq M} f_n + \int_0^T dt \int_\omega dx \int_{|v|\leq M} dv |L_1(\psi, f_n - f)|$$

$$\leq \frac{C_\psi T}{M^2} \sup_{t\in[0,T]} K(t) + \|L_1(\psi, f_n - f)\|_{L^1_{loc}(\Omega\times]0,T[)}.$$

Let $f^\varepsilon \in \mathcal{D}$ such that $\|f^\varepsilon - f\|_{L^1} \leq \varepsilon$,

$$\int_{\mathbb{R}^d} L_1(\psi, f)(f_n - f) = \int_{\mathbb{R}^d} L_1(\psi, f^\varepsilon)(f_n - f) + \int_{\mathbb{R}^d} L_1(\psi, f - f^\varepsilon)(f_n - f), \tag{12.65}$$

$L_1(\psi, f^\varepsilon)$ is regular, and then the convergence of the first term of (12.65) in $L^1(\omega\times]0, T[)$ is a consequence of the averaging lemma. The second term is equivalent to $C\|f - f^\varepsilon\|_{L^1(\Omega\times]0,T[)}$.

To obtain the convergence of $f_n L_3(f_n, f_n)$, we write

$$\int_0^T \int_\Omega f_n L_3(f_n, f_n)\psi dx dv dt = \int_0^T \int_\Omega L_1(f_n\psi, f_n) f_n dx dv dt.$$

The sequence $(f_n)$ is bounded in $L^\infty$, and $L_1(f_n\psi, f_n)$ converges in $L^1(\Omega\times]0, T[)$. Then,

$$\int\int\int f_n L_1(f_n\psi, f_n) dx dv dt \to \int\int\int f L_1(f\psi, f) dx dv dt.$$

This ends the proof of Lemma 12.5.15. □

For the trilinear term $L_2$, we have

$$\int_{\mathbb{R}^d} L_2(f)\varphi(v)dv = \int_{\mathbb{R}^d} f(v) \left\{ \int_{\mathbb{R}^d\times S^{d-1}} B(v - v_1, w) f(v') f(v_1')\varphi(v_1) dv_1 dw \right\} dv.$$

The expression between the square brackets is similar to $L_1(f, f)$ by changing the operator $T$ by $T_\varphi$ :

$$T_\varphi\psi(z) = \int_{S^{d-1}} \varphi(z + (z.w)w) B(z, w)\psi((z.w)w) dw.$$

Then, for all $f_n \in F^\alpha$,

$$\int_0^T \int_\Omega (L_2(f_n) - L_2(f))\psi dx dv dt \longrightarrow 0 \quad \forall\psi \in \mathcal{D}.$$

This completes the proof of the stability result (12.49). □

*Remark 12.5.3* The previous stability results are dependent only on estimates of lemma (12.4.7) that will be essential for the study of the existence of a weak solution of the self-consistent problem. ∎

## 12.6 The Boltzmann–Poisson System

### 12.6.1 Main Result

**Theorem 12.6.10** *We assume that $d \leq 3$. Then, under the assumptions (H1)–(H3), the Boltzmann–Poisson system (BP)*[1] *has a weak solution $(f, E)$ such that*

$$f \in L^\infty(\mathbb{R}^+;\ L^1 \cap L^\infty(\Omega)), \quad 0 \leq f \leq 1, \quad \|f(t)\|_{L^1(\Omega)} = \|f_0\|_{L^1(\Omega)},\ p.p.$$

$$E \in L^\infty_{loc}(\mathbb{R}^+;\ [W^{1,\frac{d+2}{d}}(\omega)]^d),$$

$$\int_\Omega |v|^2 f(t, x, v)dxdv + \int_\omega |E(t, x)|^2 dx + 2\int_\omega \varrho(t, x)\varphi_0(t, x)dx \in L^\infty_{loc}(\mathbb{R}^+).$$

□

To show the existence of a solution of the Boltzmann–Poisson system, we use a fixed-point procedure. $E \in L^2_{loc}(\mathbb{R}^+;\ W^{1,\infty})$, and $f(E)$ is a solution of the Boltzmann equation. This solution is uniformly bounded in $L^\infty(\mathbb{R}^+;\ L^1 \cap L^\infty(\Omega))$; its density $\varrho(E) = \int_{\mathbb{R}^d} f(E)dv$ is bounded in $L^\infty(\mathbb{R}^+; L^{\frac{d+2}{d}}(\Omega))$. The solution of Poisson equation: $-\Delta_x \Phi(E) = \varrho(E)$ in $L^\infty(0, T;\ W^{2,\frac{d+2}{d}}(\omega))$. Then, the new value of the electric field $E^* = -\nabla_x \Phi(E)$ is in $L^\infty(0, T;\ W^{1,\frac{d+2}{d}}(\omega))$ and not necessarily in $W^{1,\infty}$. A possible solution to this difficulty is to regularize the Poisson equation in order to get more regularity for the electric field.

### 12.6.2 The Modified Boltzmann–Poisson System

We consider the following modified Boltzmann–Poisson system:

**Theorem 12.6.11** *Assume that (H1)–(H3) hold. Then, the system*

[1] see page 298.

$$
(BP^\varepsilon)\qquad \begin{cases}
\partial_t f^\varepsilon + v.\nabla_x f^\varepsilon + (E^\varepsilon + E_0).\nabla_v f^\varepsilon = (Q_0 + Q_1)(f^\varepsilon) = Q(f^\varepsilon),\\
E^\varepsilon = -\nabla_x \Phi^\varepsilon,\\
-(1-\varepsilon\Delta_x)^2 \Delta_x \Phi^\varepsilon = \varrho^\varepsilon = \int_{\mathbb{R}^d} f^\varepsilon dv,\\
f^\varepsilon(x, v, t=0) = f_0(x, v),\\
f^\varepsilon(x, \bar{v}, t) = f^\varepsilon(x, v, t), \quad (x, v) \in \Gamma^-,\\
\Phi^\varepsilon(x, t) = \Delta\Phi^\varepsilon = \Delta^2\Phi^\varepsilon = 0, \quad x \in \partial\omega,
\end{cases}
$$

*where* $\bar{v} = v - 2(v.n(x))n(x)$ *has a weak solution* $(f^\varepsilon, E^\varepsilon)$ *satisfying*

$$f^\varepsilon \in L^\infty_{loc}(\mathbb{R}^+,\ L^1 \cap L^\infty(\Omega)), \quad 0 \le f \le 1, \quad \|f(t)\|_{L^1(\Omega)} = \|f_0\|_{L^1(\Omega)}, \ p.p.$$

$$E^\varepsilon \in L^\infty_{loc}(\mathbb{R}^+,\ [W^{5,r_0}(\omega)]^d), \quad r_0 = \frac{(d+2)^2}{(d+2)^2-1}.$$

□

To prove this theorem, we use a second regularization of the electric field by considering

$$\xi_\alpha(x, t) = \frac{1}{\alpha^{d+1}}\xi\left(\frac{x}{\alpha}, \frac{t}{\alpha}\right),$$

where $\xi \in \mathcal{D}(\mathbb{R}^+ \times \omega)$, $\xi \ge 0$, and $\int_{\mathbb{R}^+\times\omega} \xi dx dt = 1$.

We define

$$F_\alpha = \xi_\alpha * (\bar{E} + \bar{E}_0),$$

where $\bar{E}$ and $\bar{E}_0$ are, respectively, the extensions by zero outside $[0, T] \times \omega$ of $E$ and $E_0$ and $*$ is the convolution with respect to the variable $(t, x)$.

We consider the following Boltzmann–Poisson system:

$$
(BP_\alpha^\varepsilon) \quad \begin{cases} \partial_t f + v \cdot \nabla_x f + F_\alpha \cdot \nabla_v f = (Q_0 + Q_1)(f) = Q(f), \\ E = -\nabla_x \Phi, \\ -(1 - \varepsilon\Delta_x)^2 \Delta_x \Phi = \varrho = \int_{\mathbb{R}^d} f dv, \\ f(x, v, t = 0) = f_0(x, v), \\ f(x, \bar{v}, t) = f(x, v, t), \qquad (x, v) \in \Gamma^- \\ \Phi(x, t) = \Delta\Phi = \Delta^2\Phi = 0, \qquad x \in \partial\omega. \end{cases}
$$

**Theorem 12.6.12** *The system $(BP_\alpha^\varepsilon)$ has a weak $(f_\alpha^\varepsilon, E_\alpha^\varepsilon)$ satisfying*

$$
f_\alpha^\varepsilon \in L^\infty_{loc}(\mathbb{R}^+;\ L^1 \cap L^\infty(\Omega)), \quad 0 \le f_\alpha^\varepsilon \le 1, \quad \|f_\alpha^\varepsilon(t)\|_{L^1(\Omega)} = \|f_0\|_{L^1(\Omega)},
$$

$$
E_\alpha^\varepsilon \in L^\infty_{loc}(\mathbb{R}^+;\ [W^{5,\frac{d+2}{d}}(\omega)]^d),
$$

$$
\varrho_\alpha^\varepsilon \in L^\infty_{loc}(\mathbb{R}^+;\ L^{\frac{d+2}{d}}(\omega)), \quad j_\alpha^\varepsilon \in L^\infty_{loc}(\mathbb{R}^+;\ [L^{\frac{d+2}{d+1}}(\omega)]^d).
$$

*Moreover,* $\|E_\alpha^\varepsilon\|_{L^\infty(0,T;\ [W^{5,\frac{d+2}{d}}(\omega)]^d)}$, $\|\varrho_\alpha^\varepsilon\|_{L^\infty(0,T;\ L^{\frac{d+2}{d}}(\omega))}$, *and* $\|j_\alpha^\varepsilon\|_{L^\infty(0,T;[L^{\frac{d+2}{d+1}}(\omega)]^d)}$ *are uniformly bounded with a uniform bound independent of the parameter* $\alpha$. □

The proof of this result uses several lemmata.
We define

$$
\Lambda \ :\ L^2(0, T;\ [W^{4,r_0}(\omega)]^d) \longrightarrow L^2(0, T;\ [W^{4,r_0}(\omega)]^d)
$$

$$
E \longmapsto E^*,
$$

where $r_0$ is given by Theorem 12.6.11 and $E^*$ is defined as follows.

We regularize $E + E_0$ by $\xi_\alpha$; we define $f_\alpha(E)$ as the unique solution of the Boltzmann equation associated with $F_\alpha$.

Then, we consider the modified Poisson equation

$$
\begin{cases} -(1 - \varepsilon\Delta)^2 \Delta \Phi_\alpha^\varepsilon = \varrho_\alpha(E) = \int_{\mathbb{R}^d} f_\alpha(E) dv, \\ \Phi_\alpha^\varepsilon = \Delta\Phi_\alpha^\varepsilon = \Delta^2\Phi_\alpha^\varepsilon = 0, \quad x \in \partial\omega, \end{cases} \tag{12.66}
$$

and finally, we take $\Lambda(E) = E^* := -\nabla_x \Phi_\alpha^\varepsilon$.

The proof of the previous theorem consists of proving that $\Lambda$ has a fixed point. We start with the following result on elliptic operators.

**Lemma 12.6.16 ([1, 12])** *Let $\varrho \in L^r(\omega)$, $r \in ]1, +\infty[$, $m \in \mathbf{N}$. Then, the solution of*

$$\begin{cases} -(1-\varepsilon\Delta_x)^{2m}\Delta_x\Phi = \varrho, \\ \Phi(x,t) = \Delta_x\Phi = \cdots = \Delta_x^{2m}\Phi = 0, \qquad x \in \partial\omega, \end{cases}$$

*belongs to $W^{4m+2,r}(\omega)$ and satisfies*

$$\|\Phi^\varepsilon\|_{W^{4m+2,r}(\omega)} \le C(r,\varepsilon)\|\varrho\|_{L^r(\omega)} \tag{12.67}$$

$$\|\Phi^\varepsilon\|_{W^{2,r}(\omega)} \le C(r)\|\varrho\|_{L^r(\omega)}, \tag{12.68}$$

*where $C(r)$ is independent of $\varepsilon$. Moreover, if $\varrho \ge 0$, then $\Phi^\varepsilon \ge 0$.* □

Now, we shall prove the existence of a weak solution of $(BP_\alpha^\varepsilon)$.

**Lemma 12.6.17** *Let $T > 0$. Then, there exists $C(\varepsilon)$ dependent on $T$ and $\varepsilon$, such that*

$$\|E^*(t)\|_{W^{5,r_0}(\omega)} \le C(\varepsilon)(1 + \|E(s)\|_{L^2(0,T;\ W^{4,r_0}(\omega))})^{1/2}. \tag{12.69}$$

□

***Proof of Lemma 12.6.17*** Let $r \in ]1, \frac{d+2}{d}]$, $E_0 \in L^2(0,T;\ L^{d+2}(\omega))$, and $E \in L^2(0,T;\ W^{4,r_0}(\omega))$. Then, for $t \le T$, $\varrho_\alpha(E(t)) \in L^r(\omega)$ and

$$\|E^*(t)\|_{W^{5,r}} \le C(r,\varepsilon)\|\varrho(E(t))\|_{L^r(\omega)}.$$

By applying the $L^p$-interpolation properties and Lemma 12.4.7, we get

$$\|E^*(t)\|_{W^{5,r}} \le C(r,\varepsilon)\|\varrho_0\|_{L^1(\omega)}^\beta \left\{1 + \int_0^T \|F_\alpha(s)\|_{L^{d+2}(\omega)} ds\right\}^{(1-\beta)(d+2)},$$

where $\beta$ vérifiant $\frac{1}{r} = \beta + \frac{(1-\beta)d}{d+2}$.

We remark that the choice $r = r_0$ gives $(1-\beta)(d+2) = 1/2$. As a consequence, the previous estimate becomes

$$\begin{aligned} \|E^*(t)\|_{W^{5,r_0}} &\le C(\varepsilon)\left\{1 + \int_0^T \|F_\alpha(s)\|_{L^{d+2}(\omega)} ds\right\}^{1/2} \\ &\le C(\varepsilon)\left\{1 + \sqrt{T}\|E + E_0\|_{L^2(0,T;\ L^{d+2}(\omega))}\right\}^{1/2}. \end{aligned}$$

The fact that $E_0$ belongs to $L^2_{loc}(\mathbb{R}^+;\ L^{d+2}(\omega))$ and the fact that $W^{4,r_0}$ is continuously embedded in $L^{d+2}(\omega)$ for $d \leq 3$ lead to

$$\|E^*(t)\|_{W^{5,r_0}} \leq C(\varepsilon)\left\{1 + \|E\|_{L^2(0,T;\ W^{4,r_0}(\omega))}\right\}^{1/2},$$

where $C(\varepsilon)$ depends only on $\varepsilon$ and $T$.

As a consequence, there exists $R > 0$ such that $\bar{\mathcal{B}}_R$ of $L^2(0, T;\ [W^{4,r_0}(\omega)]^d)$ is invariant by $\Lambda$. Indeed, $W^{5,r_0}(\omega) \hookrightarrow W^{4,r_0}(\omega)$,

$$\|E^*(t)\|^2_{W^{4,r_0}} \leq C(\varepsilon)\left\{1 + \|E(s)\|_{L^2(0,T;\ [W^{4,r_0}(\omega)]^d)}\right\},$$

and $\Lambda(\bar{\mathcal{B}}_r) \subset \bar{\mathcal{B}}_R$ for all $R$ such that $R^2 = C(\varepsilon)(1 + R)$. □

**Lemma 12.6.18** *$\Lambda(\bar{\mathcal{B}}_R)$ is a precompact subset of $L^2(0, T;\ [W^{4,r_0}(\omega)]^d)$.* □

***Proof of Lemma 12.6.18*** This lemma is a consequence of the compactness results of evolution operators used to prove Lemme 12.5.10. Let $E_n$ a sequence in $\bar{\mathcal{B}}_R$. The previous inequality gives

$$\Lambda(E_n) \textit{ is bouneded in } L^2(0, T;\ [W^{5,r_0}(\omega)]^d). \tag{12.70}$$

Lemma 12.4.7 implies that $\varrho_\alpha(E_n)$ (respectively, $j_\alpha(E_n)$)[2] are uniformly bounded in $L^\infty(0, T;\ L^{\frac{d+2}{d}}(\omega))$ (respectively, in $L^\infty(0, T;\ L^{\frac{d+2}{d+1}}(\omega))$). One can deduce using

$$\partial_t \varrho_\alpha(E_n) + \nabla_x \cdot j_\alpha(E_n) = 0$$

that $\partial_t \varrho_\alpha(E_n)$ is bounded in $L^\infty(0, T;\ W^{-1,r_0}(\omega))$ (since $r_0 < \frac{d+2}{d+1}$) and $\partial_t \Lambda(E_n)$ is the solution of

$$\begin{cases} \partial_t \Lambda(E_n) = -\nabla_x \psi, \\ -(1 - \varepsilon \Delta_x)^2 \Delta_x \psi = \partial_t \varrho_\alpha(E_n), \\ \psi = \Delta \psi = \Delta^2 \psi = 0 \quad x \in \partial\omega. \end{cases}$$

Then

$$\partial_t \Lambda(E_n) \textit{ est bornée dans } L^\infty(0, T;\ [W^{4,r_0}(\omega)]^d). \tag{12.71}$$

The uniform bounds (12.70) and (12.71) imply that $\Lambda(\bar{\mathcal{B}}_R)$ is precompact in $L^2(0, T;\ [W^{4,r_0}(\omega)]^d)$. □

---

[2] $\varrho_\alpha(E_n) = \int_{\mathbb{R}^d} f_\alpha(E_n) dv$ and $j_\alpha(E_n) = \int_{\mathbb{R}^d} v f_\alpha(E_n) dv$.

**Lemma 12.6.19** $\Lambda$ *is continuous.* □

***Proof of Lemma 12.6.19*** Let $E_n$ a sequence in $\bar{\mathcal{B}}_R$ converging towards $E$ in $L^2(0, T,\ [W^{4,r_0}]^d)$ strongly. Let $F_\alpha(E_n)$ the regularization of $E_n + E_0$ by $\xi_\alpha$. We consider $f_\alpha(E_n)$ the solution of the Boltzmann equation associated with $F_\alpha(E_n)$; $\varrho_\alpha(E_n) = \int_{\mathbb{R}^d} f_\alpha(E_n)dv$ and $E_n^* = \Lambda(E_n)$.

The subset $\Lambda(\bar{\mathcal{B}}_R)$ is precompact. There exists a sub-sequence, also denoted $E_n^*$ that converges towards $E^* \in \bar{\mathcal{B}}_R$. To prove the convergence of $(\Lambda(E_n))_n$, it is sufficient to verify that $E^* = \Lambda(E)$.

The sequence $f_\alpha(E_n)$ is uniformly bounded, $0 \le f_\alpha(E_n) \le 1$, and $\varrho_\alpha(E_n)$ is bounded in $L^\infty(0, T;\ L^{\frac{d+2}{d}}(\omega))$.

Let $f_\alpha$ and $\varrho_\alpha$ be their limits, respectively :

$$f_\alpha(E_n) \overset{*}{\rightharpoonup} f_\alpha \quad in \quad L^\infty(0, T;\ L^\infty(\Omega)),$$

$$\varrho_\alpha(E_n) \rightharpoonup \varrho_\alpha \quad in \quad L^2(0, T;\ L^{\frac{d+2}{d}}(\omega)).$$

The sequence $(F_\alpha(E_n))_n$ is bounded in $L^2(0, T;\ L^{d+2}(\Omega))$. One can deduce that the assumptions of Lemma 12.4.7 are satisfied. The properties of compactness of the collision operator $Q$ imply: there exists a sub-sequence $(f_\alpha(E_n))_n$ such that

$$Q(f_\alpha(E_n)) \rightharpoonup Q(f_\alpha) \quad in \quad L^1(\Omega \times ]0, T[)\ weakly.$$

Moreover,

$$F_\alpha(E_n) \longrightarrow F_\alpha = \xi_\alpha * (\bar{E} + \bar{E}_0) \quad in \quad L^2(0, T;\ L^{d+2}(\omega))\ strong.$$

We can pass to the limit in the Green formula (12.29) for all $\psi \in C_c^1(]0, T[\times\bar{\Omega})$ satisfying (12.26):

$$\int_0^T \int_\Omega (\partial_t\psi + v\nabla_x\psi + F_\alpha.\nabla_v\psi) f_\alpha + \int_0^T \int_\Omega Q(f_\alpha)\psi + \int_\Omega f_0\psi(t = 0)dxdv = 0.$$

By passing to the limit in the modified Poisson equation, we get $E^*$ verifies

$$\begin{cases} E^* = -\nabla_x\Phi^*, \\ (1 - \varepsilon\Delta)^2\Delta\Phi^* = \varrho_\alpha, \\ \Phi^* = \Delta\Phi^* = \Delta^2\Phi = 0, \quad x \in \partial\omega. \end{cases}$$

To prove that $\varrho^\alpha = \int f_\alpha dv$, we remark that $f_\alpha(E_n)$ belongs to a compact subset of $L^1(0, T;\ L^1(\Omega)) - weak$; the density $\varrho_\alpha(E_n)$ is bounded in $L^\infty(0, T;\ L^{\frac{d+2}{d}}(\omega))$, and for all $\psi \in \mathcal{D}(]0, T[\times\omega)$,

$$\lim_{n\to\infty}\int_0^T\int_\omega\left[\varrho_\alpha(E_n)-\int_{\mathbb{R}^d}f_\alpha dv\right]\psi(t,x)dxdt=\lim_{n\to\infty}\langle f_\alpha(E_n)-f,\psi\rangle_{L^1,L^\infty}=0.$$

The sequence $f_\alpha(E_n)$ converges to $f_\alpha$ in $L^1$ weakly, and its limit in $\mathcal{D}'$ of $\varrho_\alpha(E_n)$ is equal to $\int f_\alpha dv$. Then, $E^*=\Lambda(E)$. □

***Proof of Theorem 12.6.12*** As a conclusion, the application $\Lambda$ verifies the assumptions of the Schauder fixed-point theorem. This yields the existence of $E^\varepsilon_\alpha\in\bar{\mathcal{B}}_R$. In addition, it satisfies (12.69). The estimates on the charge density $\varrho^\varepsilon_\alpha$ and the current $j^\varepsilon_\alpha$ (due to Lemma 12.4.7) are uniformly bounded with respect to $R$ and depend only on $\varepsilon$ and $T$. □

*Remark 12.6.4* We proved the existence of a weak solution (local in time: on $[0,T]$) of the modified Boltzmann–Poisson system. To prove a global solution in time, we consider a sequence of intervals $I_n=[n,n+1]$ to extend the solution on $I_{n+1}$ by a solution associated with an initial data $f=f(t=n)$ (in a weak sense). ■

***Proof of Theorem 12.6.11*** Now, we can proceed to the proof of Theorem 12.6.11 by considering a sequence of weak solutions of $(BP^\varepsilon_\alpha)$ $(f^\varepsilon_\alpha,\varrho^\varepsilon_\alpha,j^\varepsilon_\alpha,E^\varepsilon_\alpha)_\alpha$ belonging to $(BP^\varepsilon_\alpha)$. Up to the extraction of a sub-sequence, there exists $f_\alpha,\varrho_\alpha,j_\alpha,E_\alpha$ such that :

$$f^\varepsilon_\alpha\overset{*}{\rightharpoonup}f^\varepsilon\in L^\infty_{loc}(\mathbb{R}^+;\ L^1\cap L^\infty(\Omega)),$$

$$\varrho^\varepsilon_\alpha\rightharpoonup\varrho^\varepsilon\in L^\infty_{loc}(\mathbb{R}^+;\ L^{\frac{d+2}{d}}(\omega)),$$

$$j^\varepsilon_\alpha\rightharpoonup j^\varepsilon\in L^\infty_{loc}(\mathbb{R}^+;\ [L^{\frac{d+2}{d+1}}(\omega)]^d).$$

The parameter $\varepsilon$ is fixed. $E^\varepsilon_\alpha$ is bounded in $L^\infty_{loc}(\mathbb{R}^+,\ [W^{5,r_0}(\omega)]^d)$, and $\partial_t E^\varepsilon_\alpha$ is bounded in $L^\infty_{loc}(\mathbb{R}^+;\ [W^{4,\frac{d+2}{d+1}}(\omega)]^d)$; we can deduce

$$E^\varepsilon_\alpha\rightharpoonup E^\varepsilon\quad in\quad L^\infty_{loc}(\mathbb{R}^+,\ [W^{5,r_0}(\omega)]^d)\ weakly,$$

$$E^\varepsilon_\alpha\longrightarrow E^\varepsilon\quad in\quad L^2(0,T,\ [W^{4,r_0}(\omega)]^d)\ strongly.$$

Furthermore, $(\varrho^\varepsilon,E^\varepsilon)$ verifies

$$\begin{cases}E^\varepsilon=-\nabla_x\Phi^\varepsilon,\\[2ex]-(1-\varepsilon\Delta_x)^2\Delta_x\Phi^\varepsilon=\varrho^\varepsilon,\\[2ex]\Phi^\varepsilon=\Delta_x\Phi^\varepsilon=\Delta_x^2\Phi^\varepsilon=0,\quad x\in\partial\omega.\end{cases}$$

Then,

$$E^\varepsilon \in L^\infty_{loc}(\mathbb{R}^+;\ [W^{5,\frac{d+2}{d}}(\omega)]^d).$$

By passing to the limit in $\mathcal{D}'$ on the continuity equation $\partial_t \varrho^\varepsilon_\alpha + \nabla_x \cdot j^\varepsilon_\alpha = 0$, implying $\partial_t \varrho^\varepsilon + \nabla_x \cdot j^\varepsilon = 0$,

$$\partial_t E \in L^\infty_{loc}(\mathbb{R}^+;\ [W^{4,\frac{d+2}{d+1}}(\omega)]^d)$$

and

$$E^\varepsilon \in [W^{1,\infty}(\omega \times ]0, T[)]^d.$$

The strong convergence in $L^2(]0, T[\times\omega)$ of $E^\varepsilon_\alpha$ towards $E$ and $F^\varepsilon_\alpha$ to $E^\varepsilon + E_0$ and the compactness properties of $Q$ yield: the weak limit $f^\varepsilon$ of $f^\varepsilon_\alpha$ verifies the formula (12.29):

$$\int_0^T\int_\Omega [(\partial_t \psi + v \cdot \nabla_x \psi + (E^\varepsilon + E_0) \cdot \nabla_v \psi) f^\varepsilon + Q(f^\varepsilon)\psi] + \int_\Omega f_0 \psi(0, x, v) = 0$$

for all $\psi$ belonging to $\mathcal{D}([0, T] \times \mathbb{R}^{2d})$ and satisfying (12.26).

Furthermore, $f^\varepsilon_\alpha$ belongs to a weakly compact subset of $L^1(]0, T[\times\Omega)$. Note that one can easily prove that $\varrho^\varepsilon_\alpha$ converges in $\mathcal{D}'$ towards $\varrho^\varepsilon = \int_{\mathbb{R}^d} f^\varepsilon dv$. In conclusion, the limit $(f^\varepsilon, E^\varepsilon)$ is a weak solution of $(BP^\varepsilon)$. □

### 12.6.3 Unmodified Boltzmann–Poisson System

Let $(f^\varepsilon, E^\varepsilon)$ be a weak solution $(BP^\varepsilon)$. The estimates on $f^\varepsilon$ are uniform. It remains to provide estimates independent of $\varepsilon$, giving compactness for an $E^\varepsilon$. This estimate can be obtained if we have a uniform $\varrho^\varepsilon$ in $L^{\frac{d+2}{d}}$. To do this, we propose to establish a uniform kinetic energy. Using Lemma 12.4.7, we obtain a $L^p$-estimate on the density.

Let

$$K^\varepsilon(t) = \int_\Omega |v|^2 f^\varepsilon(t, x, v)dxdv, \quad V^\varepsilon(t) = \int_\omega \varrho^\varepsilon(t, x)\Phi^\varepsilon(t, x)dx,$$

$$\mathcal{E}^\varepsilon(t) = K^\varepsilon(t) + V^\varepsilon(t) + 2\int_\omega \varrho^\varepsilon(t, x)\Phi_0(t, x)dx.$$

The energy $\mathcal{E}^\varepsilon$ verifies the following lemma.

**Lemma 12.6.20** *$K^\varepsilon$, $V^\varepsilon$, and $\mathcal{E}^\varepsilon$ belong to $W^{1,1}_{loc}(\mathbb{R}^+)$. Moreover,*

$$K^\varepsilon(0) = \int_\Omega |v|^2 f_0 dx dv, \quad V^\varepsilon(0) = \int_\Omega f_0(x,v)\Phi^\varepsilon(0,x) dx dv \tag{12.72}$$

$$\mathcal{E}^\varepsilon(0) = K^\varepsilon(0) + V^\varepsilon(0) + 2\int_\Omega f_0(x,v)\Phi_0(0,x) dx dv \tag{12.73}$$

$$\frac{d}{dt}K^\varepsilon(t) = 2\int_\omega j^\varepsilon \cdot (E^\varepsilon + E_0) dx \tag{12.74}$$

$$\frac{d}{dt}V^\varepsilon(t) = 2\int_\omega \varrho^\varepsilon(t,x)\partial_t\Phi^\varepsilon(t,x) dx = 2\int_\omega j^\varepsilon \cdot \nabla_x \Phi^\varepsilon dx \tag{12.75}$$

$$[\mathcal{E}^\varepsilon]_0^t = 2\int_0^t \int_\omega \varrho^\varepsilon(s,x)\partial_t\Phi_0(s,x) dx ds. \tag{12.76}$$

□

***Proof of Lemma 12.6.20*** Let $\psi = |v|^2\chi_R(v)\chi_{\bar\omega}(x)h(t)$, where $h \in C^1_c([0,T[)$ and $\chi_R$ is defined by

$$\chi_R(v) = \begin{cases} 1 & if \quad |v| \le R, \\ 0 & if \quad |v| \ge R, \\ R+1-|v| & if \quad R \le |v| \le R+1. \end{cases}$$

The function $\chi_R$ is uniformly bounded in $W^{1,\infty}$ and converges to 1 a.e., and its gradient converges towards zero. The Green formula is

$$\begin{aligned}
&\int_0^t \int_\Omega |v|^2 f^\varepsilon \chi_R(v)\frac{dh}{ds} + \int_0^t \int_\Omega |v|^2 f^\varepsilon \nabla_v \chi_R(v) \cdot (E^\varepsilon + E_0) h(s) \\
&+2\int_0^t \int_\Omega v.(E^\varepsilon + E_0) f^\varepsilon \chi_R h(s) = \int_0^t \int_\Omega Q(f^\varepsilon)|v|^2 \chi_R(v) h(s) \\
&+\int_0^t \int_{\partial\Omega} |v|^2 f^\varepsilon \chi_R(v) h(t)(v \cdot n(x)) - h(0)\int_\Omega |v|^2 f_0 \chi_R(v).
\end{aligned} \tag{12.77}$$

$f^\varepsilon$ verifies the specular reflection condition (in the weak sense ). So that,

$$\int_0^t \int_{\partial\Omega} |v|^2 f^\varepsilon \chi_R(v) h(t)(v \cdot n(x)) = 0.$$

By passing to the limit ($R \to +\infty$) and using the properties of $Q$, we obtain

$$\int_0^t \frac{d}{ds}\left\{\int_\Omega |v|^2 f^\varepsilon(s)\right\} h(s)ds = 2\int_0^t \left[\int_\omega j^\varepsilon \cdot (E^\varepsilon + E_0)(s,x)d\right] h(s)ds$$

for all $h \in \mathcal{D}(\mathbb{R}_*^+)$.

Or also

$$\frac{d}{dt}K^\varepsilon(t) = 2\int_\omega j^\varepsilon \cdot (E^\varepsilon + E_0)dx \quad in \quad \mathcal{D}'(]0,\infty[).$$

The second term belongs to $L^1_{loc}(]0,+\infty[)$, leading to: $K^\varepsilon \in C(\mathbb{R}^+)$, and by passing to the limit on $R$ in (12.77), we get

$$K(0) = \int_\Omega |v|^2 f_0 dx dv$$

and

$$K^\varepsilon(t) = K(0) + 2\int_0^t \int_\omega j^\varepsilon \cdot (E^\varepsilon + E_0)dxds. \tag{12.78}$$

Furthermore,

$$\begin{aligned} 2\int_0^t \int_\omega j^\varepsilon \cdot (E^\varepsilon + E_0)dxds &= -2\int_0^t \int_\omega j^\varepsilon \cdot \nabla_x(\Phi^\varepsilon + \Phi_0)dxds \\ &= -2\int_0^t \int_\omega \partial_t \varrho^\varepsilon (\Phi^\varepsilon + \Phi_0)dxds \\ &= 2\int_0^t \int_\omega \varrho^\varepsilon (\partial_t \Phi^\varepsilon + \partial_t \Phi_0)dxds \\ &\quad +2\left[\int_\omega \varrho^\varepsilon(\Phi_0 + \Phi^\varepsilon)dx\right]_t^0. \end{aligned}$$

By passing to the limit ($R \to \infty$) with a test function $\psi_R = \chi_R(v)\Phi^\varepsilon(t,x)h(t)$, we get

$$\frac{d}{dt}V^\varepsilon(t) = \int_\omega \varrho^\varepsilon(t,x)\frac{\partial \Phi^\varepsilon}{\partial t}(t,x)dx + \int_\omega j^\varepsilon \cdot \nabla_x \Phi^\varepsilon(t,x)dx \in L^1_{loc}(]0,+\infty[).$$

As a consequence, $V^\varepsilon(0) = \int_\omega f_0(x,v)\Phi^\varepsilon(0,x)dx$, and thanks to the modified Poisson equation (12.66) that

$$V^\varepsilon(t) = \int_\omega \|(1 - \varepsilon\Delta)\nabla(\Phi^\varepsilon(t))\|^2 dx.$$

Moreover, the continuity equation leads to

$$\frac{dV^\varepsilon}{dt}(t) = 2\int_\omega (1-\varepsilon\Delta)\nabla\Phi^\varepsilon \cdot (1-\varepsilon\Delta)\nabla\partial_t\Phi^\varepsilon dx$$

$$= 2\int_\omega \varrho^\varepsilon(t,x)\partial_t\Phi^\varepsilon(t,x)dx = 2\int_\omega j^\varepsilon \cdot \nabla_x\Phi^\varepsilon dx.$$

Identity (12.78) becomes

$$\int_0^t\int_\omega j^\varepsilon \cdot (E^\varepsilon + E_0)(s,x) = [V^\varepsilon(s)]_t^0 + \int_0^t\int_\omega \varrho^\varepsilon\partial_t\varphi_0(s,x)dxds + \left[\int_\omega \varrho^\varepsilon\varphi_0 dx\right]_t^0.$$

Finally,

$$K^\varepsilon(t) + V^\varepsilon(t) = K(0) + V^\varepsilon(0) + 2\int_0^t\int_\omega \varrho^\varepsilon\partial_t\varphi_0 dx + 2\left[\int_\omega \varrho^\varepsilon\varphi_0 dx\right]_t^0$$

and

$$[\mathcal{E}^\varepsilon]_0^t = 2\int_0^t\int_\omega \varrho^\varepsilon\partial_t\varphi_0(s,x)dxds.$$

**Lemma 12.6.21** *There exists $C_T$ such that*

$$K^\varepsilon(t) \le C_T, \qquad \forall\varepsilon \le 1.$$

***Proof of Lemma* 12.6.21** By applying (12.76), we have

$$\mathcal{E}^\varepsilon(t) \le \mathcal{E}^\varepsilon(0) + 2\int_0^t\int_\omega \varrho^\varepsilon(s,x)\partial_t\varphi_0(s,x)dxds$$

$$\le C_T\left(1 + \int_0^T \|\varrho^\varepsilon(t)\|_{L^1(\omega)}\|\partial_t\varphi_0(t)\|_{L^\infty(\omega)}dt\right).$$

Using (H3), the energy $\mathcal{E}^\varepsilon$ is uniformly bounded in $C([0,T])$.

As a consequence,

**Lemma 12.6.22** $(\varrho^\varepsilon, j^\varepsilon, \Phi^\varepsilon)$ *satisfies:*

1. $\varrho^\varepsilon$ *is uniformly bounded in* $L^\infty(0,T;\ L^{\frac{d+2}{d}}(\omega))$.
2. $j^\varepsilon$ *is uniformly bounded in* $L^\infty(0,T;\ [L^{\frac{d+2}{d+1}}(\omega)]^d)$.
3. $\Phi^\varepsilon$ *is uniformly bounded in* $L^\infty(0,T;\ W^{2,\frac{d+2}{d}}(\omega))$.
4. $\partial_t\Phi^\varepsilon$ *is uniformly bounded in* $L^\infty(0,T;\ W^{1,\frac{d+2}{d+1}}(\omega))$. □

***Proof of Lemma** 12.6.22* $i)$ and $ii)$ are a consequence of Lemma (12.4.7), by replacing $K$ by $K^\varepsilon$. The property $iii)$ is a consequence (12.67), and $iv)$ can be deduced from $\partial_t\varrho^\varepsilon + \nabla_x \cdot j^\varepsilon = 0$, which implies a uniform bound $\partial_t\varrho^\varepsilon$ in $L^\infty(0,T;\ W^{-1,\frac{d+2}{d+1}}(\omega))$. □

Using this lemma, we can extract a sub-sequence satisfying

$$f^\varepsilon \overset{*}{\rightharpoonup} f \quad in \quad L^\infty(0,T;\ L^\infty(\Omega)) \quad |v|^2 f^\varepsilon \overset{*}{\rightharpoonup} F \quad in \quad L^\infty(0,T;\ \mathcal{M}_b(\Omega)),$$

$$\varrho^\varepsilon \rightharpoonup \varrho \quad in \quad L^\infty(0,T;\ L^{\frac{d+2}{d}}(\omega)), \quad j^\varepsilon \rightharpoonup j, \quad in \quad L^\infty(0,T;\ [L^{\frac{d+2}{d+1}}(\omega)]^d),$$

$$E^\varepsilon \rightharpoonup E \quad in \quad L^\infty(0,T;\ [W^{1,\frac{d+2}{d}}(\omega)]^d),$$

and

$$E^\varepsilon \longrightarrow E \quad in \quad L^\infty(0,T;\ [L^{\frac{d+2}{d}}(\omega)]^d).$$

Moreover, we have:

**Lemma 12.6.23** *The functions $F$, $\varrho$, and $j$ are:*

1. $F = |v|^2 f$.
2. $\varrho = \int_{\mathbb{R}^d} f dv$.
3. $j = \int_{\mathbb{R}^d} v f dv$. □

***Proof of Lemma** 12.6.23* The proof can be carried out by passing to the limit in the weak formulation using a test function Soit $\psi \in \mathcal{D}$, leading to

$$\int |v|^2 f^\varepsilon \psi \longrightarrow \langle F, \psi\rangle_{\mathcal{D}',\mathcal{D}}.$$

Using the fact that $|v|^2\psi \in \mathcal{D}$ and $f^\varepsilon \overset{*}{\rightharpoonup} f$, then

$$\int |v|^2 f^\varepsilon \psi \longrightarrow \langle f, |v|^2\psi\rangle = \langle |v|^2 f, \psi\rangle,$$

which implies that $F = |v|^2 f \in L^\infty(0,T;\ \mathcal{M}_b(\Omega))$. Moreover, $|v|^2 f^\varepsilon$ is uniformly bounded in $L^\infty_{loc}$ and then $F \in \mathcal{M}_b \cap L^\infty_{loc} = L^1 \cap L^\infty_{loc}$.
(ii) Let $\psi \in D(\bar{\omega} \times [0,T])$,

$$\int_0^T \int_\omega (\varrho^\varepsilon - \int_{\mathbb{R}^d} f dv)\psi dx dv dt = \int_0^T \int_\Omega (f^\varepsilon - f)\psi dx dv dt.dx dv.$$

The function $\psi \in L^\infty(\Omega\times]0,T[)$ and $f^\varepsilon \rightharpoonup f$ in $L^1$-weak.

$$\varrho = \int_{\mathbb{R}^d} f dv.$$

(iii)

$$\left|\int_0^T\int_\omega (j^\varepsilon - \int_{\mathbb{R}^d} vf dv)\psi dx dv\right| \leq \left|\int_0^T\int_\omega\int_{|v|\leq R} (vf^\varepsilon - vf)\psi dx dv dt\right|$$
$$+\frac{C\psi}{R}\int_0^T (K^\varepsilon(t) + K(t))dt$$
$$\leq |\langle f^\varepsilon - f, v\chi_{|v|\leq R}(v)\psi(x,t)\rangle_{L^1,L^\infty}| + \frac{C}{R}.$$

We pass to the limit in the first term of the right-hand side for a fixed $R$ using the weak convergence of $f^\varepsilon$. Then, we let $R$ going to infinity, and we obtain iii). □

**Lemma 12.6.24** *The weak limit $(f, E)$ is a solution of the Boltzmann–Poisson system (BP).* □

***Proof of Lemma** 12.6.24* Let $\psi \in \mathcal{D}$ and satisfy (12.26). Then,

$$\int_0^T\int_\Omega f^\varepsilon[\partial_t\psi + v.\nabla_x\psi + (E^\varepsilon + E_0).\nabla_v\psi]dx dv dt$$
$$+\int_0^T\int_\Omega Q(f^\varepsilon)\psi dx dv dt + \int_\Omega f_0\psi(x,v,0) = 0.$$

The convergence of the first integral is a consequence of the weak convergence of $f^\varepsilon$ and the strong convergence of $E^\varepsilon$. Moreover, $W^{1,\frac{d+2}{d}}(\omega) \hookrightarrow L^{d+2}(\omega)$ giving $E^\varepsilon$ is bounded in $L^2(0,T;\ L^{d+2}(\omega))$. As a consequence, Lemma (12.4.7) implies

$$\int_0^T\int_\Omega Q(f^\varepsilon)\psi dx dv dt \longrightarrow \int_0^T\int_\Omega Q(f)\psi dx dv dt.$$

Finally, $f$ is a weak solution of the Boltzmann equation associated with $E$ and satisfies (12.7) and (12.8). It remains to justify that it is a gradient of the potential, solution of the Poisson equation. Indeed, $\Phi^\varepsilon$ is uniformly bounded in $L^\infty(0,T;\ W^{1,\frac{d+2}{d}}(\omega))$, solution of

$$\int_\omega (1-\varepsilon\Delta)\nabla\Phi^\varepsilon(x,t)(1-\varepsilon\Delta)\nabla\theta(x)dx = \int_\omega \varrho^\varepsilon(t,x)\theta(x)dx$$

for a.e. $t \in ]0, T[$ and all $\theta \in \mathcal{D}(\omega)$. By passing to the limit on $\varepsilon$, we deduce that $\Phi$ verifies

$$\int_\omega \nabla\Phi^\varepsilon(x,t)\nabla\theta(x)dx = \int_\omega \varrho(x,t)\theta(x)dx,$$

and this means that $\Phi$ is the solution of the homogeneous Dirichlet problem $-\Delta\Phi = \varrho$ in $\mathcal{D}'$. To finish, the weak convergence of $f^\varepsilon$ and the strong convergence of $E^\varepsilon$ prove, by passing to the limit in (12.76), that

$$\mathcal{E}(t) = \int_\Omega |v|^2 f(x,v,t)dxdv + \int_\omega |E(t,x)|^2 dx \\ +2\int_\omega \varrho(x,t)\Phi(t,x)dx \in L^\infty_{loc}(\mathbb{R}^+).$$

## References

1. S. Agmon, A. Douglis, L. Nirenberg, Estimates near the boundary for solutions of elliptic partial differential equations satisfying general boundary conditions. II. Commun. Pure Appl. Math. **17**, 35–92 (1964)
2. V.I. Agoshkov, The existence of traces of functions in spaces used in problems of transport theory. (Russian) Dokl. Akad. Nauk SSSR **288**(2), 265–269 (1986)
3. R. Alexandre, Weak solutions of the Vlasov-Poisson initial-boundary value problem. Math. Methods Appl. Sci. **16**(8), 587–607 (1993)
4. L. Arkeryd, C. Cercignani, A global existence theorem for the initial-boundary value problem for the Boltzmann equation when the boundaries are not isothermal. Arch. Ration. Mech. Anal. **125**(3), 271–287 (1993)
5. A. Arseniev, Global existence of weak solution of the Vlasov system of equations. U.S.S.R. Comp. and Math. Phys. **15**, 131–143 (1975)
6. A.W. Ashcroft, N.D. Mermin, *Solid-State Physics* (Saunders College, 1976)
7. C. Bardos, Problèmes aux limites pour les équations aux dérivées partielles du premier ordre à coefficients réels; théorèmes d'approximation; application à l'équation de transport. Ann. Sci. de l'Ec. Norm. Sup., 4e série **3**, 185–233 (1970)
8. C. Bardos, R. Caffisch, B. Nicolaenko, Different aspects of the Milne Problem (based on energy estimates). Transp. Theory Stat. Phys. **16**, 561–585 (1987)
9. C. Bardos, P. Degond, Global existence for the Vlasov-Poisson equation in three space variable with small initial data. Ann. Inst. H. Poincaré, Anal. non linéaire **2**, 101–118 (1985)
10. C. Bardos, J. Rauch, Maximal positive boundary value problems as limits of singular perturbation problems. Trans. Am. Math. Soc. **270**(2), 377–408 (1982)
11. C. Bardos, R. Santos, R. Sentis, Diffusion approximation and computation of the critical size. Trans. Am. Math. Soc. **284**(2), 617–649 (1984)
12. N. Ben Abdallah, Weak solutions of the initial-boundary value problem for the Vlasov-Poisson system. Math. Methods Appl. Sci. **17**, 451–476 (1994)
13. N. Ben Abdallah, P. Degond, On a hierarchy of macroscopic models for semiconductors. J. Math. Phys. **37**(7), 3306–3333 (1996)
14. N. Ben Abdallah, P. Degond, S. Génieys, An energy-transport model for semiconductors derived from the Boltzmann equation. J. Stat. Phys. **84**(1–2), 205–231 (1996)

15. N. Ben Abdallah, J. Dolbeault, Relative entropies for kinetic equations in bounded domains (irreversibility, uniqueness and stationary solutions). en préparation
16. N. Ben Abdallah, L. Desvillettes, S. Génieys, On the convergence of the Boltzmann equation for semiconductors towards the energy transport model. J. Stat. Phys. **98**(3–4), 835–870 (2000)
17. F. Bouchut, L. Desvillettes, Averaging lemmas without time Fourier transform and application to discretized kinetic equations. Proc. R. Soc. Edinb. Sect. A **129**(1), 19–36 (1999)
18. H. Brézis, *Analyse fonctionnelle; théorie and applications* (Masson, Paris, 1992)
19. M. Cannone, C. Cercignani, A trace theorem in kinetic theory. Appl. Math. Lett. **4**(6), 63–67 (1991)
20. S. Chapman, T.G. Cowling, *The Mathematical Theory of Nonuniform Gases* (Cambridge University Press, Cambridge, 1990)
21. C. Cercignani, *The Theory and Application of the Boltzmann Equation* (Springer, New York, 1988)
22. M. Cessenat, Théorèmes de trace $L^p$ pour les espaces de fonctions de la neutronique. C.R. Acad. Sci. Paris Sér. I Math. **299**(16), 831–834 (1984)
23. P. Degond, S. Schmeiser, Macroscopic models for semiconductor heterostructures. J. Math. Phys. **39**(9), 4634–4663 (1998)
24. R.J. Diperna, P.L. Lions, On the Cauchy problem for the Boltzmann equation; global existence and weak stability. Ann. Math. **130**, 321–366 (1989)
25. R.J. Diperna, P.L. Lions, Ordinary differential equations, transport theory and Sobolev spaces. Invent. Math. **98**, 707–741 (1989)
26. R.J. Diperna, P.L. Lions, Global weak solutions of Vlasov-Maxwell systems. Commun. Pure Appl. Math. **42**, 729–757 (1989)
27. R.J. Diperna, P.L. Lions, Solutions globales d'équations de type Vlasov-Poisson. Note C. R. Acad. Sci. Paris, Série I **307**, 655–658 (1988)
28. R.J. Diperna, P.L. Lions, Y, Meyer, $L^p$ regularity of velocity averages. Ann. Inst. H. Poincaré Anal. Non Linéaire **8**(3–4), 271–287 (1991)
29. J. Dolbeault, Stationary states in plasma physics: Maxwellian solutions of the Vlasov-Poisson system. Math. Models Methods Appl. Sci. **1**(2), 183–208 (1991)
30. J. Dolbeault, Free energy and solutions of the Vlasov-Poisson-Fokker-Planck system: external potential and confinement (large time behavior and steady states). J. Math. Pures Appl. (9) **78**(2), 121–157 (1999)
31. J. Dolbeault, Kinetic models and quantum effects: a modified Boltzmann equation for Fermi-Dirac particles. Arch. Ration. Mech. Anal. **127**(2), 101–131 (1994)
32. H. Federer, *Geometric Measure Theory* (Springer, Berlin, 1969)
33. D. Gilbarg, N.S.Trudinger, *Elliptic Partial Differential Equations of Second Order* (Springer Verlag, Berlin Heidelberg, New York 1977)
34. F. Golse, Particle transport in nonhomogeneous media, in *Mathematical Aspects of Fluid and Plasma Dynamics (Salice Terme, 1988)*. Lecture Notes in Math., 1460 (Springer, Berlin, 1991), pp. 152–170
35. F. Golse, P.L. Lions, B. Perthame, R. Sentis, Regularity of the moments of the solution of a transport equation. J. Funct. Anal. **76**(1), 110–125 (1988)
36. H. Grad, Asymptotic theory of the Boltzmann equation. Phys. Fluids **6**, 147–181 (1963)
37. H. Grad, Principles of the kinetic theory of gases. Hand. Phys. **12**, 205–294 (1958)
38. C. Greengard, P.-A. Raviart, A boundary-value problem for the stationary Vlasov-Poisson equations: the plane diode. Commun. Pure Appl. Math. **43**(4), 473–507 (1990)
39. D. Hilbert, Math Ann. **72**, 562 (1912)
40. E. Horst, On the classical solutions of the initial value problem for the unmodified nonlinear Vlasov equation. II. Special cases. Math. Methods Appl. Sci. **4**(1), 19–32 (1982)
41. E. Horst, R. Hunze, Weak solutions of the initial value problem for the unmodified nonlinear Vlasov equation. Math. Methods Appl. Sci. **6**(2), 262–279 (1984)
42. T. Kato, *Perturbation Theory for Linear Operators* (Springer, New York, 1966)
43. C. Kittel, *Introduction to Solid State Physics* (Wiley, New York, 1968)

44. J.L. Lions, *Quelques méthodes de résolution des problèmes aux limites non linéaires* (Dunod, Paris, 1969)
45. P.L. Lions, Compactness in Boltzmann's equation via Fourier integral operator and applications I-III. J. Math. Kyoto Univ. **34**, 391-427; 429–461; 539–584 (1994)
46. P.L. Lions, B. Perthame, E. Tadmor, A kinetic formulation of multidimensional scalar conservation laws and related equations. J. Am. Math. Soc. **7**(1), 169–191 (1994)
47. P.L. Lions, B. Perthame, Propagation of moments and regularity for the 3-dimensional Vlasov-Poisson system. Invent. Math. **105**(2), 415–430 (1991)
48. P.A. Markowich, C.A, Ringhofer, C. Schmeiser, *Semiconductor Equations* (Springer, New York, 1989)
49. S. Mischler, On the trace problem for solutions of the Vlasov equation. Commun. Partial Differ. Equ. **25**(7–8), 1415–1443 (2000)
50. S. Mischler, On the initial boundary value problem for the Vlasov-Poisson-Boltzmann system. Commun. Math. Phys. **210**(2), 447–466 (2000)
51. S. Mischler, B. Perthame, Boltzmann equation with infinite energy: renormalized solutions and distributional solutions for small initial data and initial data close to a Maxwellian. SIAM J. Math. Anal. **28**(5), 1015–1027 (1997)
52. A. Pazy, *Semigroups of Linear Operators and Applications to Partial Differential Equations*. Applied Mathematical Sciences, vol. 44 (Springer, New York-Berlin, 1983)
53. K. Pfaffelmoser, Global classical solutions of the Vlasov-Poisson system in three dimensions for general initial data. J. Differ. Equ. **95**(2), 281–303 (1992)
54. J. Schaeffer, Global existence of smooth solutions to the Vlasov-Poisson system in three dimensions. Commun. Partial Differ. Equ. **16**(8–9), 1313–1335 (1991)
55. C. Schmeiser, Macroscopic models for almost elastic nonlinear electron-phonon interaction in semiconductors. SIAM J. Appl. Math. **60**(1), 109–120 (2000)
56. W. Shockley, Problems related to $p-n$ junctions in silicon. Solid State Electr. **2**, 35–67 (1961)
57. W. Shockley, W.T. Read, Statistics of recombinations of Holes and Electrons. Phys. Rev. **87**, 835–842 (1952)
58. S.M. Sze, *Physics of Semiconductor Devices*, 2nd edn. (Wiley, New York, 1981)